D0970899

BIOACTIVE CARBOHYDRATES:
In Chemistry, Biochemistry and Biology

BIOACTIVE CARBOHYDRATES:
In Chemistry, Biochemistry and Biology

JOHN F. KENNEDY, B.Sc., Ph.D., D.Sc., C.Chem., F.R.S.C., F.I.Biol.
Director of the Research Laboratory for the
Chemistry of Bioactive Carbohydrates and Proteins
Department of Chemistry, University of Birmingham, Birmingham, U.K.

and

CHARLES A. WHITE, B.Sc., Ph.D., C.Chem., M.R.S.C.
Development Manager, Vincent Kennedy Ltd.
Sutton Coldfield, West Midlands, U.K.

ELLIS HORWOOD LIMITED
Publishers · Chichester

Halsted Press: a division of
JOHN WILEY & SONS
New York · Brisbane · Chichester · Toronto

First published in 1983 by

ELLIS HORWOOD LIMITED
Market Cross House, Cooper Street, Chichester, West Sussex, PO19 1EB, England

The publisher's colophon is reproduced from James Gillison's drawing of the ancient Market Cross, Chichester.

Distributors:

Australia, New Zealand, South-east Asia:
Jacaranda-Wiley Ltd., Jacaranda Press,
JOHN WILEY & SONS INC.,
G.P.O. Box 859, Brisbane, Queensland 40001, Australia

Canada:
JOHN WILEY & SONS CANADA LIMITED
22 Worcester Road, Rexdale, Ontario, Canada.

Europe, Africa:
JOHN WILEY & SONS LIMITED
Baffins Lane, Chichester, West Sussex, England.

North and South America and the rest of the world:
Halsted Press: a division of
JOHN WILEY & SONS
605 Third Avenue, New York, N.Y. 10016, U.S.A.

British Library Cataloguing in Publication Data
Kennedy, John F.
Bioactive carbohydrates in chemistry, biochemistry and biology.
1. Carbohydrates
I. Title II. White, Charles A.
547.7'8 QD321

Library of Congress Card No. 82-9286 AACR2

ISBN 0-85312-201-6 (Ellis Horwood Ltd., Publishers – Library Edn.)
ISBN 0-85312-467-1 (Ellis Horwood Ltd., Publishers – Student Edn.)
ISBN 0-470-27527-8 (Halsted Press)

Typeset in Press Roman by Ellis Horwood Ltd.
Printed in Great Britain by R. J. Acford, Chichester.

Table of Contents

CHAPTER 4 ANALYTICAL METHODS FOR IDENTIFICATION AND DETERMINATION OF MACROMOLECULAR STRUCTURE

CHAPTER 5 CHEMICAL AND BIOCHEMICAL SYNTHESES

CHAPTER 6 MONOSACCHARIDES

Foreword

The last 10–15 years have seen substantial progress in many areas of carbohydrate chemistry, biochemistry and biology. For example, 'the new organic chemistry' which uses understanding of stereochemistry and reaction mechanisms and aims for predictive and specific synthetic routes, has been fully developed and exploited with carbohydrates; biosynthetic pathways of glycoconjugates have now been largely elucidated: the stereochemistry of glycan chains has become much better understood and powerful physical methodologies are now available for investigations in this area: there has been much activity with new technological exploitations such as oilfield chemicals and graft copolymers; carbohydrates have been significant in the emergence of practical biotechnology, for example in enzymic routes to sweeteners and as substrates for and products of fermentation.

If progress has been made, challenges also remain. The long recognised biological functions of carbohydrates are in the storage and mobilisation of energy and in contributing to the integrity of cells and tissues through their structural and water binding properties; many proposals have recently been made however, that they could have biological importance of additional and general types, perhaps especially in intra- and inter-cellular recognition events. This has some support from newly discovered phenomena such as the influence of carbohydrate sequence on clearance of serum glycoproteins from circulation but the generality of such a function remains to be established. It might turn out to be an oversimplification to suppose that the function of a carbohydrate moiety can be dissected away from that of the larger molecule of which it may be a part. Nevertheless, the conserved nature of carbohydrate covalent structures, the substantial energy that is expended, and the considerable amount of genetic information that is stored for their synthesis would all suggest that glycoconjugates in particular have biological importance beyond our present level of understanding. Such questions will guarantee that a lively interest in carbohydrates will continue and indeed that knowledge of carbohydrates will be necessary background for diverse investigations in years to come.

Despite the considerable progress made and the importance of problems that remain, it is surprising and indeed unfortunate that no advanced and up-to-date text book of general scope has been available in this area for many years. Drs Kennedy and White are to be congratulated for deciding to write the present book to fill an important gap.

Dr. D. A. REES, D.Sc., F.R.S.
Director, National Institute for Medical Research
Medical Research Council
Mill Hill, London, UK

October 1982

Preface

The study of carbohydrates has always been of secondary importance in comparison to the study of other biological macromolecules such as proteins and nucleic acids, but the recent interest in biotechnology and renewable sources of raw materials has brought about a general upsurge of interest. For those of us who have been involved in carbohydrates for many years this second rate status has always seemed unjust and irrational and due, in part, to the lack of suitable texts which cover the whole range of interest from chemistry through biochemistry to biology of all aspects of carbohydrates from the simple building blocks (monosaccharides) through the carbohydrate polymers (oligosaccharides and polysaccharides) to carbohydrate-containing macromolecules (glycoproteins, proteoglycans, nucleic acids, lipids, antibiotics, etc.). In fact, this lack of a suitable text for students was felt personally by one of us and was the instigation of the many ideas over almost 10 years which have culminated in the production of this text.

It has been our principle aim to provide, under one cover, a full description of the chemistry, biochemistry, and biology of all groups of carbohydrates and provide an introduction to their technological aspects suitable for undergraduate use. Whilst we appreciate that the resulting text is in no way comprehensive in depth or coverage, we believe that this book will not only provide a sound basis for those reading for an Honours degree course but also provide a valuable aid to those embarking on, or involved in, carbohydrate research in all its aspects and to those academics and industrialists who are not specialists in, but require a working knowledge of, the many aspects of carbohydrates. For these reasons we have assumed no previous knowledge of carbohydrates, but a general knowledge of organic chemistry is required.

In order to teach new students, and re-educate existing specialists, we have based our nomenclature on that which is now being recommended by the International Union of Pure and Applied Chemistry and the International Union of Biochemistry and provided many illustrations and tables to allow those not familiar with the new accepted nomenclature to convert to it whilst at the same time allow new students to easily understand much of the older literature which is characterised by its use of trivial rather than systematic nomenclature. From our experiences of teaching students we have based much of our discussions on the pictorial approach to enable the reader to visualise the subtle stereochemical relationships which are the most important aspects of carbohydrate structures. For this pictorial approach we have preferred to use conformational structures rather than the traditional Haworth structures since the former provide a better understanding of the stereochemistry involved.

The authors wish to thank many friends and colleagues for their assistance in making it possible for us to write this book. Professors M. Stacey, C.B.E., F.R.S. and S. A. Barker are thanked for their help and interest in the early stages of our careers, Dr. R. S. Tipson for his valuable help in reading a substantial part of the manuscript and his advice on many aspects of nomenclature, Drs. E. Morris, R. Gigg, E. Tarelli and V. A. McKusick for their helpful suggestions and provision of copious amounts of information, the various members of the IUPAC/IUB committees who provided preprints of many documents on carbohydrate nomenclature and the large number of colleagues and students, especially Meriam H. Adam Hussain, Dr. Dave P. Atkins, A. Jane Griffiths and David L. Stevenson who have encouraged us and assisted with reading and correction of manuscripts and proofs. Finally, the authors thank our parents and friends and in particular Sandra, Helen and Elizabeth for their co-operation, understanding and encouragement during the preparation of this book.

John F. Kennedy
Charles A. White

October 1982

CHAPTER 1

Introduction

GENERAL

Carbohydrates comprise the most abundant group of natural compounds. They can be looked upon as prime biological substances, produced by the processes of photosynthesis from the atmosphere by plants for their own nutritional and physiological needs. They must not, however, be regarded solely as a phenomenon of the plant world as they are constituents of a wide range of biological systems. They are, in many cases, utilized as food by man, other animals or micro-organisms and frequently become converted into more complex compounds. They serve as sources of energy (monosaccharides) and as major devices for the storage of solar energy, and as the principle source of metabolic energy for the human body (starch and glycogen). They form the major constituents of the shells of insects, crabs, and lobsters (chitin), and the supporting tissue of plants (cellulose), and are present in the cell walls of plants and bacteria, and the soft coats of animal cells. They form compounds with proteins and lipids, and, in such combinations, are essential to many biological reactions.

Carbohydrates, so-called because they were originally believed to be hydrates of carbon having the general formula $C_n(H_2O)_n$, are multifunctional compounds containing a number of hydroxyl groups of similar or equal reactivity, and at least one asymmetric carbon atom. With accumulation of knowledge, the definition has been expanded to include polyhydroxy aldehydes, ketones, alcohols, and acids and simple derivatives thereof, but the chemistry is largely that of simple functional groups, with extensive stereochemical variations superimposed. However, the manipulation of a single, selected hydroxyl group is a difficult problem and requires complex series of reactions. The condensation reaction between two molecules of a single monosaccharide to form a dimer results in not one exclusive product, but a possibility of up to 25 different compounds, and condensation to form a trimer yields up to 176 different trisaccharides. Fortunately, nature's catalysts (enzymes) are very selective and will only produce the required disaccharide, unlike chemical synthesis which produces a mixture of products. It is this balance between simplicity and complexity that has caused some scientists to avoid a study of carbohydrates.

Most modern textbooks on chemistry or biochemistry give little mention, if any, of carbohydrates and are usually restricted to the stereochemistry of the simple monosaccharides and the structure of the simple structural materials such as starch and cellulose, often with a very historical approach. Very little mention is made of the essential roles that carbohydrates undertake. This lack of interest is by no means a recent phenomenon because as long ago as the 1930s Bell was discouraged from embarking on a study of carbohydrates "as the field had now been fully worked out". Whelan in 1957 and Rees in 1971 were also led to believe that carbohydrate chemistry and biochemistry were running down. During the past decade, however, attitudes towards carbohydrates have changed markedly.

There are many reasons for this upsurge in interest. The development of new chemical methods and chromatographic techniques has allowed achievement of syntheses that earlier carbohydrate chemists could only carry out by carefully controlled chemical transformations and crude separation techniques. Now, much more is being discovered about the structure of carbohydrates because analyses are being performed on smaller and smaller quantities of material. The advance of carbohydrate chemistry has also been helped by studies on the action of enzymes on carbohydrates and has resulted in a new understanding of enzyme action and its applications both at the large-scale industrial level and at the detailed, small-scale level, as in the biochemical processes of life. Such overall developments have given impetus to a study of the role of carbohydrates in the processes of health and disease.

The monosaccharides D-glucose (Greek γλυκυσ, meaning sweet) and D-fructose have come to the front in industrial sweetening and bulking agents and, in the form of a mixture, as a serious competitor to the conventional carbohydrate sweetener, namely cane sugar (sucrose). Such polysaccharides as xanthan gum and alginic acid are much in vogue as food additives for gelling etc., and in other industries such as oil-drilling. In combination with protein, carbohydrates are involved in the defence mechanisms of the living human cell and play a part in successful organ transplantations, etc. It is now obvious that carbohydrate chemistry and biochemistry as we know them to-day hold much in store for far-reaching discoveries of Nobel Prize-winning standard and importance. We present a balanced approach to the chemistry and biochemistry of the whole spectrum of carbohydrates from simple monosaccharides to complex polysaccharides and carbohydrate-containing macromolecules, and hope to encourage further interest in a rapidly expanding, exciting field.

NOMENCLATURE

Throughout this book, we use nomenclature that is recommended by the International Union of Pure and Applied Chemistry (IUPAC) and the International Union of Biochemistry (IUB) Joint Commission on Biochemical Nomenclature

(JCBN) and have, where possible, included those recommendations which have not yet been published. We gratefully acknowledge the assistance of members of the JCBN for the provision of the draft final documents. We have also used the internationally accepted system of chemical nomenclature in preference to the system which the Association for Science Education (ASE) proposes for teaching in British schools, since the former is that found in the original literature and used by carbohydrate chemists worldwide.

A number of abbreviations for the more common biological compounds are used throughout this book. The names of these compounds are given in Table 1.1.

In many instances a trivial name is added, in parentheses, for a sugar or sugar derivative in order to help the reader understand the older chemical literature and current biochemical and biological literature. Most of these trivial names are 'not recommended' in the nomenclature rules and recommendations and are not used in carbohydrate chemistry. However, some of these trivial names are still acceptable in biochemistry and biology, whilst the chemical and pharmaceutical industries continue to use many of the 'not recommended' and uninformative names.

Table 1.1 Abbreviations used for the common biological compounds

ADP	adenosine 5′-diphosphate
AMP	adenosine 5′-monophosphate
ATP	adenosine 5′-triphosphate
dATP	deoxyadenosine 5′-triphosphate
CDP	cytidine 5′-diphosphate
CMP	cytidine 5′-monophosphate
dCMP	deoxycytidine 5′-monophosphate
CoA	coenzyme A
CTP	cytidine 5′-triphosphate
dCTP	deoxycytidine 5′-triphosphate
DFP	diisopropyl fluorophosphate
DNA	deoxyribonucleic acid
FAD	flavin-adenine dinucleotide
FMN	flavin mononucleotide (riboflavin 5′-monophosphate)
GDP	guanosine 5′-diphosphate
dGDP	deoxyguanosine 5′-diphosphate
GMP	guanosine 5′-monophosphate
dGMP	deoxyguanosine 5′-monophosphate
GTP	guanosine 5′-triphosphate
dGTP	deoxyguanosine 5′-triphosphate
IDP	inosine 5′-diphosphate
IMP	inosine 5′-monophosphate
ITP	inosine 5′-triphosphate

Table 1.1 (*continued*)

NAD^+	oxidized nicotinamide-adenine dinucleotide
NADH	reduced nicotinamide-adenine dinucleotide
$NADP^+$	oxidized nicotinamide-adenine dinucleotide phosphate
$NAD(P)^+$	indicates either NAD^+ or $NADP^+$
NADPH	reduced nicotinamide-adenine dinucleotide phosphate
NAD(P)H	indicates either NADH or NADPH
NDP	nucleoside 5′-diphosphate
NMN	nicotinamide mononucleotide
NMP	nucleoside 5′-monophosphate
dNMP	deoxynucleoside 5′-monophosphate
NTP	nucleoside 5′-triphosphate
poly(C)	synthetic polynucleotide composed of cytidylate residues
poly(G)	synthetic polynucleotide composed of guanylate residues
PPi	inorganic diphosphate
RNA	ribonucleic acid
tRNA	transfer ribonucleic acid
TDP	ribothymidine 5′-diphosphate
dTDP	thymidine 5′-diphosphate
TMP	ribothymidine 5′-monophosphate
dTMP	thymidine 5′-monophosphate
TTP	ribothymidine 5′-triphosphate
dTTP	thymidine 5′-triphosphate
UDP	uridine 5′-diphosphate
UMP	uridine 5′-monophosphate
dUMP	deoxyuridine 5′-monophosphate
UTP	uridine 5′-triphosphate
dUTP	deoxyuridine 5′-triphosphate

CHAPTER 2

Classification

Carbohydrates may be classified into three groups: monosaccharides, oligosaccharides and polysaccharides. Monosaccharides are the simple sugars which cannot be hydrolysed to smaller sugar molecules. Oligosaccharides are simple polymers of monosaccharides joined by glycosidic linkages; by definition, they contain from 2 to 10 monosaccharide units to which they can be hydrolysed. Polysaccharides are polymers of high molecular weight consisting of more than 10 units, but the division between oligosaccharides and polysaccharides is made easier since carbohydrates containing between 5 and 15 units rarely occur in nature. The majority of naturally occurring polysaccharides contain 80 to 100 units, with only a few in the range 25 to 75 units. There are some polysaccharides which contain more than 100 units, for example, native cellulose contains an average of 3000 units, or, more accurately, contains a series of polymers having a molecular weight distribution about a mean value equivalent to 3000 units. Such polysaccharides rarely exist as collections of discrete macromolecules of identical molecular weight.

MONOSACCHARIDES

Monosaccharides are classified according to the number of carbon atoms they contain. Trioses have the formula ($C_3H_6O_3$), tetroses ($C_4H_5O_4$), pentoses ($C_5H_{10}O_5$) and hexoses ($C_6H_{12}O_6$), etc.

Configuration

D-Glucose, a hexose and the most common monosaccharide, occurs in the juices of fruits and in honey and is a common hydrolysis product of polysaccharides. The structures of D-glucose and some of the other hexoses and pentoses were determined, and their stereochemical relationships proved by the work of Fischer, for which he was awarded the Nobel Prize for chemistry in 1907. Each family of monosaccharides (hexoses, pentoses, etc.) was found to comprise a collection of molecular forms (stereoisomers) which have identical structures but different three-dimensional arrangements of the atoms and functional groups. Those stereoisomers which are not mirror images of each other are referred to as diastereoisomers.

D-Glucose forms a penta-acetate (1), not a hexa-acetate, showing that there are only 5 hydroxyl groups present; since the presence of 2 hydroxyl groups on the same carbon atom is an unstable situation, there are 5 carbon atoms carrying the hydroxyl groups. The remaining carbon atom has been shown to be part of an aldehyde group (by the reaction with mild oxidizing agents to afford a monocarboxylic acid). Reactions which destroy stereochemistry by formation of a double bond (for example, phenylosazone formation) at one carbon atom have shown that certain pairs of sugars give the same product, for example, D-glucose and D-mannose react with loss of stereochemistry at carbon atom 2 (C-2) to give the same phenylosazone (Scheme 2.1) and must therefore have the same configuration at the other carbon atoms. They are said to be epimeric at C-2. Such reactions as these, and reactions which add or remove a carbon atom have shown that the structures of monosaccharides are related as in Fig. 2.1(a) and (b), where structures are depicted in straight chain, Fischer projection formulae.

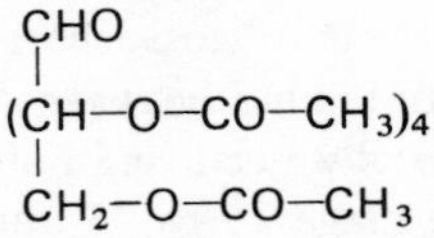

$$\begin{array}{l} CHO \\ | \\ (CH{-}O{-}CO{-}CH_3)_4 \\ | \\ CH_2{-}O{-}CO{-}CH_3 \end{array}$$

An *aldehydo*-aldohexose penta-acetate

(1)

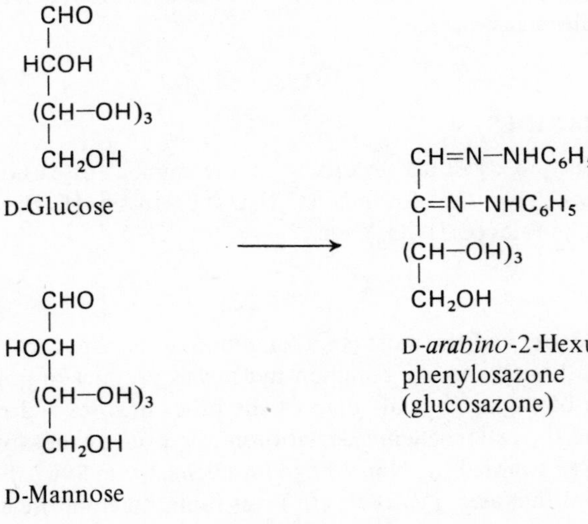

$$\begin{array}{l} CHO \\ | \\ HCOH \\ | \\ (CH{-}OH)_3 \\ | \\ CH_2OH \end{array}$$

D-Glucose

$$\begin{array}{l} CHO \\ | \\ HOCH \\ | \\ (CH{-}OH)_3 \\ | \\ CH_2OH \end{array}$$

D-Mannose

$$\longrightarrow \quad \begin{array}{l} CH{=}N{-}NHC_6H_5 \\ | \\ C{=}N{-}NHC_6H_5 \\ | \\ (CH{-}OH)_3 \\ | \\ CH_2OH \end{array}$$

D-*arabino*-2-Hexulose phenylosazone (glucosazone)

Scheme 2.1

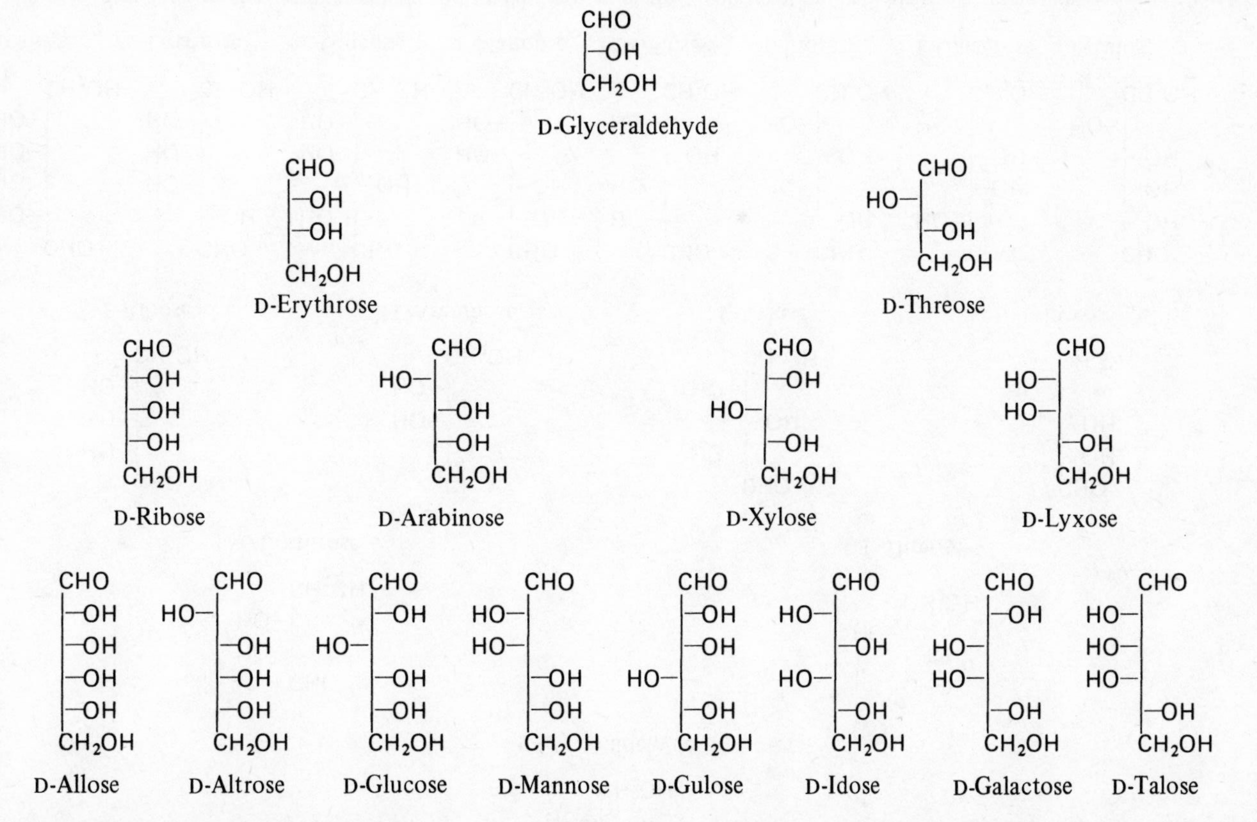

Fig. 2.1(a) – Structures of the *aldehydo*-aldoses, up to aldohexoses, of the D configuration, with hydrogen atoms omitted for clarity (for the full structure of D-Glucose see Fig. 2.2).

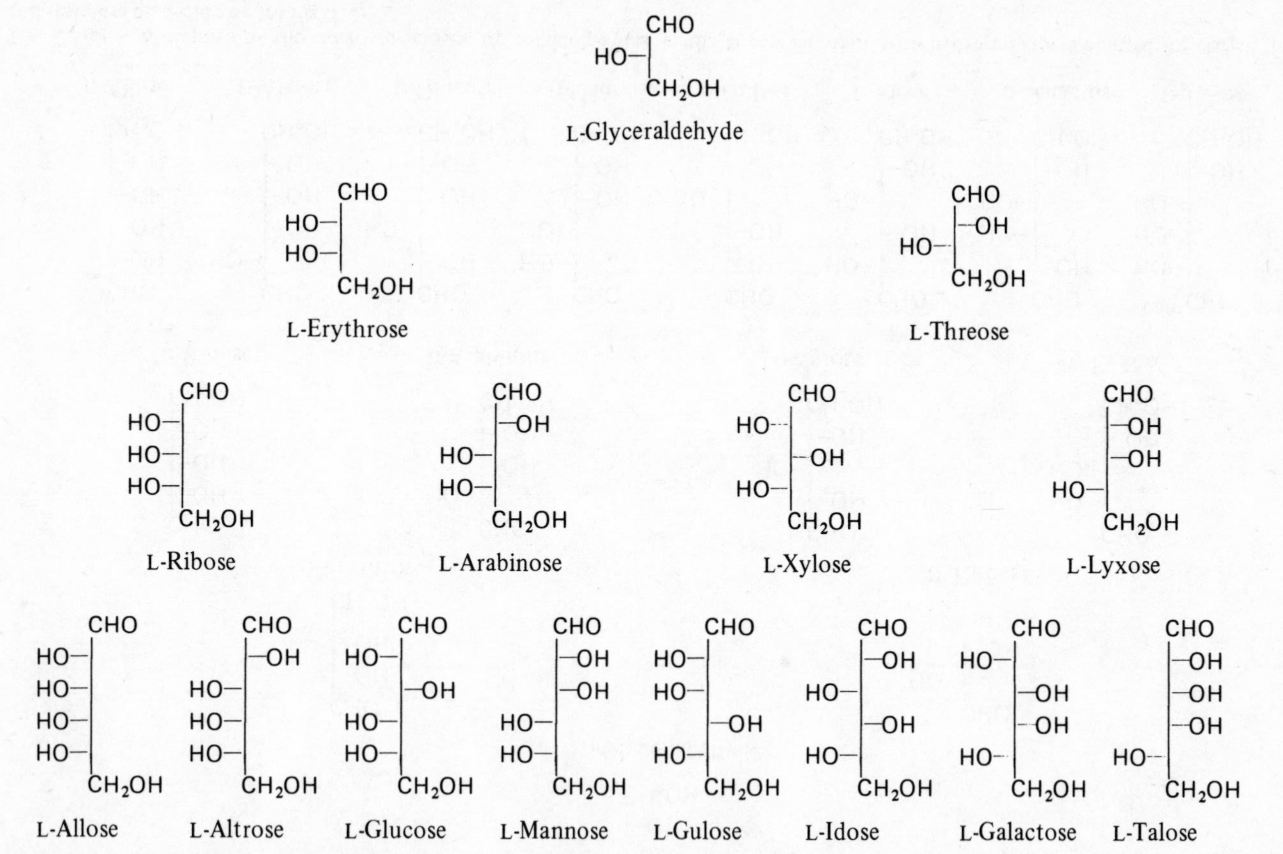

Fig. 2.1(b) – Structures of the *aldehydo*-aldoses, up to aldohexoses, of the L configuration, with hydrogen atoms omitted for clarity (for the full structure of L-Glucose see Fig. 2.2).

The stereochemistry at the penultimate carbon atom in the chain, counting the carbon atom of the aldehyde group as the first (C-1), determines the series to which the sugar belongs (that is, which enantiomeric form or absolute configuration it has). For glucose, the stereochemistry at C-5 determines whether it is D-glucose or L-glucose, D-glucose being related to D-glyceraldehyde (Fig. 2.2). To build up to D-series of D-aldopentoses, the 'initials' RAXL give the order ribose, arabinose, xylose, and lyxose which have (see Fig. 2.1(a)):

C-4 hydroxyl groups all to the right
C-3 hydroxyl groups first pair right, second pair left
C-2 hydroxyl groups alternating right then left.

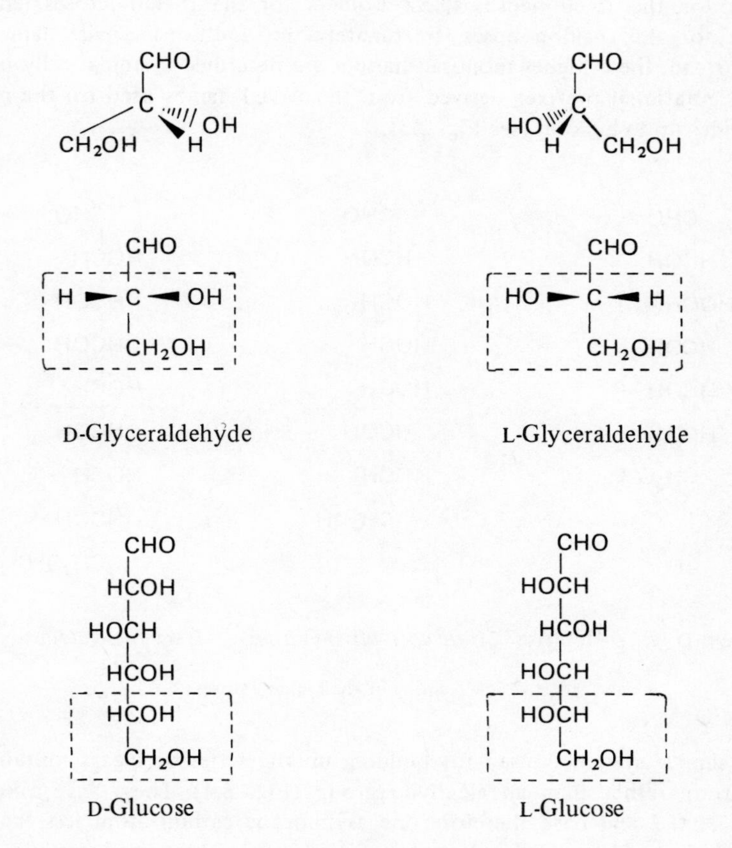

Fig. 2.2 – Relationship of the configuration at C-5 in D- and L-glucose to the configuration of D- and L-glyceraldehyde.

The mnemonic: '*All altruists gladly make gum in gallon tanks*' gives the correct sequences for the D-aldohexoses which can be built up as follows:

C-5 hydroxyl groups all to the right
C-4 hydroxyl groups, first four right, second four left
C-3 hydroxyl groups, alternating pairs to right then left
C-2 hydroxyl groups, alternating right then left.

In both the above sequences it is worthwhile noting that the number of hydroxyl groups, which run first to the right and then to the left, is reduced by a half on descent of the relevant carbon atom number.

For the monosaccharides containing more than 6 carbon atoms the number of isomers doubles for every extra carbon atom. Thus, there are 16 possible isomers for the D-aldoheptoses, 32 isomers for the D-aldo-octoses, and 64 isomers for the D-aldononoses. Fortunately, no additional trivial names are used; instead, these higher monosaccharides are described systematically by use of configurational prefixes derived from the trivial names used for the monosaccharides up to hexoses (see Fig. 2.3).

D-*glycero*-D-*gluco*-Heptose	L-*threo*-L-*altro*-Octose	D-*xylo*-L-*galacto*-Nonose
CHO	CHO	CHO
HCOH	HCOH	HOCH
HOCH	HOCH	HCOH
HCOH	HOCH	HCOH
HCOH	HOCH	HOCH
- - - - - -	- - - - - -	- - - - - -
HCOH	HCOH	HCOH
CH_2OH	HOCH	HOCH
	CH_2OH	HCOH
		CH_2OH

Fig. 2.3 – Naming of the higher aldoses.

A similar system is used for building up the series of sugars containing a keto group rather than an aldehydo group (Fig. 2.4). These have a ketonic carbon at C-2 and have therefore one asymmetric carbon atom less than the corresponding aldose. Higher ketoses, such as heptuloses, octuloses and nonuloses are named as derivatives of the lower ketoses, using the same system as applied to the higher aldoses (see Fig. 2.3).

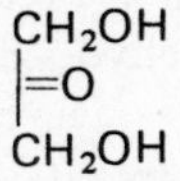

CH_2OH
=O
CH_2OH

1,3-Dihydroxy-2-propanone†
(Dihydroxyacetone)
(Glycerone)

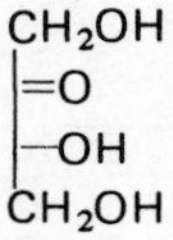

CH_2OH
=O
—OH
CH_2OH

D-*glycero*-Tetrulose
(D-Erythrulose)
(D-Threulose*)

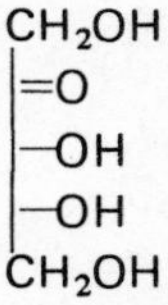

CH_2OH
=O
—OH
—OH
CH_2OH

D-*erythro*-Pentulose
(D-Ribulose)
(D-Adonose*)

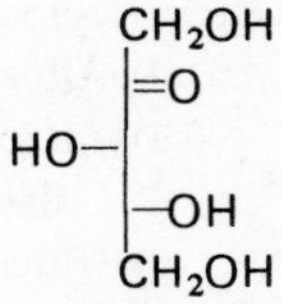

CH_2OH
=O
HO—
—OH
CH_2OH

D-*threo*-Pentulose
(D-Xylulose)
(D-Lyxulose*)

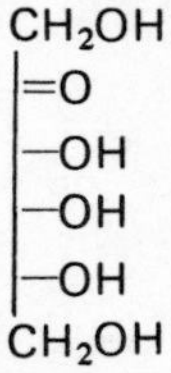

CH_2OH
=O
—OH
—OH
—OH
CH_2OH

D-*ribo*-Hexulose
(D-Psicose)
(D-Allulose*)

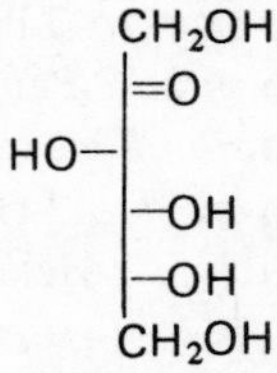

CH_2OH
=O
HO—
—OH
—OH
CH_2OH

D-*arabino*-Hexulose
(D-Fructose)
(D-Levulose*)

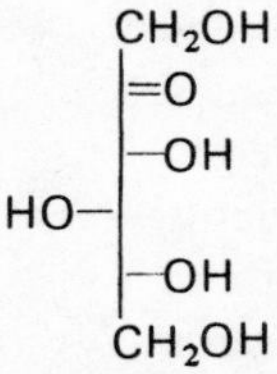

CH_2OH
=O
—OH
HO—
—OH
CH_2OH

D-*xylo*-Hexulose
(D-Sorbose)

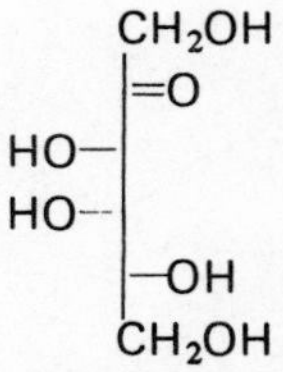

CH_2OH
=O
HO—
HO—
—OH
CH_2OH

D-*lyxo*-Hexulose
(D-Tagatose)

Fig. 2.4 – Structures of the ketoses, up to ketohexoses, of the D configuration (trivial names in brackets; * not recommended trivial name), with hydrogen atoms omitted for clarity.
† Not regarded as being a sugar, due to absence of asymmetric carbon atom.

The convention used for Fischer projection formulae is that the vertical bonds (carbon–carbon) lie behind the plane of the page (except the one under consideration) and the horizontal bonds (carbon–hydrogen and carbon–oxygen) project in front of the page. The aldehydo group must be placed at the top (C-1). For ketoses, the keto group is placed next to the top (C-2). Frequently, the hydrogen atoms are omitted for simplicity as in Fig. 2.1. The trivial names used in Figs. 2.1 and 2.4 are preferred for simple sugars and have corresponding systematic names (Table 2.1) which are used to describe more-complex structures.

Table 2.1 Systematic names for neutral monosaccharides

Trivial	Systematic
D-Glyceraldehyde	D-*glycero*-Triose
D-Erythrose	D-*erythro*-Tetrose
D-Threose	D-*threo*-Tetrose
D-Arabinose	D-*arabino*-Pentose
D-Lyxose	D-*lyxo*-Pentose
D-Ribose	D-*ribo*-Pentose
D Xylose	D-*xylo*-Pentose
D-Allose	D-*allo*-Hexose
D-Altrose	D-*altro*-Hexose
D-Galactose	D-*galacto*-Hexose
D-Glucose	D-*gluco*-Hexose
D-Gulose	D-*gulo*-Hexose
D-Idose	D-*ido*-Hexose
D-Mannose	D-*manno*-Hexose
D-Talose	D-*talo*-Hexose
D-Erythrulose*	D-*glycero*-2-Tetrulose
D-Ribulose*	D-*erythro*-2-Pentulose
D-Xylulose*	D-*threo*-2-Pentulose
D-Psicose	D-*ribo*-2-Hexulose
D-Fructose	D-*arabino*-2-Hexulose
D-Sorbose	D-*xylo*-2-Hexulose
D-Tagatose	D-*lyxo*-2-Hexulose

* less-preferred trivial names which are not recommended for current usage.

As may be seen from Figs. 2.1 and 2.4, the carbon atoms not at the end of the chain have four different atoms or groups attached to them and are said to be asymmetric. This characteristic gives rise to optical activity and standard solutions of different sugars have different optical rotations, not all in the same direction (Table 2.2).

Table 2.2 Optical rotations of some naturally occurring monosaccharides

Pyranose		Mutaform
D-Ribose (α)	− 23	− 24
L-Arabinose (β)	+ 191	+ 105
D-Xylose (α)	+ 94	+ 19
D-Lyxose (α)	+ 6	− 14
(β)	− 73	− 14
D-Glucose (α)	+ 113	+ 53
(β)	+ 19	+ 53
D-Galactose (α)	+ 150	+ 80
(β)	+ 53	+ 80
D-Mannose (α)	+ 29	+ 14
(β)	− 17	+ 14
L-Rhamnose	− 9	− 8
D-Quinovose	+ 73	+ 30
L-Fucose (α)	− 153	− 76
D-Fructose (β)	− 132	− 94
L-Sorbose	− 44	− 43
D-Tagatose	− 3	− 4
D-Glucuronic acid	+ 12	+ 36
D-Galacturonic acid (α)	+ 107	+ 52
(β)	+ 31	+ 52
D-Mannuronic acid (α)	+ 16	− 24
(β)	− 48	− 24
2-Amino-2-deoxy-D-glucose (α)	+ 100	+ 48
(D-Glucosamine) (β)	+ 14	+ 48
2-Amino-2-deoxy-D-glucose hydrochloride (α)	+ 100	+ 73
(D-Glucosamine hydrochloride) (β)	+ 25	+ 73
2-Acetamido-2-deoxy-D-glucose (*N*-Acetyl-D-glucosamine)	+ 56	+ 41
2-Amino-2-deoxy-D-galactose (α)	+ 121	+ 95
(D-Galactosamine hydrochloride) (β)	+ 45	+ 95
2-Acetamido-2-deoxy-D-galactose (*N*-Acetyl-D-galactosamine)	+ 131	+ 98
5-Acetamido-3,5-dideoxy-D-*glycero*-D-*galacto*-2-nonulosonic acid (*N*-Acetylneuraminic acid)	− 32	− 32

Ring Structures

These projections do not completely represent the structure of monosaccharides; for example, the aldehydo group does not react as a normal aldehydo group, as it will not colour Schiff's reagent and produces a mixture of glycosides on reaction with methanolic hydrogen chloride and not a dimethyl acetal. This indicates that the aldehydo group is in some way masked. D-Glucose can cry-

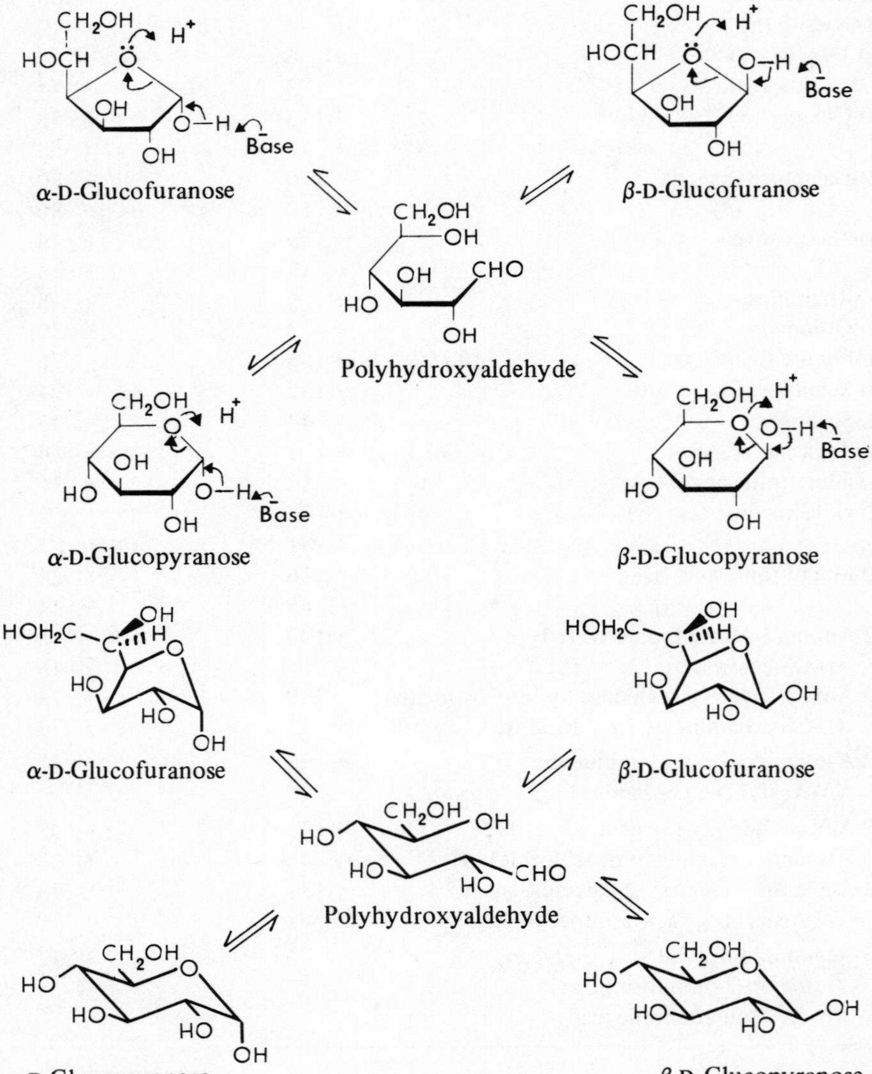

α-D-Glucofuranose β-D-Glucofuranose

Polyhydroxyaldehyde

α-D-Glucopyranose β-D-Glucopyranose

α-D-Glucofuranose β-D-Glucofuranose

Polyhydroxyaldehyde

α-D-Glucopyranose β-D-Glucopyranose

Fig. 2.5 – Mutarotation of D-glucose.

stallize, depending on conditions, in two forms that have different melting points (146° and 149°) and optical rotations (+113° and +19°) and are known as α- and β-D-pyranose forms. If these crystalline forms are dissolved in water, the optical rotations change, and both solutions reach a constant value of +53°, showing that the two forms are interconvertible – this process is called mutarotation, and is exhibited by all mono- and most reducing di-saccharides. The different forms are known as anomers and the C-1 atom of aldoses is the anomeric carbon. These observations were rationalized by Haworth, who devised the ring structures for monosaccharides on the basis of their existence as cyclic hemiacetals. The two forms of D-glucose are thereby shown to be due to the arrangement of the hydroxyl group at C-1 (Fig. 2.5) due to the formation of another asymmetric carbon atom (mutarotation). Haworth also established that the favoured ring size of the aldehexoses is the six-membered (pyranose) ring, which is denoted by the symbol *p* after the three-letter symbol for the monosaccharide (for example, Glc*p*). The five-membered (furanose) ring can and does exist in nature, and is denoted by an *f* (for example, Glc*f*).

Conformation

In order to visualize the six-membered ring the so-called 'Haworth formulae' are inadequate and the Haworth 'conformational structures' are used, which show puckered rings not planar rings. 'Conformation' is the word, introduced by Haworth, used to describe the arrangement in space of the atoms of a single chemical structure (configuration), the various arrangements being produced only by rotation about single bonds. There are two types of strainless pyranose rings possible, the boat (*B*) (2a) and the chair (*C*) (2b) forms, and of these, the chair is usually preferred as there are usually fewer interactions across the ring between substituents (that is, non-bonded interactions). Other strained rings exist but are only found when other constraints, such as double bonds, distort the ring. These forms include the skew (*S*) (3) and half-chair (*H*) (4) conformations.

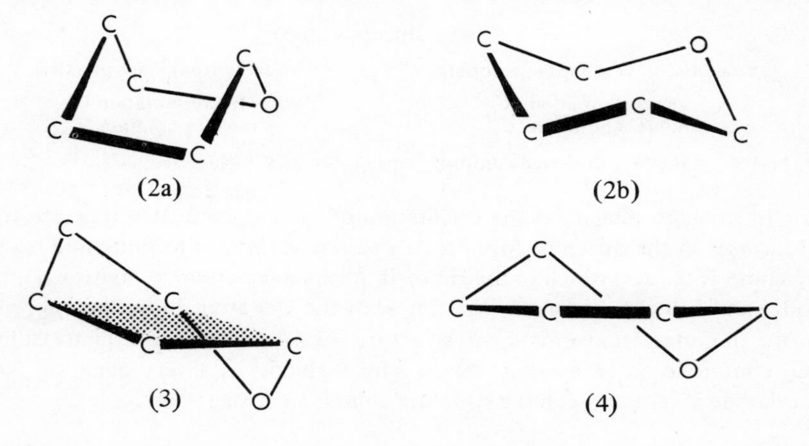

(2a) (2b)

(3) (4)

Reeves uses the terms *C1* and *1C* to define the ring shapes but as may be seen from Fig. 2.6, the enantiomeric (D and L) forms of a sugar in the same conformation (that is, with the same axial-equatorial arrangement) have different symbols, and must therefore always be linked to the absolute configuration of the sugar. Therefore, β-D-glucopyranose in the *C1(D)* conformation has all its substituents in equatorial positions, at the greatest distance from other sub stituents. In order to overcome these problems a revised and improved system of conformational nomenclature has been devised that is based on Reeve's system, but which defines the ring without resort to the substituents (Recommendations, 1980a). Thus, using the plane described by the oxygen atom and carbon atoms 2, 3 and 5, the rings can be defined as $^4C_1$ if C-4 is above the plane and C-1 is below it, or as $^1C_4$ for the alternative ring. (The numbering of carbon atoms appears clockwise if viewed from above.)

β-D-Glucopyranose

(all hydroxyl groups equatorial)
Reeves notation *C1*
modern notation $^4C_1$

(all hydroxyl groups axial)
Reeves notation *1C*
modern notation $^1C_4$

mirror plane

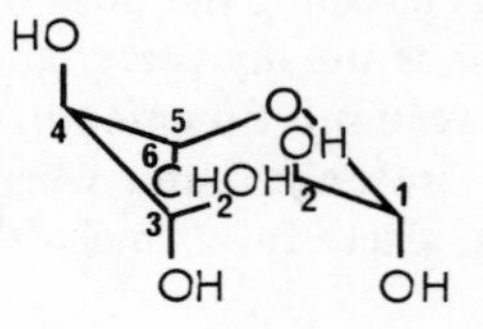

β-L-Glucopyranose

(all hydroxyl groups equatorial)
Reeves notation *1C*
modern notation $^1C_4$

(all hydroxyl groups axial)
Reeves notation *C1*
modern notation $^4C_1$

Fig. 2.6 – Preferred conformations of β-D- and -L-glucopyranose.

In order to remember the orientation of the groups in the pyranose forms of hexoses in the different formula, the easiest method is to think of β-D-glucopyranose (Fig. 2.7) which in the Haworth formula has alternate hydroxyl groups above and below the plane of the ring with the C-6 atom above the ring, whilst in the preferred 'conformational' structure, all substituents are equatorial in the $^4C_1$ conformation (a possible reason why D-glucose is a very common monosaccharide in nature – it has a structure containing little strain).

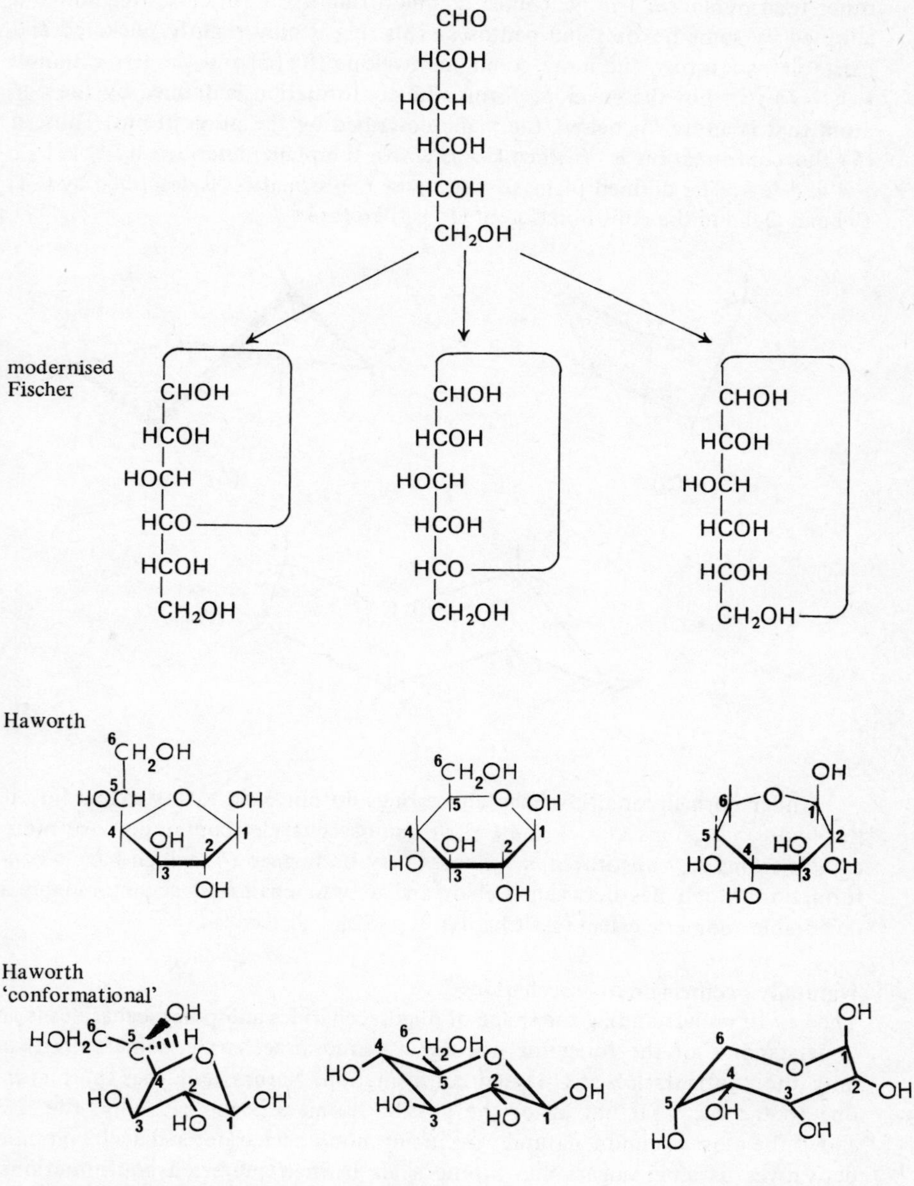

modernised Fischer

Haworth

Haworth 'conformational'

β-D-Glucofuranose-E_3 β-D-Glucopyranose $^4C_1$ β-D-Glucoseptanose

Fig. 2.7 – Conformations of β-D-glucose (hydrogen atoms omitted from Haworth structures for clarity).

As indicated earlier, hexoses can, under certain conditions, exist in rings other than pyranose. The next most common ring is the furanose ring which is adopted by some hexoses and pentoses. This ring is only slightly puckered and exists in two forms, the more common envelope (*E*) (5) and the less common twist (*T*) (6). For the envelope forms, the conformation is defined by the one atom that is above, or below, the plane described by the other atoms. Thus, in (5) the conformation is ^3E since C-3 is above the plane described by C-1, C-2, C-4 and O-4. The defined plane for the twist conformation is described by C-1, C-4 and O-4 and the conformation of (6) is therefore $^3T_2$.

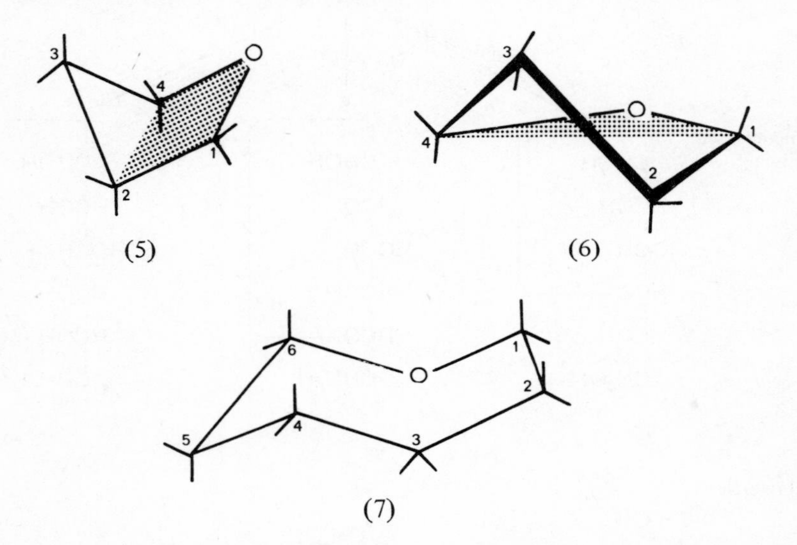

(5) (6)

(7)

Under normal conditions, septanose rings do not exist to any extent but, if the hydroxyl groups at C-4 and C-5 of monosaccharides containing 7 or more carbon atoms are substituted, septanoses may be formed (7) which have a conformation which lies between a chair and a twist chair and accommodates a favourable anomeric effect (see Chapter 3, p. 62).

Naturally occurring monosaccharides

The key to understanding the shape of oligosaccharides and polysaccharides is an understanding of the foregoing discussion of monosaccharide isomers (particularly the configuration at C-1) and conformation. Nature simplifies the picture due to the fact that not all of the possible isomers occur naturally. Fig. 2.8 shows the most common, naturally occurring monosaccharides and such common derivatives as aminosugars and uronic acids in their preferred conformations. Structures of less common monosaccharides and other derivatives will be given in subsequent Chapters, for example, complex aminosugars in antibiotics (Chapter 12) and in microbial polysaccharides (Chapter 8).

Pentoses

β-D-Ribofuranose-^{3}E

2-Deoxy-β-D-*erythro*-pentafuranose-^{3}E (2-Deoxy-D-ribose)

β-D-Xylopyranose-$^{4}C_1$

α-L-Arabinopyranose-$^{4}C_1$

α-L-Arabinofuranose-E_3

Hexoses

β-D-Glucopyranose-$^{4}C_1$

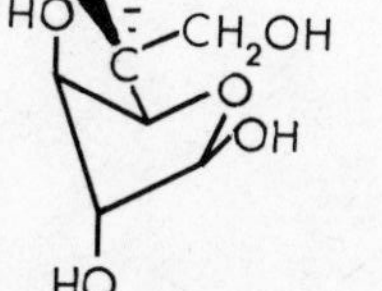
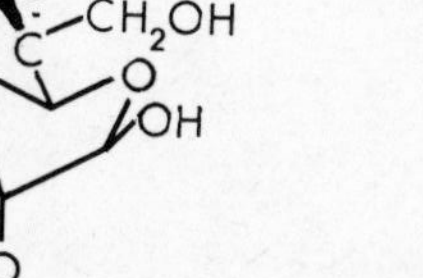

β-D-Glucofuranose-^{3}E

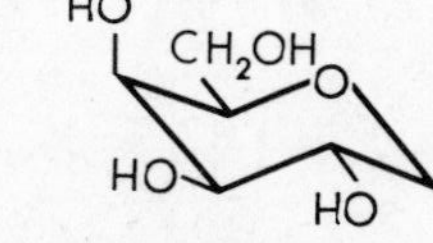

β-D-Galactopyranose-$^{4}C_1$

β-L-Galactopyranose-$^{1}C_4$

Fig. 2.8 – Structures of common monosaccharides (shown in one anomeric form only) (*continued on next page*).

Hexoses (*continued*)

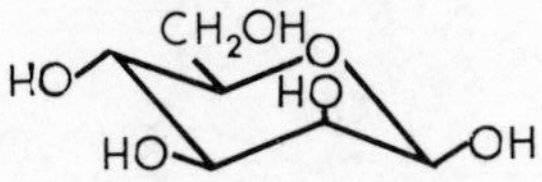

β-D-Mannopyranose-$^4C_1$

Ketohexoses

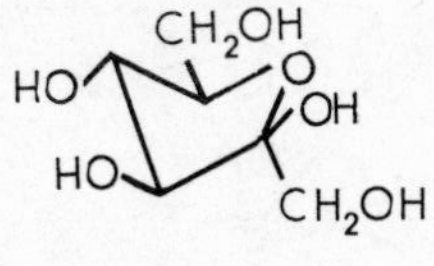

β-D-Fructofuranose-^3E

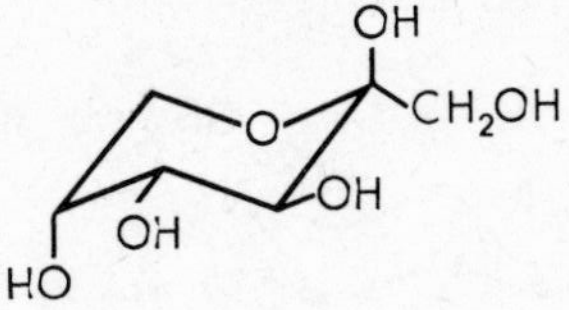

α-D-Fructopyranose-$^1C_4$

Deoxyhexoses

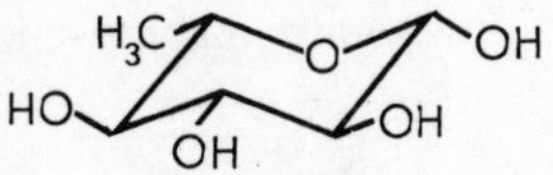

6-Deoxy-β-L-mannopyranose-$^1C_4$
(L-Rhamnopyranose)

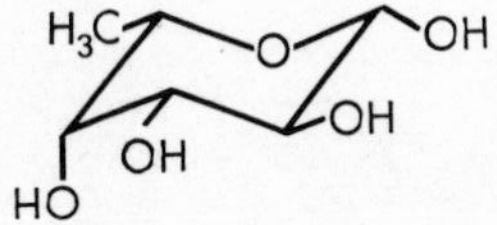

6-Deoxy-β-L-galactopyranose-$^1C_4$
(L-Fucopyranose)

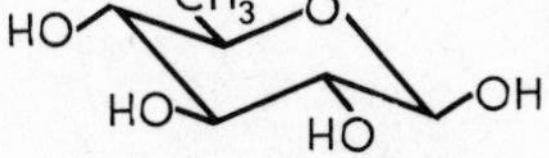

6-Deoxy-β-D-glucopyranose-$^4C_1$
(D-Quinovopyranose)

Hexuronic acids

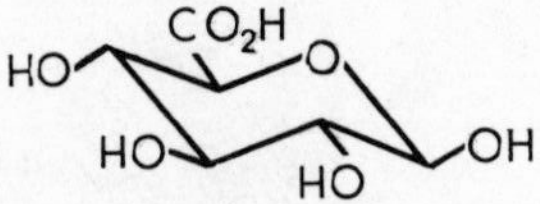

β-D-Glucopyranuronic acid-$^4C_1$

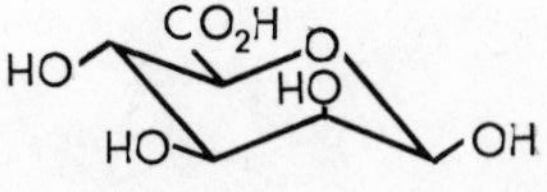

β-D-Mannopyranuronic acid-$^4C_1$

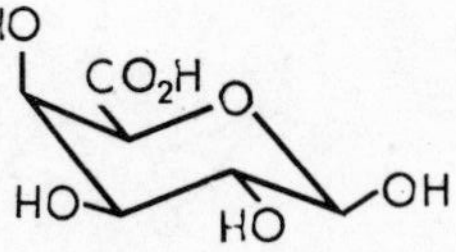

β-D-Galactopyranuronic acid-$^4C_1$

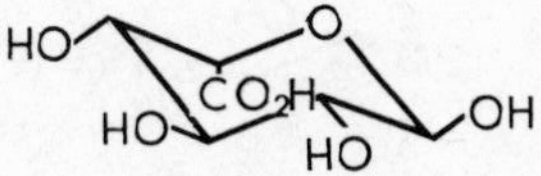

α-L-Idopyranuronic acid-$^4C_1$

2-Amino-2-deoxyhexoses

2-Amino-2-deoxy-β-D-glucopyranose-$^4C_1$ (Glucosamine)*

2-Amino-2-deoxy-β-D-galactopyranose-$^4C_1$ (Galactosamine)*

Neuraminic acids

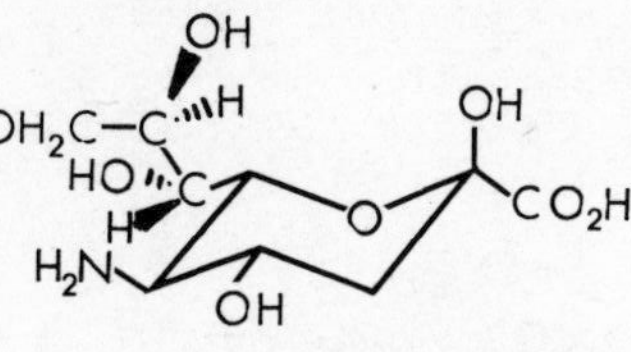

5-Amino-3,5-dideoxy-D-*glycero*-α-D-*galacto*-2-nonulopyranonic acid-$^1C_4$ (Neuraminic acid)*

5-Acetamido-3,5-dideoxy-D-*glycero*-α-D-*galacto*-2-nonulopyranonic acid-$^1C_4$ (*N*-Acetylneuraminic acid)*

Fig. 2.8 – Structures of common monosaccharides (shown in one anomeric form only). See Figs. 8.9 and 12.1 for structures of less-common monosaccharides.

* These trivial names are accepted as standard biochemical and biological nomenclature but not for chemical usage.

OLIGOSACCHARIDES

When two or more monosaccharides are joined together, the bond that holds them together is known as the glycosidic bond. This is formed between the hydroxyl group on the anomeric carbon atom of one monosaccharide and any hydroxyl group on another monosaccharide through formation of an acetal (Scheme 2.2). Formation of a disaccharide by the condensation reaction (see also Chapter 5) between two identical hexopyranose ring structures can result in 11 different isomers if the identical hexopyanose residues belong to the D-series. There will of course be another series of 11 different isomers if the residues belong to the L-series and additional isomers can occur if furanose forms are also considered. The series of 11 isomers is composed of eight isomers having glycosidic linkages between C-1 of one residue, in either anomeric configuration, and C-2, C-3, C-4 or C-6 of the other pyranose residue. These are termed (1 → 2)-α-D-, (1 → 3)-β-D- linkages etc., where α- and β- refer to the anomeric con-

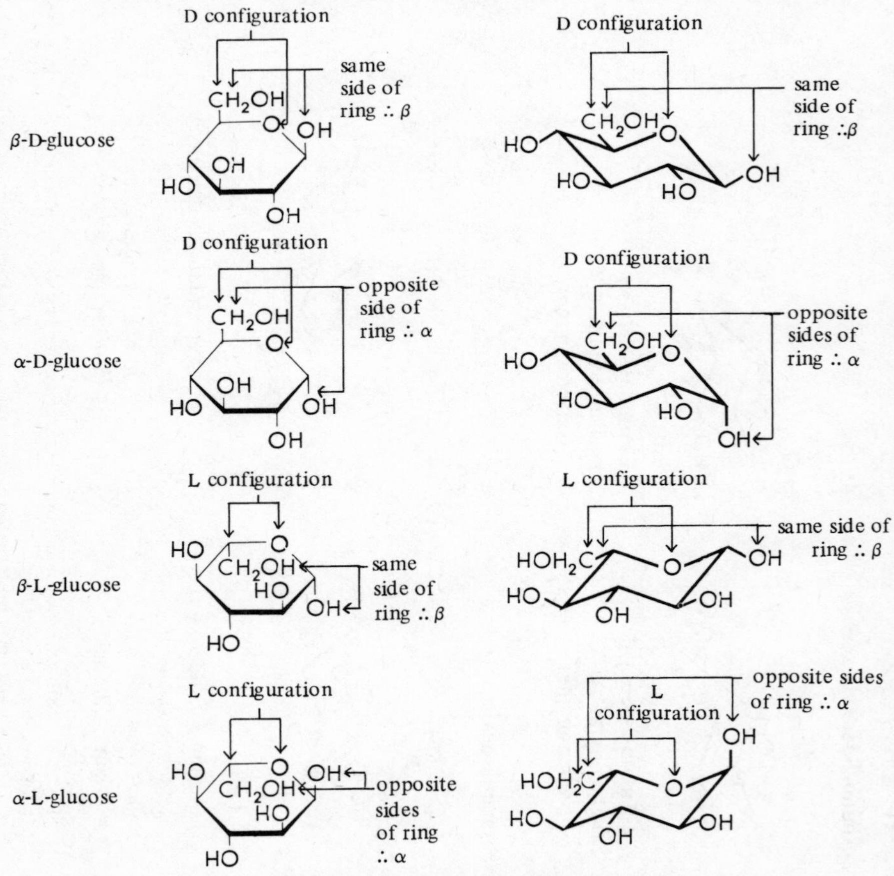

Fig. 2.9 – Definition of anomeric configuration at carbon atom 1.

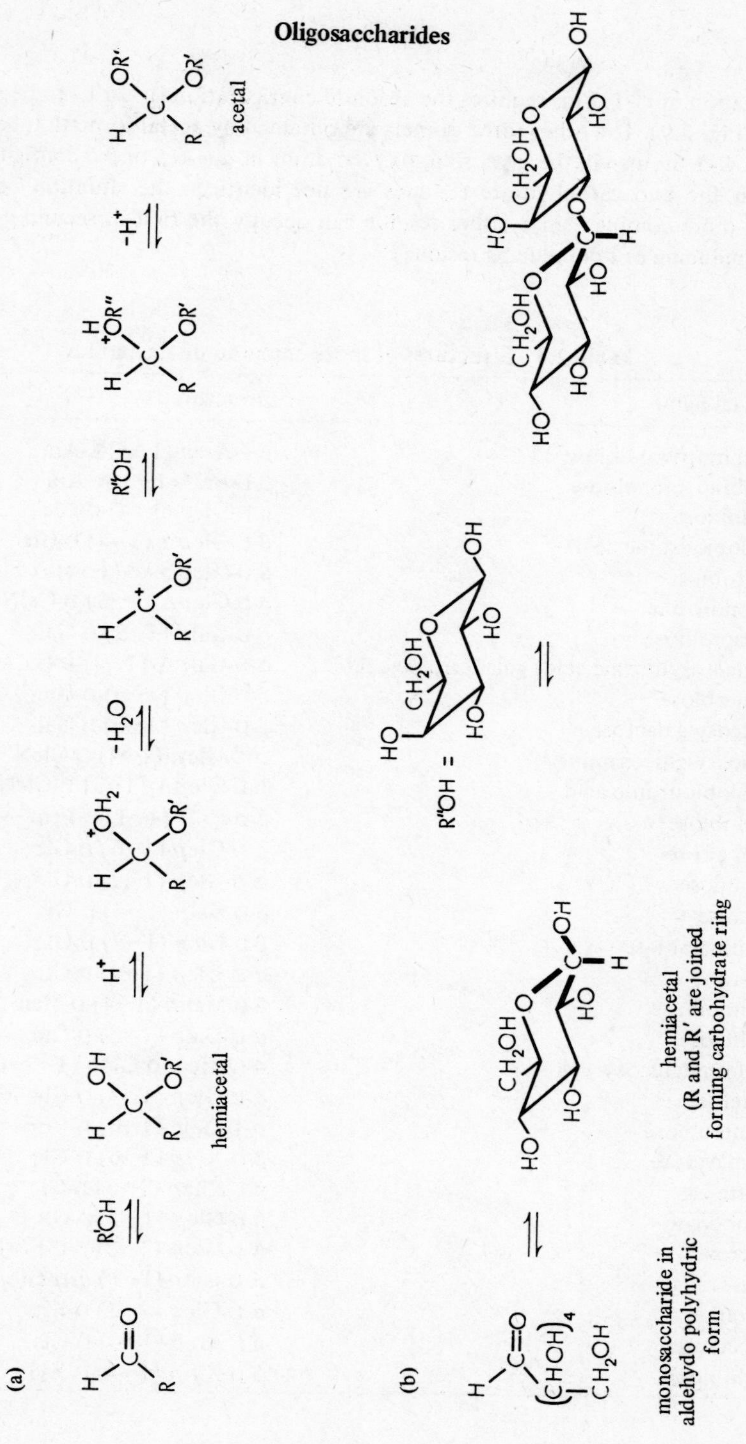

(a)

$R'OH$ ⇌ hemiacetal ⇌ H^+ ⇌ $-H_2O$ ⇌ $R''OH$ ⇌ $-H^+$ ⇌ acetal

hemiacetal

acetal

(b)

$R''OH$ =

monosaccharide in aldehydo polyhydric form

hemiacetal (R and R′ are joined forming carbohydrate ring

Scheme 2.2

figuration at C-1. This requires the absolute configuration (D- or L- to be stated) (see Fig. 2.9). The other three isomers are obtained by acetal formation between both C-1 atoms via the glycosidic oxygen atom in α,α, α,β or β,β configuration. When the two carbohydrate residues are not identical, the situation becomes even more complex since either residue can occupy the first or second position (i.e. reducing or nonreducing residue).

Table 2.3 Structures of more-common disaccharides

Trivial name	Structure
Arabinopyranobiose	β-L-Ara*p*-(1→3)-L-Ara
Arabinofuranobiose	β-L-Ara*f*-(1→3)-L-Ara
Cellobiose	β-D-Glc*p*-(1→4)-D-Glc
Cellobiouronic acid	β-D-Glc*p*A-(1→4)-D-Glc
Chitobiose	β-D-Glc*p*NAc-(1→4)-D-GlcNAc
Chondrosine	β-D-Glc*p*A-(1→3)-D-GalN
Galactobiose	β-D-Gal*p*-(1→3)-D-Gal
(Galactosyluronic acid) galacturonic acid	α-D-Gal*p*A-(1→4)-D-GalA
Gentobiose	β-D-Glc*p*-(1→6)-D-Glc
Glucosylgalactose	β-D-Glc*p*-(1→6)-D-Gal
Glucosylglucosamine	α-D-Glc*p*-(1→4)-D-GlcN
Hyalobiouronic acid	β-D-Glc*p*A-(1→3)-D-GlcN
Inulobiose	β-D-Fru*f*-(1→1)-D-Fru
Isomaltose	α-D-Glc*p*-(1→6)-D-Glc
Kojibiose	α-D-Glc*p*-(1→2)-D-Glc
Lactose	β-D-Gal*p*-(1→4)-D-Glc
Laminarabiose	β-D-Glc*p*-(1→3)-D-Glc
Maltose	α-D-Glc*p*-(1→4)-D-Glc
Mannobiose	β-D-Man*p*-(1→4)-D-Man
Melibiose	α-D-Gal*p*-(1→6)-D-Glc
4-Methylglucosylxylose	4-*O*-Me-α-D-Glc*p*-(1→2)-D-Xyl
Nigerose	α-D-Glc*p*-(1→3)-D-Glc
Planteobiose	α-D-Gal*p*-(1→6)-D-Fru
Primaverose	β-D-Xyl*p*-(1→6)-D-Glc
Rutinose	β-L-Rha*p*-(1→6)-D-Glc
Sophorose	β-D-Glc*p*-(1→2)-D-Glc
Sucrose	α-D-Glc*p*-(1↔2)-β-D-Fru*f*
α,α-Trehalose	α-D-Glc*p*-(1↔1)-α-D-Glc*p*
Turanose	α-D-Glc*p*-(1→3)-D-Fru
Vicianose	β-L-Ara*p*-(1→6)-D-Glc
Xylobiose	β-D-Xyl*p*-(1→4)-D-Xyl

Table 2.4 Structures of more-common higher oligosaccharides

Trivial name	Structure
Trisaccharides	
Cellotriose	β-D-Glc*p*-(1→4)-β-D-Glc*p*-(1→4)-D-Glc
Gentianose	β-D-Glc*p*-(1→6)-α-D-Glc*p*-(1↔2)-β-D-Fru*f*
6-*O*-Glucosylmaltose	α-D-Glc*p* 1 ↓ 6 α-D-Glc*p*-(1→4)-D-Glc
Isokestose	β-D-Fru*f*-(2→1)-β-D-Fru*f*-(2↔1)-α-D-Glc*p*
Isomaltotriose	α-D-Glc*p*-(1→6)-α-D-Glc*p*-(1→6)-D-Glc
Isopanose	α-D-Glc*p*-(1→4)-α-D-Glc*p*-(1→6)-D-Glc
Kestose	β-D-Fru*f*-(2→6)-β-D-Fru*f*-(2↔1)-α-D-Glc*p*
Laminaratriose	β-D-Glc*p*-(1→3)-β-D-Glc*p*-(1→3)-D-Glc
Maltotriose	α-D-Glc*p*-(1→4)-α-D-Glc*p*-(1→4)-D-Glc
Melezitose	α-D-Glc*p*-(1→3)-β-D-Fru*f*-(2↔1)-α-D-Glc*p*
Neokestose	β-D-Fru*f*-(2→6)-α-D-Glc*p*-(1↔2)-β-D-Fru*f*
Neuraminolactose	NeuNAc-(2→3)-β-D-Gal*p*-(1→4)-D-Glc
Panose	α-D-Glc*p*-(1→6)-α-D-Glc*p*-(1→4)-D-Glc
Planteose	α-D-Gal*p*-(1→6)-β-D-Fru*f*-(2→1)-α-D-Glc
Raffinose	α-D-Gal*p*-(1→6)-α-D-Glc*p*-(1↔2)-β-D-Fru*f*
Umbelliferose	α-D-Gal*p*-(1→2)-α-D-Glc*p*-(1↔2)-β-D-Fru*f*
Tetrasaccharides	
Cellotetraose	β-D-Glc*p*-(1→4)-β-D-Glc*p*-(1→4)-β-D-Glc*p*- (1→4)-D-Glc
Maltotetraose	α-D-Glc*p*-(1→4)-α-D-Glc*p*-(1→4)-α-D-Glc*p*- (1→4)-D-Glc
Stachyose	α-D-Gal*p*-(1→6)-α-D-Gal*p*-(1→6)-α-D-Glc*p*- (1↔2)-β-D-Fru*f*
Pentasaccharides	
Verbascose	α-D-Gal*p*-(1→6)-α-D-Gal*p*-(1→6)-α-D-Gal*p*- (1→6)-α-D-Glc*p*-(1↔2)-β-D-Fru*f*
Cyclic oligosaccharides*	
Cyclomaltohexaose	→4)-[α-D-Glc*p*]$_6$-(1—
Cyclomaltoheptaose	→4)-[α-D-Glc*p*]$_7$-(1—
Cyclomalto-octaose	→4)-[α-D-Glc*p*]$_8$-(1—

* Previously referred to as Schardinger dextrins, cyclodextrins, cycloamyloses.

The major classification of oligosaccharides is according to the number of monosaccharide units, into di-, tri- and tetra-saccharides etc. Each classification is further subdivided into homo- or hetero-oligosaccharides (homo-oligosaccharides consist of only one type of monosaccharide); and reducing or nonreducing oligosaccharides (depending on the presence or absence of a free hemiacetal group). For example, sucrose is a nonreducing heterodisaccharide, α,α-trehalose a nonreducing homodisaccharide, and maltotriose, a reducing homotrisaccharide (see Tables 2.3 and 2.4).

Nomenclature of oligosaccharides is undergoing a process of systematisation (in common with all carbohydrate nomenclature). The original Tentative Rules (1969) stated that reducing oligosaccharides are named as glycosylglycoses, glycosylglycosylglycoses, etc., and nonreducing oligosaccharides as glycosylglycosides, etc. Further information on anomeric configuration, configuration of monosaccharides residues and positions of linkages and substitution are included. Hence, maltose is 4-*O*-α-D-glucopyranosyl-D-glucose (8), the '4' indicating that the linkage is between C–1 of one residue and C–4 of the other, the '*O*' indicating that substitution is on the oxygen atom at C-4, the 'α' indicating the anomeric configuration of the glycosidic linkages and the 'D' indicating that both monosaccharides have stereochemistry related to D-glyceraldehyde (see Fig. 2.9). Tables 2.3 and 2.4 show the structures of the more-common oligosaccharides, some of which are naturally occurring and some of which occur only as hydrolysis products of polysaccharides. Where available, trivial names have been included since these are often used for brevity, if no ambiguity is possible. Such names as agarotetraose and nigerotriose should *not* be used to describe oligomers containing two or more different monosaccharides or types of linkage, as there is no indication as to the order of monosaccharide or linkages.

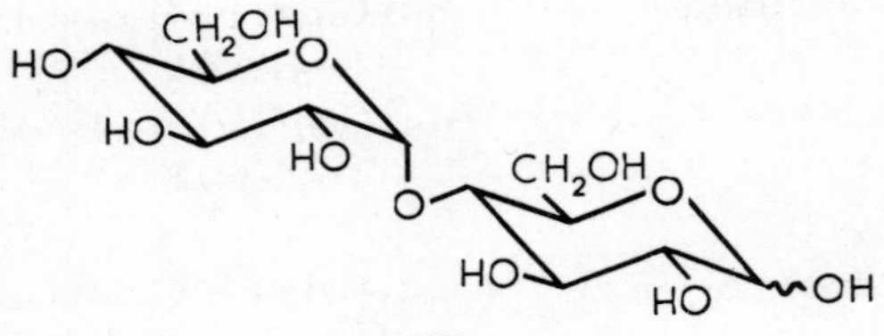

(8)

Structures of oligosaccharides may be written in abbreviated form (as in Tables 2.3 and 2.4) by using a system of symbols (Recommendations 1980b) containing three letters (usually the first three letters of the trivial name) to represent both name and structure of component monosaccharides (see Table 2.5). 2-Amino-2-deoxy sugars are represented by the symbol for the parent sugar followed by the letter N, and uronic acids by A. The symbols for less common sugars can be derived from trivial names, but it is usual to give the systematic name as well. Examples include:

3,6-Dideoxy-D-*xylo*-hexose (abequose = Abe)
6-Deoxy-D-glucose (quinovose = Qui)
3-*C*-(Hydroxymethyl)-D-*glycero*-aldotetrose (D-apiose = Api)

Configuration, ring size and anomeric configuration are added by using the usual symbols (as in Tables 2.3 and 2.4). Where a branch point occurs, the branch to the main chain is added by use of a vertical arrow to the central letter of the 3-letter symbol for the parent sugar, with appropriate numbers used to denote the position of the linkage (9).

Table 2.5 Symbols used for monosaccharides

Allose	=	All
Altrose	=	Alt
Arabinose	=	Ara
Fructose	=	Fru
Fucose	=	Fuc
Galactose	=	Gal
Glucose	=	Glc
Gulose	=	Gul
Idose	=	Ido
Lyxose	=	Lyx
Mannose	=	Man
Muramic acid	=	Mur
Neuraminic acid	=	Neu
Rhamnose	=	Rha
Ribose	=	Rib
Talose	=	Tal
Xylose	=	Xyl

α-D-Gal*p*-(1→4)-β-D-Glc*p*NAc-(1→4)-D-Glc*p*-
6
↑
1
β-D-Fru*f*

(9)

The linkage between the two residues, A and B, in the trisaccharide shown in Fig. 2.10 can be described as a (1→6)-linkage, which means that a glycosidic bond is formed between C-1 of residue A and C-6 of residue B. Since residue B is also linked to another residue C, by, in this case a (1→3)-linkage, residue B can be described as a 1,6-disubstituted residue.

The standard nomenclature for describing the various monosaccharide units in oligosaccharides (and polysaccharides, etc.) is based on the distinction between terminal locations of such units. In general, the appropriate terms are: 'glycosyl group', 'glycosyl residue' and 'glycose residue'. Thus, in the trisaccharide shown in Fig. 2.10, unit A is a D-glucopyranosyl group, unit B is a D-glucopyranosyl residue, and unit C is a D-galactopyranose residue.

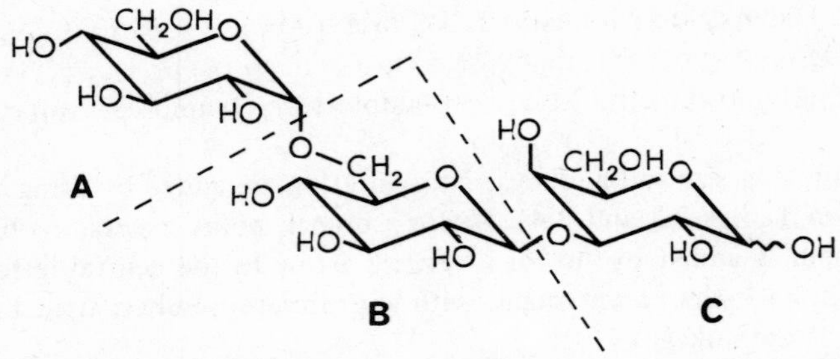

Fig. 2.10 – The trisaccharide, α-D-Glc*p*-(1→6)-β-D-Glc*p*-(1→3)-D-Gal*p*, used as an example for the definition of oligosaccharide, etc. nomenclature (see text for details).

Oligosaccharides with branched structures are described as *O*-substituted oligosaccharides using a system in which a locant is used showing the position of the hydroxyl group involved in the branch point. If the branch point occurs in the glycose residue the locant alone is used whereas if the branch point occurs in a glycosyl residue or group the locant carries a superscript denoting the position of the glycosyl residue or group in the oligosaccharide, counting from the glycose residue. Thus 6-*O*-α-D-glucosylmaltose and panose (see Table 2.4) are both D-glucosyl derivatives of maltose with the substitution being described as 6-*O*-α-D-glucopyranosylmaltose and 6^2-*O*-α-D-glucopyranosylmaltose respectively. Although the numbering of the glycosyl residues has the inconvenience of reading from right to left in the conventional representation of oligosaccharides, it has the advantage that shortening or lengthening of chains by the usual mechanism of transglycosylation at the nonreducing end (see Chapter 5) leaves the numbering of residues in the chain unaltered. Alternative nomenclature would be α-D-glucopyranosyl-(1 → 4)-*O*-[α-D-glucopyranosyl-(1 → 6)-*O*-] -D-glucose and α-D-glucopyranosyl-(1 → 6)-*O*-α-D-glucopyranosyl-(1 → 4)-*O*-D-glucose.

Further examples of oligosaccharides will be discussed subsequently in the chapters on oligosaccharides (Chapter 7) and antibiotics (Chapter 12).

POLYSACCHARIDES

Polysaccharides are natural macromolecules occurring in almost all living organisms, constituting one of the largest groups of natural compounds classified thus far, and function either as an energy source or as structural units in the morphology of the living material in which they are endogenous. Examples of polysaccharides possessing structural functions are: (a) cellulose, a polymer of D-glucose, that is probably the most abundant naturally occurring organic substance and is the structural material of plants, and (b) chitin, a polymer of 2-acetamido-2-deoxy-D-glucose, which is the major organic component of the exoskeleton (shells) of insects, crabs, lobsters etc.

As one of the main sources of energy for living organisms, certain polysaccharides form part of the central pathway of energy in most cells. The starches and glycogens, long-chain polymers of D-glucose, are the media for energy storage in plants and animals respectively.

Polysaccharides also perform more-specific roles such as being responsible for the type specificity of the pneumococcal polysaccharides. Other natural macromolecules, which are not composed entirely of sugar units, contain blocks of monosaccharide units as part of the molecular structure, and contribute extensively to the production and maintenance of living tissues of animals. The blood-group substances, for example, constitute a group of glycoproteins in which the arrangements of monosaccharide residues in the carbohydrate sub units contribute towards the blood-group specificity of the overall molecule.

Trivial names of the polysaccharides usually reflect their origin; examples include cellulose, the principal component of cell walls in plants, and dermatan, a polysaccharide normally occurring in its sulphated form and originally found in the dermal layer of skin (Greek, *derma*). The trivial names can also reflect some property of the isolated polymer, for example, starch, a name derived from the old English word *stercan*, meaning to stiffen.

The species origin of a polysaccharide leads to differences within a polysaccharide type. Thus, since, for example, starches from various plant sources are readily distinguished chemically, it is necessary to specify the origin in definitively naming the starch, for example, maize starch. The traditional names of long standing, such as cellulose, glycogen and amylose are still inevitably retained, but, with the increase in knowledge of the structure of these compounds, nomenclature and classification are now being made in terms of structure and, in the interests of systematisation, all new discoveries should be named systematically (Recommendations, 1980c). The term *glycan* derived from glycose, meaning a simple monosaccharide, is another word for polysaccharide, but a more specific term is obtained by using the configurational prefix of the parent sugar with the suffix 'an' to signify a polymer, for example mannan, for a polymer based on the monosaccharide mannose. Further specificity is achieved by inclusion of the D or L configuration, as appropriate, for example, D-glucan from D-glucose. Such nomenclature, and any classification derived therefrom, should ideally include information on chemical structure. Polysaccharides which, on hydrolysis, yield only one type of monosaccharide are called homoglycans, whereas those which can be hydrolysed to more than one type of monosaccharide are called heteroglycans, with designatory prefixes of di-, tri-, etc., for the number of monosaccharide types involved.

There is, at present, no proof of the existence of polysaccharides which contain more than about six different types of monosaccharide unit. The most common constituents are the pentose and hexose monosaccharides and monosaccharides derived from them, for example hexuronic acids, 6-deoxyhexoses, 2-amino-2-deoxyhexoses (hexosamines) and simple derivatives (including sulphates, acetates and methyl ethers), and these usually form a regular repeating-unit throughout the polymer. This repeating-unit is usually used in any chemical representation of the polysaccharide, and its structure can be written by using the same system of abbreviations and symbols used for oligosaccharides (Recommendations, 1980b). Tables 2.6 and 2.7 show the structures and sources of the common homoglycans and of some heteroglycans. These and other carbohydrate-containing compounds will be discussed more thoroughly in subsequent chapters.

Table 2.6 Structure and source of the common homopolysaccharides

Linkage	Source	Common Name
L-Arabinans		
(1→3)-α-L-, (1→5)-α-L- branched	Plant pectic substances	
D-Fructans		
(2→1)-β-D- linear	Dandelions, dahlias, Jerusalem artichokes	Inulin
(2→6)-β-D- linear	Various grasses	Levans
(2→6)-β-D-, (2→1)-β-D- branched	Plants and bacteria	Levans
L-Fucans		
(1→2)-α-L-, (1→4)-α-L- branched	Brown seaweed	Fucoidan
D-Galactans		
(1→3)-β-D-, (1→4)-α-D- linear	Red seaweeds	Carrageenan
(1→3)-β-D-, (1→6)-β-D- branched	Beef lung	
(1→4)-β-D- linear	Plant pectic substances	
(1→5)-β-D- linear	Penicillin mould	Galactocarolose
D-Galacturonans		
(1→4)-α-D- linear	Plant pectic substances	Pectic acid
D-Glucans		
(1→2)-β-D- linear	Agrobacteria	
(1→3)-α-D-, (1→4)-α-D- linear	*Aspergillus niger*, Iceland moss	Nigeran, elsinan isolichenan
(1→3)-β-D- linear	Brown seaweeds, plants, plants, algae, fungi and yeasts	Laminaran, callose, curdlan, pachyman
(1→3)-β-D-, (1→6)-β-D- branched	Fungi	Scleroglucan
(1→3)-β-D-, (1→4)-β-D- linear	Iceland moss, cereal grains	Lichenan
(1→4)-α-D- linear	Plants	Amylose
(1→4)-α-D-, (1→6)-α-D- linear	Fungi	Pullulan
(1→4)-α-D-, (1→6)-α-D- branched	Animals, plants and micro-organisms	Glycogen, amylopectin

Table 2.6 (*continued*)

Linkage	Source	Common name
(1→4)-β-D- linear	Plant cell-walls	Cellulose
(1→4)-β-D-, (1→3)-α-D- branched	Bacteria	Dextran
(1→6)-β-D- linear	Lichens	Pustulan
2-Amino-2-deoxy-D-glucans		
(1→4)-β-D- linear	Crab and lobster shells, fungi	Chitin
D-Mannans		
(1→2)-α-D-, (1→6)-α-D- branched	Yeasts	
(1→4)-β-D- linear	Seaweeds, plants	
D-Xylans		
(1→3)-β-D- linear	Green seaweed	Rhodymenan
(1→3)-β-D-, (1→4)-β-D- linear	Red seaweed	
(1→4)-β-D- linear	Plant cell-walls	

Table 2.7 Structure and source of some heteropolysaccharides

Constituent monosaccharides and chain type	Source	Common name
L-Arabinose, D-galactose, branched	Coniferous woods	
L-Arabinose, D-xylose, branched	Plant cell-walls	
DL-Galactose, linear	Red seaweeds	Agarose, porphyran
branched	Snails	
D-Galactose, 2-amino-2-deoxy-D-glucose, linear	Cornea	Keratan sulphate
D-Galactose, D-mannose, branched	Leguminous seeds, fungi	
2-Amino-2-deoxy-D-galactose, D-glucuronic acid, linear	Cornea, cartilage	Chondroitin, chondroitin sulphates
2-Amino-2-deoxy-D-galactose, L-iduronic acid, linear	Skin	Dermatan sulphate
D-Glucose, D-mannose, linear	Coniferous woods, seeds, bulbs	
2-Amino-2-deoxy-D-glucose, D-glucuronic acid, linear	Animal and mammalian tissues	Hyaluronic acid
D-Glucuronic acid, D-xylose, branched	Plant cell-walls	
L-Guluronic acid, D-mannuronic acid, linear	Bacteria, brown seaweeds	Alginic acid

CHAPTER 3

General chemistry

GENERAL STRUCTURES OF MACROMOLECULAR CARBOHYDRATES

All macromolecular carbohydrate materials have definite, three-dimensional structures, and the four aspects of these structures (primary, secondary, tertiary and quaternary structure) have essentially the same meaning for carbohydrates as for proteins and other biopolymers but the definition of each aspect is not as clear-cut as in, for example, protein chemistry. For this reason, carbohydrate structures are compared and contrasted, aspect for aspect, with protein structures.

Primary structure

For proteins this is the sequence of aminoacid residues within the linear chain. For polysaccharides, etc. the same definition applies for linear chains, but, as discussed previously (Chapters 1 and 2), the manner in which the various monosaccharide residues are joined is much more complex than for the corresponding combination of aminoacid residues due to the greater number of substitution positions possible in a monosaccharide molecule. A further complexity is that branched structures, occurring *via* disubstitution of a single residue, commonly exist for polysaccharides. Thus, in order to define the primary structure of a polysaccharide fully, it is essential to state the identity of all monosaccharide residues, the sequence of these residues, their position and anomeric configuration and the position of any other substituents. Polysaccharide primary structures frequently show simple repeating sequences and these repeating sequences have been determined for many polysaccharides and other carbohydrate-containing macromolecules; these will be discussed in subsequent chapters.

Secondary structure

The secondary structure of proteins can be described as the spatial relationship of near neighbours which is a result of interactions such as hydrogen bonding between C=O and N—H groups, producing some degree of double-bond character and, hence, local regularities. The geometry of the individual sugar rings in a polysaccharide is essentially rigid but the relative orientations of component residues (i.e. rotation) about the glycosidic linkage determines the overall conformation of the polysaccharide. Two torsion (rotational) angles are required in order to define the glycosidic bond between two carbohydrate residues, A and B, except for (1→6)-linked polysaccharides which require three angles (see Fig. 3.1). The angle ϕ is about the bond from the anomeric carbon atom to the oxygen atom that joins the two residues, the angle ψ is about the bond from the glycosylated oxygen atom of residue A to the carbon atom of residue B, and the angle ω is about the exocyclic carbon–carbon bond (see Recommendations, 1981a). The range of values obtained for the rotational angles ϕ, ψ and ω is severely restricted by steric hindrance between the adjacent rings and by non-bonding interactions between groups in adjacent residues. The restrictions are greatest for glycosidic linkages involving axial groups (as may be seen for such α-D-glucans as amylose, Fig. 3.1), and for residues containing such bulky substituents as *N*-acetyl in equatorial positions adjacent to the glycosidic linkage. This has the result that chains are relatively stiff with, for example, coil shapes predominating in solution. The overall secondary structure of polysaccharides is therefore heavily dependent on the primary structure.

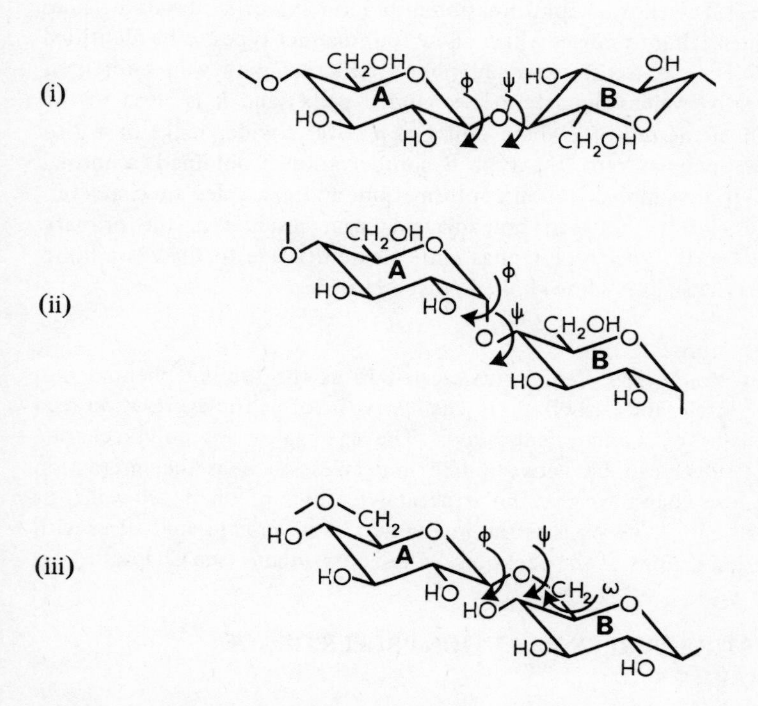

Fig. 3.1 – Rotation of individual residues about the glycosidic linkage in (i) cellulose, (ii) amylose, and (iii) dextran.

Tertiary structure

In protein structure this is the gross folding of the chain which brings together groups that are normally separated by large distances along the protein backbone, allowing such interactions as the formation of disulphide linkages, ion-pairs and hydrophobic attractions. The repeating sequences in primary structures of polysaccharides lead to regular patterns in secondary structures which lead to sterically regular gross conformations aided by favourable non-covalent interactions between hydroxyl, sulphate, amino, phosphate, carboxyl, etc. groups. Irregularities in primary and secondary structures and large branched structures inhibit tertiary structure formation whilst such external perturbations as changes in temperature and ionic concentrations can cause changes in the tertiary structure. For example, charged polysaccharides can form stable, tertiary structures by incorporation of counter-ions within the tertiary structure. A useful concept which has been used to describe the overall chain-conformation is to regard any conformation as a helix (even though it may not look like a conventional helix) and specify two parameters, namely, the number (n) of monomer residues per helix turn and this projected length (h) of each monomer residue on the helix axis. These parameters can be calculated from values of ϕ, ψ and ω (Ramachandran *et al.*, 1963). The allowed conformation of homopolysaccharides have values of n and h which fall into ranges which allow four distinct types to be identified (see Fig. 3.2). Type A is the extended ribbon structure with values for n of 2 to $\pm$ 4 (negative values indicate a left handed helix) and h is close to the absolute length of the residue. Where values of n cover a wider range (n = 2 to $\pm$ 10) and h approaches zero, the type B conformation is obtained (a normal helix). Type C is a crumpled ribbon conformation, and examples are characterized by many clusters between non-adjacent sugar residues in the primary structure. The fourth type, type D, has more flexibility due to the extra bond which separates the rings in (1→6)-linked polysaccharides.

Quaternary structure

This aspect of structure is frequently referred to as the subunit phenomenon and, in both protein and carbohydrate chemistry, involves the aggregation of a number of chains by noncovalent bonds. The aggregation of polysaccharide chains (Rees, 1977) can be between like molecules, such as the interaction between cellulose chains to give the structural features of plant cell-walls, or between unlike molecules, such as the interaction between xanthan helices with the unsubstituted regions of the backbone of galactomannans (see Chapter 8).

EFFECT OF STRUCTURE ON SOLUTION PROPERTIES OF POLYSACCHARIDES

Many of the simpler homopolysaccharides, for example, cellulose and starch, are insoluble in their natural, ordered states. This is because the adoption of an

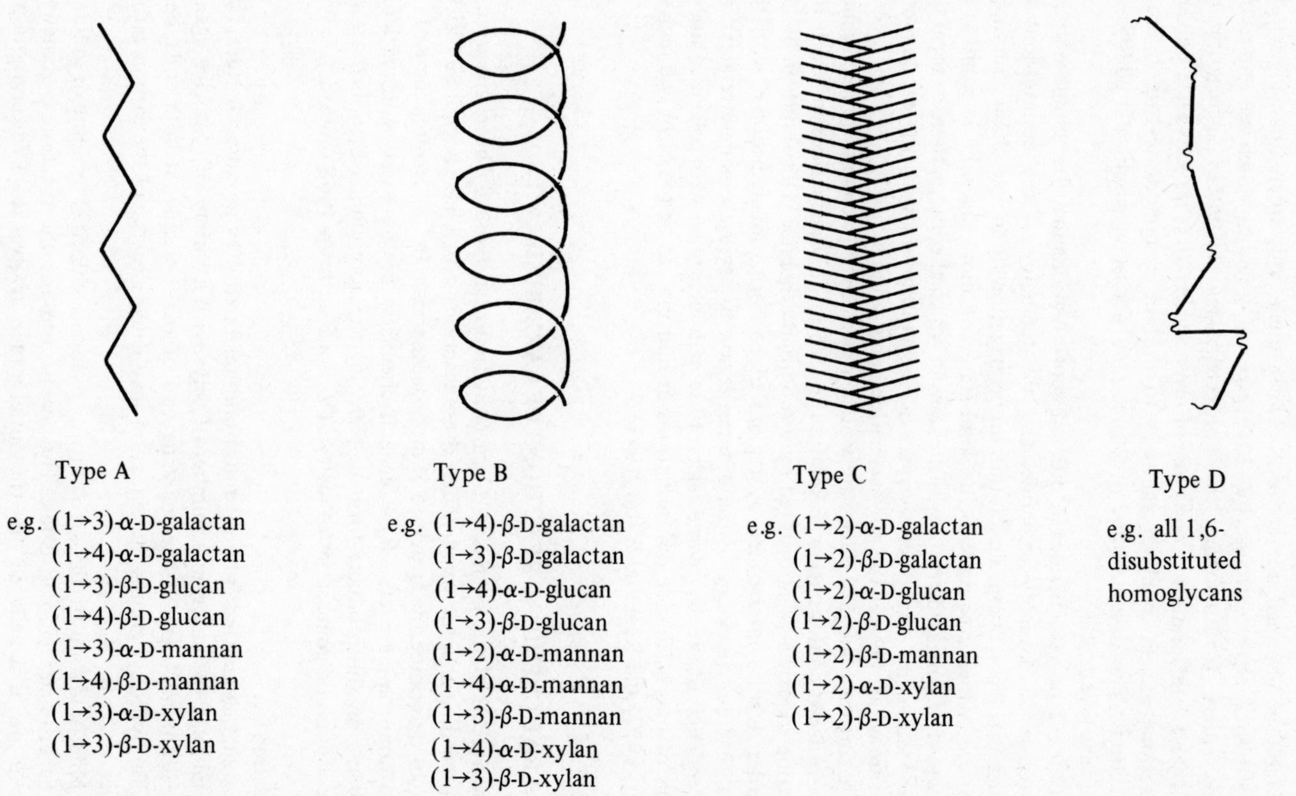

Fig. 3.2 – Tertiary structures found in homopolysaccharides. Type A, extended ribbon; Type B, flexible helix; Type C, crumpled ribbon, and Type D, flexible coil.

ordered conformation usually facilitates the ordered packing of chains with favourably non-bonded interactions between them. This ordered packing results in a stiffened chain with the loss of degrees of freedom which are important driving forces for dissolving the polysaccharides. Therefore aggregation of chains and precipitation are favoured even when the carbohydrate–solvent interactions are of similar magnitude to that of carbohydrate–carbohydrate interactions. The ordered-state structures may persist in solution if additional factors are favourable.

Charged polysaccharides in ordered conformations may be soluble because the charge on the polysaccharide, and the tendency of the counter-ions to disperse in solvent, favour dissolution not aggregation. Branched-chain structures, which are awkward to pack in the solid state, are more favoured in solution or in a weakly aggregated form, whilst chains in which the interactions to form the ordered state are weak will always exist in equilibrium with the random-coil structure which favours the polysaccharide being soluble.

A further biological device by which polysaccharide chains are maintained in an ordered conformation in solution is seen in polysaccharides in which the repeating sequence is interrupted by a modified sequence. This results in blocks of ordered state interspaced by regions of conformational disorder, with the result that any tendency of the ordered region to aggregate is counteracted by the tendency of the disordered region to remain in contact with solvent. Further details of polysaccharide conformations in solution can be found in two reviews (Rees, 1977, and Rees and Walsh, 1977).

GENERAL CHEMICAL REACTIONS OF CARBOHYDRATES

Some aspects of the chemistry of carbohydrates have already been discussed in the preceding chapter, in particular those reactions which have been used historically to determine the structure of monosaccharides. In this chapter some of the more important aspects of the general chemistry are discussed with particular emphasis on the products formed. Many of the compounds described are the subject of an on-going series of reviews (Various Authors, 1968 onwards).

Oxidation

As would be expected from the multifunctional nature of monosaccharides, they can undergo oxidation in a number of ways with a variety of oxidizing agents. Oxidation can occur at the reducing end of the molecule, at both ends, or at specific hydroxyl groups, or oxidation may cause cleavage of the carbon-carbon bonds.

Mild oxidation of aldoses with, for example, chlorine, bromine or iodine at pH 5, produces aldonic acids. The mechanism for the reaction is somewhat obscure but it is believed that the initial attack involves the hydroxyl group at

C-1 since β-D-glucopyranose is oxidized 250 times faster than the α-anomer. The aldose is oxidized directly to its corresponding lactone which is then slowly hydrolysed to the aldonic acid (Scheme 3.1). The free aldonic acid is difficult to isolate, due to its marked tendency to be dehydrated and revert to the lactone, but salts and such derivatives as amides and phenylhydrazones can be isolated. Alkaline oxidation with iodine (hypoiodite oxidation) similarly produces aldonic acids, and this reaction can be used as the basis for the quantitative determination of aldoses and for the determination of the reducing (terminal) residue of oligosaccharides (Ko and Somers, 1974). Ketoses are unaffected by mild oxidation with halogens.

β-D-Glucopyranose —(1)→ D-Glucono-1,4-lactone ⇌ D-Gluconic acid

Reagents: (1) halogens at pH 5.

Scheme 3.1

Selective oxidation of the nonreducing end of the monosaccharide (that is, C-6 of an aldohexose) with oxygen, using platinum as a catalyst, or with potassium permanganate, produces an alduronic acid. Both methods of preparation require the use of protecting groups to prevent oxidation at other positions in the molecule. Fig. 2.8 shows the structures of the naturally occurring alduronic acids. D-Glucuronic acid plays an important role in animal metabolism by aiding excretion of phenols, steroids and aromatic carboxylic acids by formation of D-glucosiduronic acids; it also occurs in heparin. D-Galacturonic acid occurs in fruit pectin; D-mannuronic acid and L-guluronic acid occur in various seaweed polysaccharides; whilst L-iduronic acid is a component of two of the glycosaminoglycans, dermatan sulphate and heparin (see Chapter 9). They are not easily isolated from their natural sources due to the drastic conditions usually required to break the glycosidic linkages. These conditions frequently result in decarboxylation and elimination reactions.

The normal form in which the alduronic acids exist is as a lactone; thus D-glucuronic acid exists as D-glucofuranurono-6,3-lactone (10), and D-galacturonic acid as D-galactopyranurono-6,3-lactone (11), because the configuration of the carboxyl and hydroxyl groups is such that a furanose ring would require a trans junction between the two rings, whereas the pyranose ring gives an unstrained structure.

Oxidation of both ends of the monosaccharide molecule leads to the formation of aldaric acids. They may be prepared by the oxidation of aldoses with nitric acid. Under the same conditions, ketoses form aldoses with one less carbon atom due to cleavage of the 1,2-bond, whereas 6-deoxyaldoses afford aldoses with one less carbon atom due to cleavage of the 5,6 bond. D-Arabinaric acid can, therefore, be formed from D-fructose, D-arabinose, D-lyxose or 6-deoxy-L-galactose (L-fucose) as shown in Scheme 3.2. D-Glucaric acid is isolated as the dilactone (12). (*meso*)-Galactaric acid (mucic acid) has very low solubility in water and can therefore be used as the basis of a gravimetric method of analysis for galactose. On heating, *meso*-galactaric acid can form a monolactone but for stereochemical reasons cannot form a dilactone.

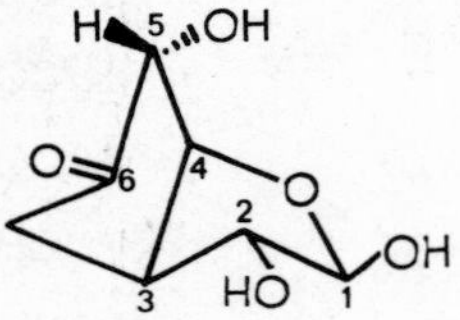

(10)

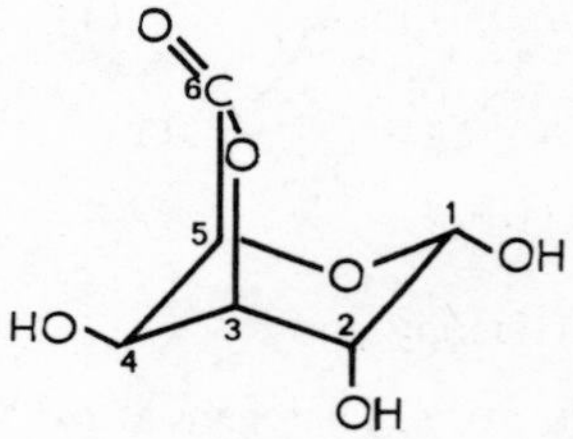

(11)

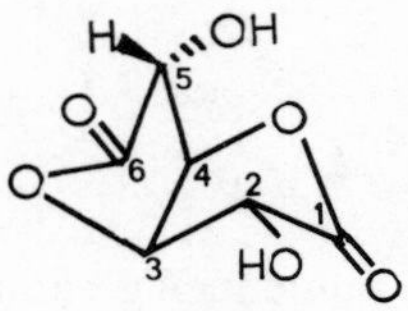

D-Glucaro-1,4;6,3-dilactone

(12)

Oxidation at the secondary hydroxyl groups is not usually accomplished directly from free sugars, but is possible using a suitably blocked derivative. Another complication is that the free aldosuloses are not usually crystalline due to the wide variety of cyclic structures which each can adopt. One possible method of preparation of aldos-2-uloses (osones) is by direct oxidation with copper(II) ions (Bayne and Fewster, 1956), aldos-3-uloses can be prepared in high yield (Angyal and James, 1970) from partially acetylated monosaccharides by using chromium(III) oxide in acetic acid (Scheme 3.3).

Cleavage of carbon-carbon bonds can be brought about by oxidation with such glycol-cleaving agents as lead(IV) acetate and periodic acid. The chemistry of these reactions is fully discussed in Chapter 4, as they are important in the structural analysis of carbohydrates.

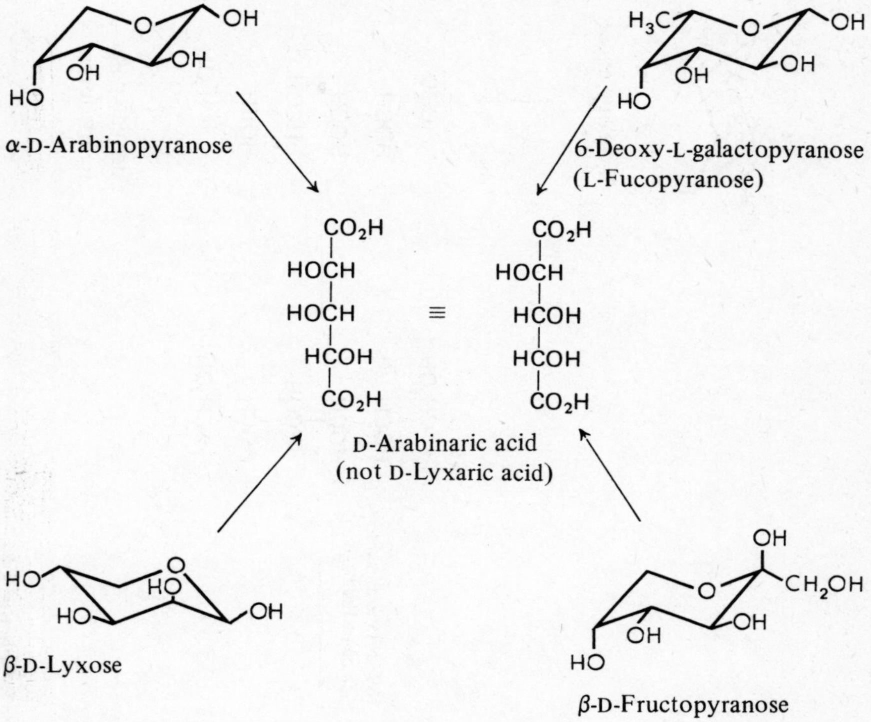

α-D-Arabinopyranose

6-Deoxy-L-galactopyranose
(L-Fucopyranose)

$$\begin{array}{c} CO_2H \\ | \\ HOCH \\ | \\ HOCH \\ | \\ HCOH \\ | \\ CO_2H \end{array} \equiv \begin{array}{c} CO_2H \\ | \\ HOCH \\ | \\ HCOH \\ | \\ HCOH \\ | \\ CO_2H \end{array}$$

D-Arabinaric acid
(not D-Lyxaric acid)

β-D-Lyxose

β-D-Fructopyranose

Scheme 3.2

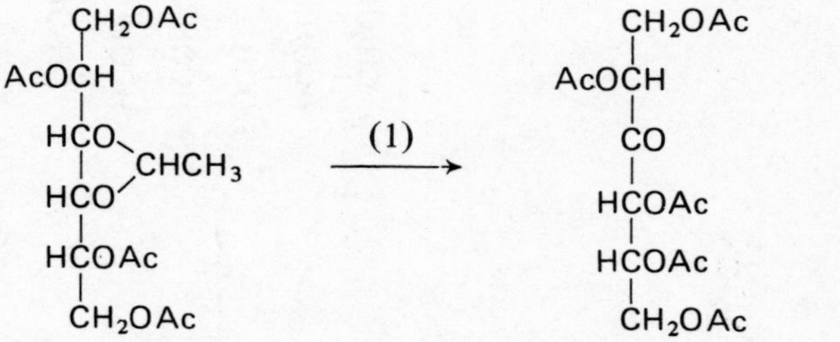

$$\begin{array}{c} CH_2OAc \\ | \\ AcOCH \\ | \\ HCO \\ | \quad\; CHCH_3 \\ HCO \\ | \\ HCOAc \\ | \\ CH_2OAc \end{array} \xrightarrow{(1)} \begin{array}{c} CH_2OAc \\ | \\ AcOCH \\ | \\ CO \\ | \\ HCOAc \\ | \\ HCOAc \\ | \\ CH_2OAc \end{array}$$

1,2,5,6-Tetra-*O*-acetyl-3,4-D-ethylidene-D-mannitol

1,2,4,5,6-Penta-*O*-acetyl-D-*arabino*-3-hexulose

Reagents: (1) Chromium(III) oxide in acetic acid

Scheme 3.3

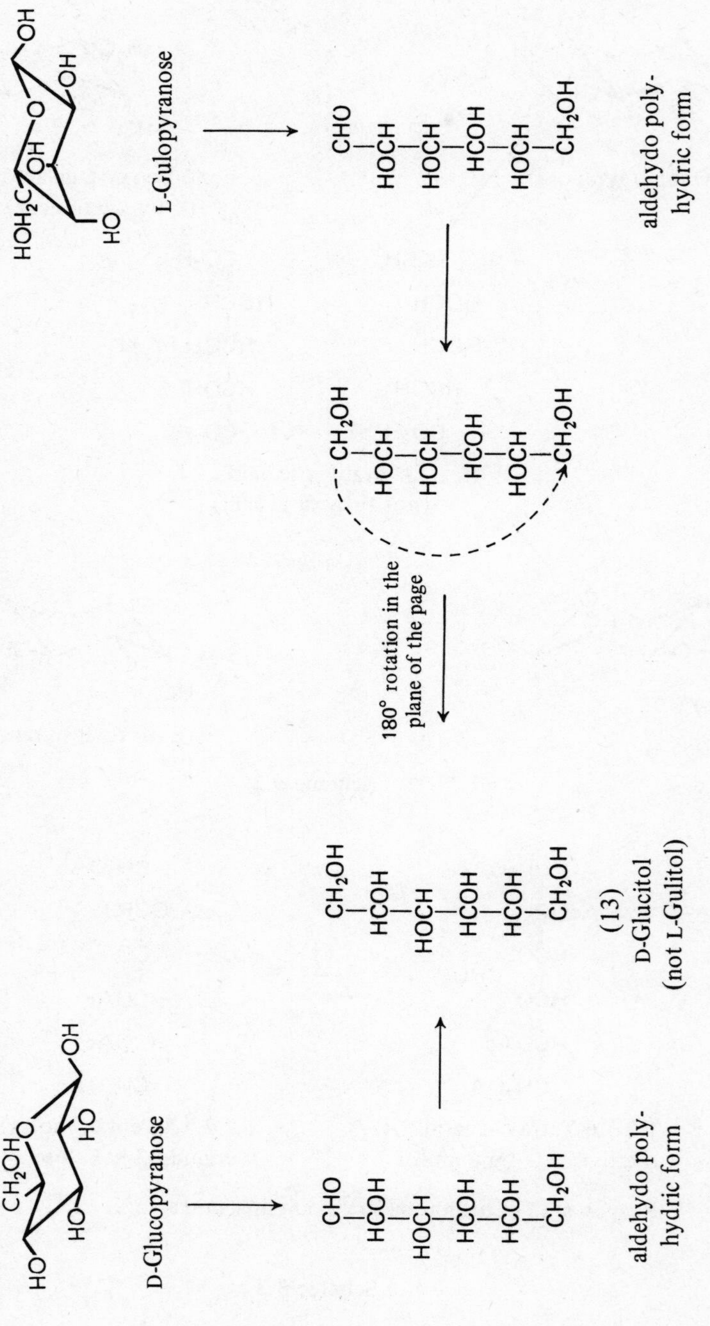

D-Glucopyranose → CHO–HCOH–HOCH–HCOH–HCOH–CH_2OH (aldehydo polyhydric form) → CH_2OH–HCOH–HOCH–HCOH–HCOH–CH_2OH

(13)
D-Glucitol
(not L-Gulitol)

L-Gulopyranose → CHO–HOCH–HOCH–HCOH–HOCH–CH_2OH (aldehydo polyhydric form) → CH_2OH–HOCH–HOCH–HCOH–HOCH–CH_2OH → (180° rotation in the plane of the page) → (13)

Scheme 3.4

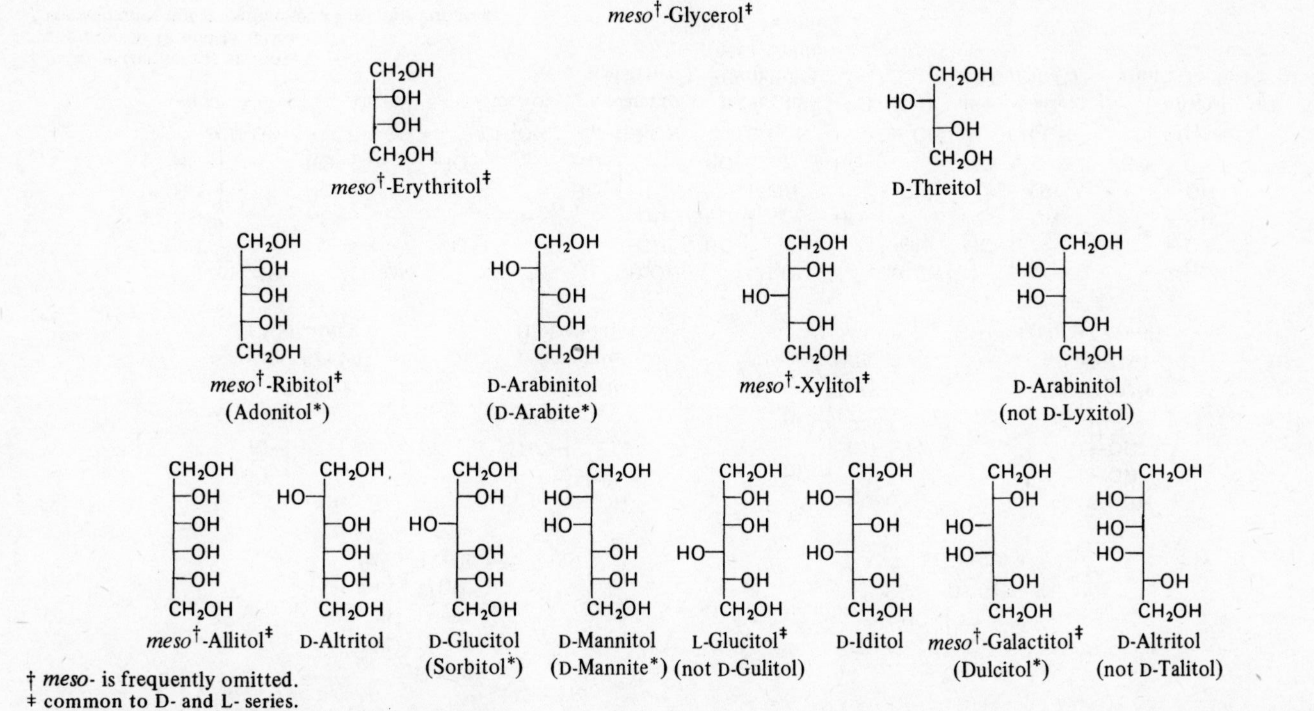

† *meso-* is frequently omitted.
‡ common to D- and L- series.
* trivial names not recommended for common usage.

Fig. 3.3a – Structures of the alditols, up to hexitols, of the D configuration with hydrogen atoms omitted for clarity (for the full structure of D-glucitol see Scheme 3.4). Rotations of 180° in the plane of the page gives the alternate structures, see Scheme 3.4.

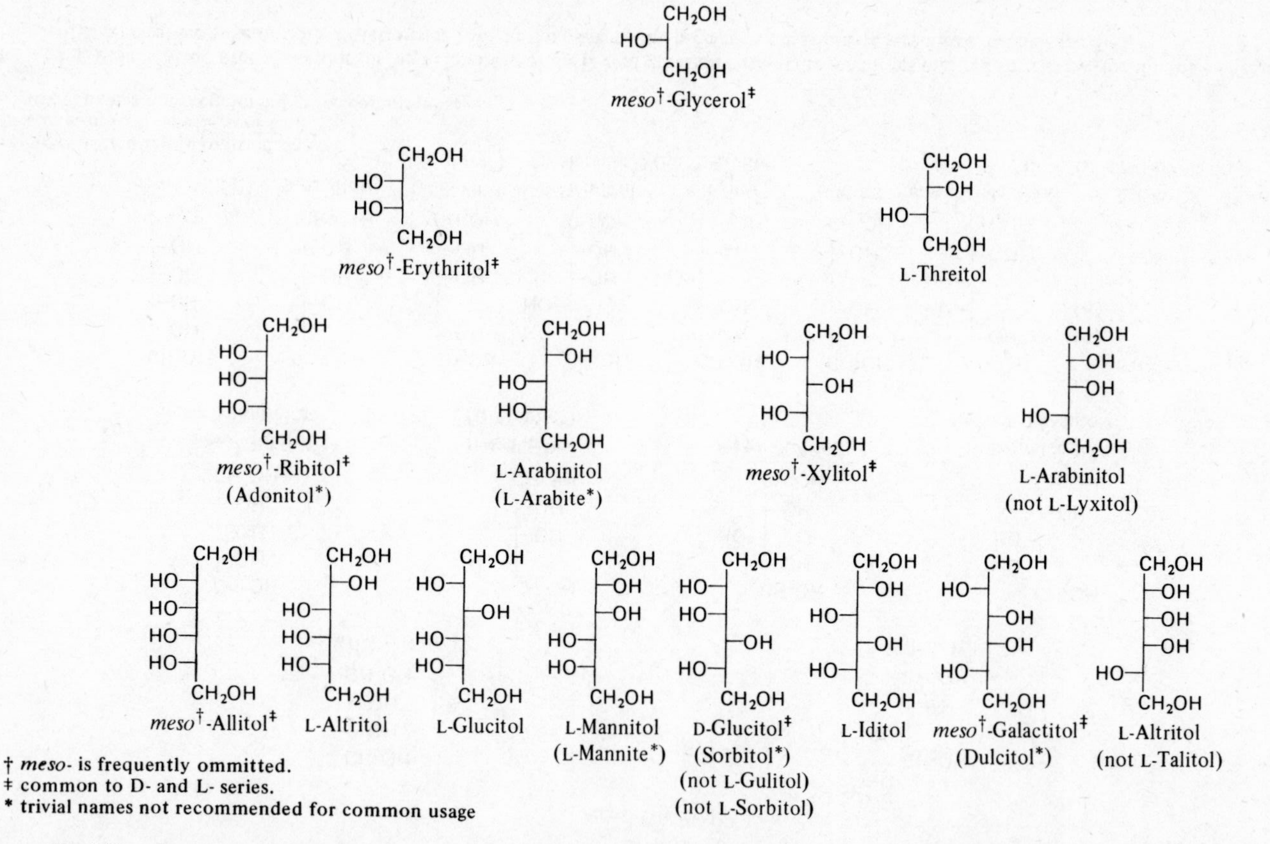

† *meso*- is frequently ommitted.
‡ common to D- and L- series.
* trivial names not recommended for common usage

Fig. 3.3b – Structures of the alditols, up to hexitols, of the L configuration with hydrogen atoms omitted for clarity (for the full structure of D-glucitol see Scheme 3.4). Rotations of 180° in the plane of the page gives the alternate structures, see Scheme 3.4.

Reduction

Reduction of the aldehyde group (potential aldehyde group if the aldose is present in a ring form) of an aldose gives a polyhydric alcohol known as an alditol and the fact that two different aldoses (for example, D-glucose and L-gulose) give the same alditol (D-glucitol (13), as shown in Scheme 3.4) was exploited by Fischer in the determination of the stereochemistry of monosaccharides (see Chapter 2). Ketoses give a mixture of products; for example, D-fructose is reduced to a mixture of D-glucitol and D-mannitol. The structures and names of the alditols, up to hexitols are given in Figs. 3.3a and 3.3b, from which it may be seen that there are only 3 tetritols, 5 pentitols and 10 hexitols derived from the 4 tetroses, 8 pentoses and 16 hexoses respectively.

The traditional method of preparation, using sodium amalgam, has been superceded by sodium borohydride for laboratory preparations; it gives a more rapid reaction, unless the aldose is substituted at C-3 thus sterically hindering this reagent. The major drawback to the use of sodium borohydride, namely the subsequent removal of inorganic materials, has led to the development of other methods such as the use of Raney nickel in boiling ethanol (Wolfrom and Schumacher, 1955) for laboratory-scale preparations, or electrolytic reduction in alkaline media or catalytic hydrogenation for industrial usage. In the last method, extensive epimerisation of the aldose occurs prior to reduction, resulting in a mixture of products; this has been exploited for the industrial preparation of D-mannitol from D-glucose rather than from D-mannose. A number of alditols exist in nature either free or in combined form, for example glycerol is an essential component of lipids, D-ribitol occurs in some microbial polysaccharides (see Chapter 8), and D-glucitol and D-mannitol in many fruits, plants and seaweeds.

The alditols are commonly used in the analysis of carbohydrates as precursors of alditol acetates, which are useful, volatile derivatives for analysis (see Chapter 4). Reduction of carbohydrates, usually monosaccharides, by reagents which produce a coloured product is used as a basis for estimation of carbohydrates using non-corrosive reagents (see review by White and Kennedy, 1981).

Action of acids and bases

Monosaccharides undergo isomerization and degradation with acids and bases to give a variety of products, but since alditols are remarkably stable to the action of acids and bases it may be assumed that the (potential) aldehyde or ketone group is responsible.

Aldoses are invariably more stable towards acid than towards alkali, although very little is known about their behaviour under mildly acidic conditions. Evaporation of a solution of an aldose in dilute mineral acid (for example, 1 mM) causes intermolecular-condensation reactions similar to glycoside formation, to give small proportions of di-, tri- and higher oligo-saccharides. This process is called reversion. Under more drastic conditions, aldoses and ketoses undergo more-extensive dehydration reactions, to give furan derivatives (Scheme 3.5). These derivatives react with a number of phenols, aromatic amines and certain aliphatic aminoacids, in particular, L-cysteine, to give coloured products. Such reactions can be used as methods for detection and determination of carbohydrates which, in some cases, allow different types of monosaccharides to be detected in mixtures. For example, it is possible to determine pentoses in the presence of hexoses by using L-cysteine, due to the different ease of formation of the furan derivatives (pentoses react under milder conditions than those required for hexoses), and to distinguish a number of hexoses from each other due to secondary reactions taking place to differing degrees. One secondary reaction which can occur is a ring-opening reaction producing 4-oxo-pentanoic acid (laevulinic acid, 14) and formic acid. For a review of the methods used for the analysis of carbohydrates based on furan derivatives, see White and Kennedy (1981).

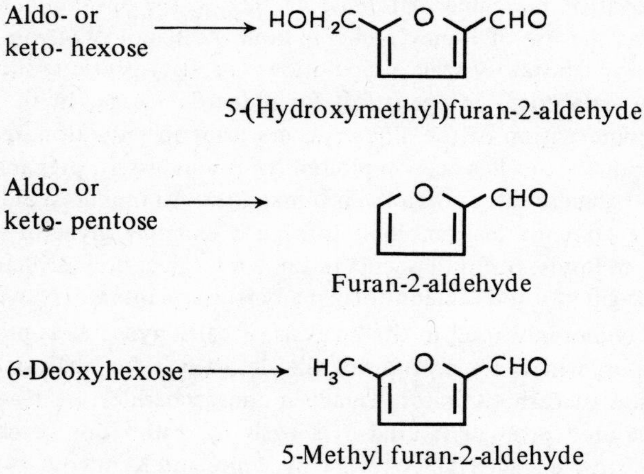

Aldo- or keto- hexose ⟶ HOH_2C–(furan)–CHO

5-(Hydroxymethyl)furan-2-aldehyde

Aldo- or keto- pentose ⟶ (furan)–CHO

Furan-2-aldehyde

6-Deoxyhexose ⟶ H_3C–(furan)–CHO

5-Methyl furan-2-aldehyde

Scheme 3.5

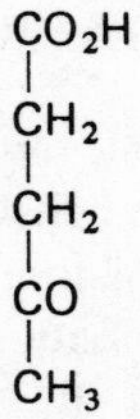

CO_2H
|
CH_2
|
CH_2
|
CO
|
CH_3

(14)

Under mildly basic conditions, such as in the presence of pyridine or aqueous calcium hydroxide, aldose ⇌ ketose isomerisation, which also results in epimerisation at C-2 (Scheme 3.6), occurs. These are known as the Lobry de Bruyn-Alberda van Eckenstein rearrangements. In conditions of 10 mM sodium hydroxide the proportions of product formed are shown in Scheme 3.6. Under more-drastic conditions, such as longer reaction times, increased temperature, or increased concentration of alkali, rearrangements take place through processes of β-elimination and benzilic acid type rearrangements (Scheme 3.7) with the formation of aldonic acid derivatives (trivial name, not accepted for current usage, saccharinic acids) (Scheme 3.8). Even more strongly basic conditions result in complex, reverse alditol reactions which give fragments containing three carbon atoms; those include 2-hydroxypropanaldehyde (15), pyruvic acid (16) and lactic acid (17).

D-Glucose (66%)	⇌	1,2-enediol	⇌	D-Fructose (30%)	⇌	2,3-enediol
CHO		CHOH		CH_2OH		CH_2OH
HCOH		‖ COH		CO		COH
HOCH		HOCH		HOCH		‖ HOC
HCOH		HCOH		HCOH		HCOH
HCOH		HCOH		HCOH		HCOH
CH_2OH		CH_2OH		CH_2OH		CH_2OH

1,2-enediol ⇌ D-Mannose (1%): CHO, HOCH, HOCH, HCOH, HCOH, CH_2OH

2,3-enediol ⇌ D-*ribo*-2-Pentulose (trace): CH_2OH, CO, HCOH, HCOH, HCOH, CH_2OH

Scheme 3.6

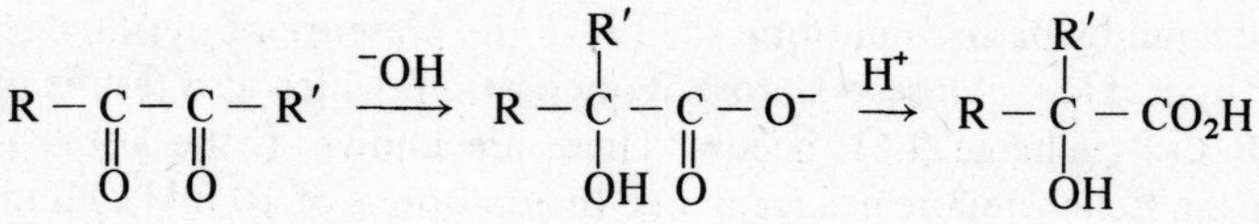

$$R-\underset{\underset{O}{\|}}{C}-\underset{\underset{O}{\|}}{C}-R' \xrightarrow{^{-}OH} R-\underset{OH}{\overset{R'}{C}}-\underset{\underset{O}{\|}}{C}-O^{-} \xrightarrow{H^{+}} R-\underset{OH}{\overset{R'}{C}}-CO_2H$$

Scheme 3.7

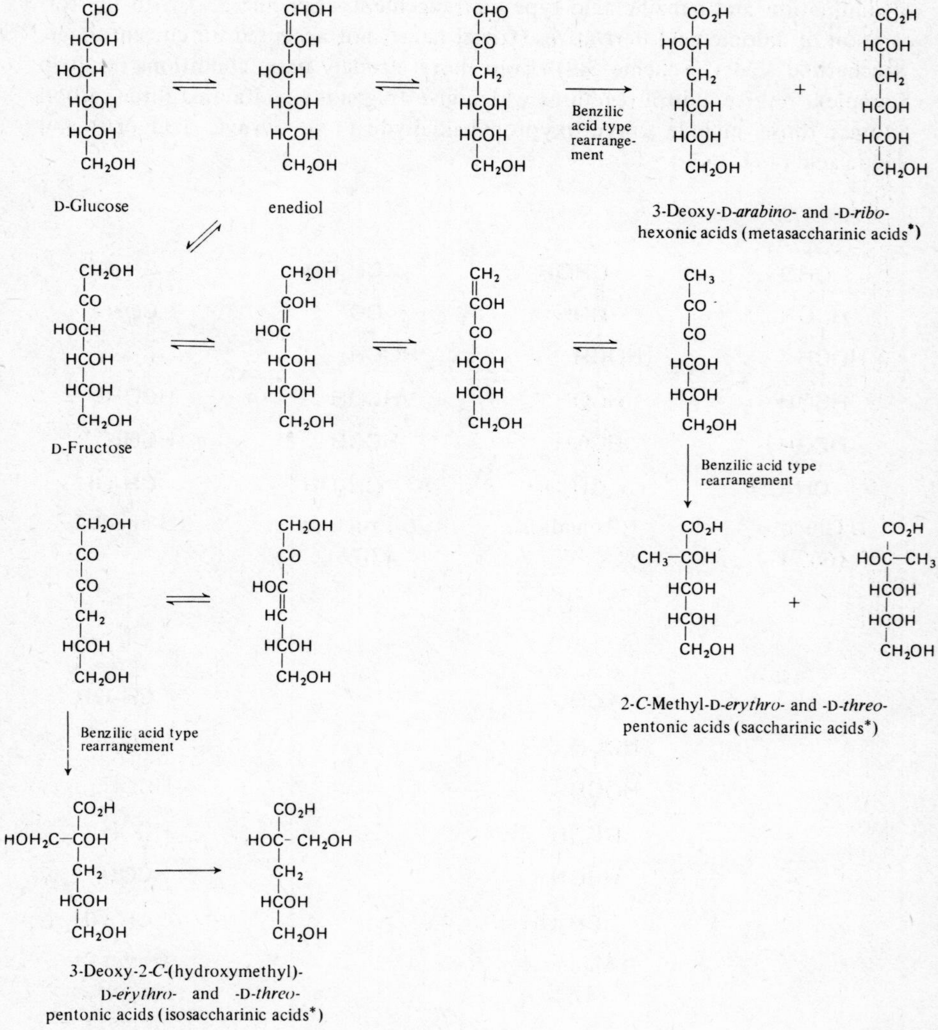

D-Glucose ⇌ enediol ⇌ Benzilic acid type rearrangement → 3-Deoxy-D-*arabino*- and -D-*ribo*-hexonic acids (metasaccharinic acids*)

D-Fructose ⇌ … Benzilic acid type rearrangement → 2-*C*-Methyl-D-*erythro*- and -D-*threo*-pentonic acids (saccharinic acids*)

Benzilic acid type rearrangement → 3-Deoxy-2-*C*-(hydroxymethyl)-D-*erythro*- and -D-*threo*-pentonic acids (isosaccharinic acids*)

* trivial names which are not accepted for current usage.

Scheme 3.8

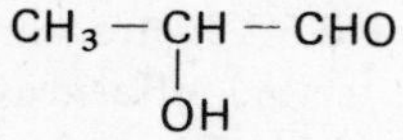

$CH_3-CH(OH)-CHO$ (15)

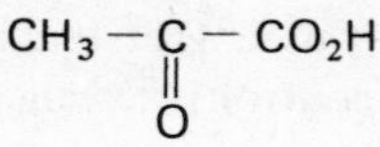

$CH_3-C(=O)-CO_2H$ (16)

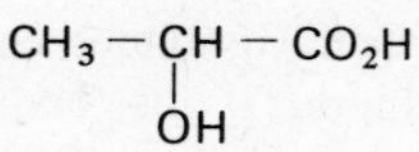

$CH_3-CH(OH)-CO_2H$ (17)

DERIVATIVES

Due to the polyfunctional nature of carbohydrates, development of the chemistry and synthesis of such compounds has depended upon the development of methods of derivatisation to react with specific hydroxyl or carbonyl groups, to protect or 'block' the groups and prevent subsequent reactions thereof. A discussion of a number of suitable derivatives will be given, but the reader is directed to reviews which deal with the subject in much greater detail (for example, Dutton, 1973 and Kennedy, 1974a) and a series which deals with the practical details of derivatisation (Whistler and Others, 1962 onwards).

It must be remembered that when monosaccharides, from pentoses upwards, are derivatised, the possibility exists that a number of isomers will be produced as a result of mutarotation. With some reactions (for example, trimethylsilylation) the rate of reaction is very much higher than that of mutarotation, with the result that essentially only one isomer is obtained from crystalline monosaccharides. On the other hand, if the monosaccharides were obtained by acid hydrolysis of polysaccharides (see Chapter 4), they will normally exist as mutarotated equilibrium mixtures and so, if the derivative is required for quantitative analysis (see Chapter 4), it is imperative that the method be standardised in order to predict exactly the ratio of isomers formed by any monosaccharide. One method used to ensure standardisation is to pre-equilibrate the monosaccharides prior to derivatisation by, for example, treating the pyridine solution with lithium perchlorate prior to addition of the silylating reagents. It is, of course, possible to study the mutarotation of carbohydrates under various conditions by trapping the isomers present in the solution using a derivatisation reaction with a reaction rate much higher than that of mutarotation.

Glycosides

As has been discussed earlier (Chapter 2), the bulk of carbohydrate found in nature exists in polysaccharides in which monosaccharides are joined together by glycosidic linkages to give full acetals (see Scheme 2.2). In a similar manner aldoses and ketoses react with simple alcohols in the presence of acid catalysts to form glycosides. Thus, the reaction of D-glucose, in the presence of anhydrous methanol containing dissolved hydrogen chloride (1–4%), produces a mixture of methyl α- and β-D-glucosides. Since this reaction (Fischer glycosidation) is thermodynamically controlled, the product obtained preferentially is the α-D-pyranoside (66%) due to the less-polar solvent having a more pronounced effect on the alignment of dipoles (see Fig. 3.4). This effect is known as the anomeric effect. If milder conditions are used (for example, less than 0.5% hydrogen chloride), this thermodynamic control is lessened and kinetic factors influence the reaction such that some of the furanosides are produced. Table 3.1 shows the products formed from D-galactose under differing conditions.

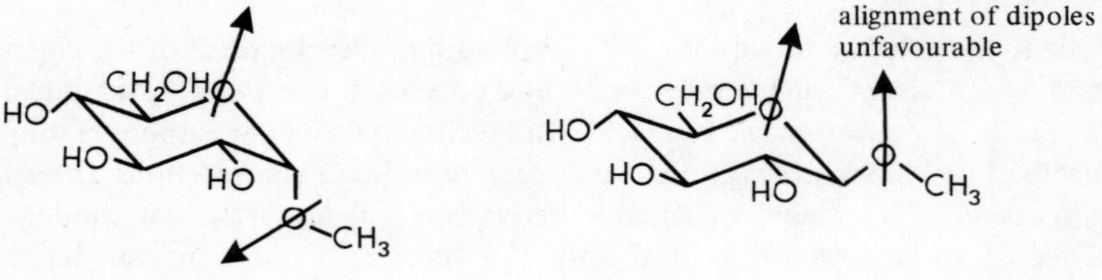

alignment of dipoles unfavourable

Methyl α-D-glucopyranoside

Methyl β-D-glucopyranoside

Fig 3.4 – The anomeric effect.

Table 3.1 Ratios of isomeric methyl D-galactosides produced under different conditions

Concentration of hydrogen chloride in dry methanol (%)	Temperature (°C)	Time (hours)	Furanoside α	Furanoside β	Pyranoside α	Pyranoside β
0.5	25	6	17	33	0	50
0.5	25	940	30	7	49	14
4.0	64	3	73	4	9	14
4.0	64	20	39	5	40	16

Where the alcohol is rare, and cannot be used in large excess, an alternative method of preparation is the Koenigs-Knorr synthesis in which the appropriate glycosyl halide reacts with an alcohol in the presence of silver oxide. The preparation and reaction of the glycosyl halide is shown in Scheme 3.9.

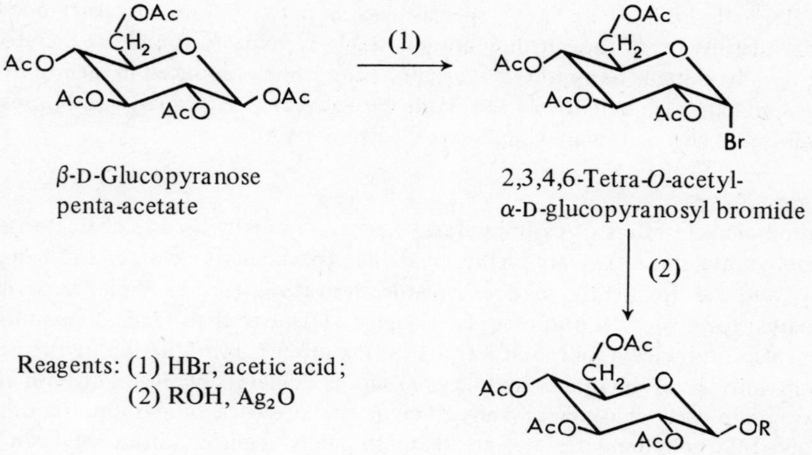

β-D-Glucopyranose penta-acetate

2,3,4,6-Tetra-*O*-acetyl-α-D-glucopyranosyl bromide

Reagents: (1) HBr, acetic acid;
(2) ROH, Ag_2O

β-anomer (not α anomer) formed.

Scheme 3.9

Ethers

Ether derivatives of carbohydrates are stable under a variety of conditions, and are, therefore, ideal for analytical purposes, but, for synthetic purposes their stability may be a disadvantage due to the lack of reliable reactions for removal of the blocking group. Methyl ethers were (and still are) a major feature in structural analysis of many oligo- and poly-saccharides (see Chapter 4) and in determination of the ring structures of monosaccharides. They may be formed either by the Purdie (methyl iodide in the presence of silver oxide or carbonate) or the Haworth (dimethyl sulphate in aqueous alkali) procedures. Both of these procedures may require several applications before complete methylation of complex molecules is achieved and many variations thereof exist. The preferred method today is that of Hakomori (1964), which uses the dimethylsulphinyl (dimsyl) anion in dimethyl sulphoxide with subsequent treatment with methyl iodide; but there are exceptions to its successful use.

Benzyl ethers, formed by action of alkali and benzyl chloride are stable to acids and alkalis, and, because they can be removed by mild hydrogenolysis, are more widely used than methyl ethers as protecting groups. Allyl ethers, produced from allyl bromide, are also common protecting groups, since they are stable to acids, but, in the presence of a base, form vinyl ethers which are readily removed by mild acid (conditions under which isopropylidene groups are retained; see later). Certain other ethers, such as triphenylmethyl (trityl) ethers, are used due to their special properties in synthetic schemes. Triphenylmethyl ethers which, due to their size, are usually attached to the O-6 position can be removed by mild acid in the presence of ester, ether and glycosidic linkages, and, if by hydrogenolysis, also in the presence of acetals.

Trimethylsilyl ethers have a special place in carbohydrate chemistry due to their volatility, even though they are not stable to hydrolysis and are therefore not usually suitable as synthetic reagents. They can be produced in quantitative yields and are used extensively as volatile derivatives for such analytical purposes as gas-liquid chromatography and mass spectrometry.

Esters

Carboxylic acid esters of carbohydrates were traditionally used to characterise a carbohydrate since they are highly crystalline (particularly acetates and benzoates) and are frequently used as volatile derivatives (for example, as alditol acetates) for analytical purposes (see Chapter 4). Due to their facile preparation, they also find wide applicability as protecting groups. Introduction of the acyl group into a carbohydrate hydroxyl group is achieved by reaction with the appropriate acid chloride or anhydride in the presence of pyridine or other catalyst. Carbohydrate acetates are stable to mildly acidic conditions and can be deacetylated under basic conditions (usually using sodium methoxide), but, under mildly basic conditions, the acetate shows a tendency to migrate to free hydroxyl groups. Benzoates, are, however more stable to mildly basic conditions, showing little tendency to migrate, and they will also withstand acid conditions better. Other carboxylic acid esters used include chloroacetates and trifluoroacetates, which are more labile and can be readily and selectively removed.

Sulphonate esters are more versatile than carboxylic acid esters due to their stability to acids and bases. They are removed by disruption of the alkyl–oxygen rather than the oxygen–sulphur bond, and are therefore very useful in the synthesis of carbohydrate derivatives (see Chapter 5). They are prepared from the sulphonic acid chloride, usually methanesulphonyl (mesyl) chloride or 4-toluenesulphonyl (tosyl) chloride, in the presence of pyridine and, usually, the carbohydrate glycoside at low temperatures. Free aldoses react rapidly with sulphonic acid chlorides, giving the sulphonylated glycosyl chloride and degradation products therefrom. Sulphonate groups can be removed by nucleophilic displacement with lithium aluminium hydride.

Other esters are used for specific purposes (for example, nitrates for protecting groups for methylation, phosphates for biological intermediates, and cyclic carbonates for preparation of immobilised derivatives – see Chapter 13).

Cyclic acetals

The polyhydroxylated nature of carbohydrates means that pairs of hydroxyl groups, if suitably placed, can form cyclic acetals by reaction with aldehydes or ketones in the presence of acidic catalysts (Scheme 3.10). In some cases, they react with the anomeric hydroxyl group and so fix the monosaccharide in a furanose or pyranose ring. They are stable to alkali but readily removed by acid (90% trifluoroacetic acid is recommended) or in the case of the benzylidene group, by catalytic hydrogenolysis in the presence of platinum.

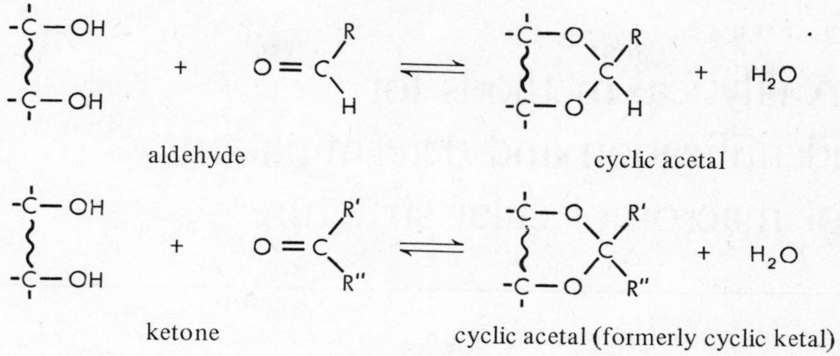

-C—OH + O=C(R)(H) ⇌ cyclic acetal + H_2O

aldehyde — cyclic acetal

-C—OH + O=C(R')(R'') ⇌ cyclic acetal + H_2O

ketone — cyclic acetal (formerly cyclic ketal)

Scheme 3.10

The most commonly used carbonyl compounds are acetaldehyde, benzaldehyde and acetone but a number of others have been used. The solvent for the reaction is frequently the carbonyl compound, and the catalysts include anhydrous sulphuric acid (up to 50%), hydrogen chloride (up to 1%), and such Lewis acids as zinc chloride and boron trifluoride. In the case of isopropylidene formation from acetone anhydrous copper(II) sulphate is often used to maintain anhydrous conditions.

A naturally occurring cyclic acetal (pyruvic acid acetal) is found in certain polysaccharides (for example, xanthan gum, see Chapter 8, p. 171, which contains the *S*-isomer). This derivative is formed from the α-keto acid, 2-oxopropanoic acid (pyruvic acid) (16) (Scheme 3.11), and is frequently, but erroneously, referred to as a 'pyruvate', implying that it is an ester derivative.

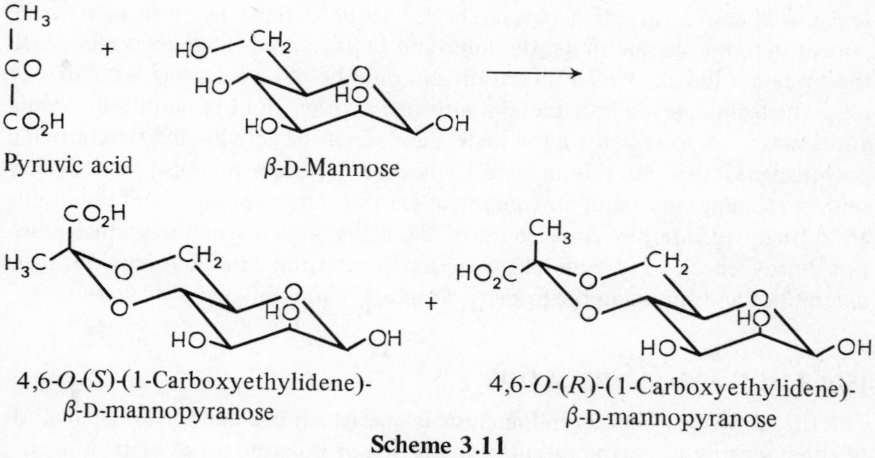

CH_3–CO–CO_2H + β-D-Mannose ⟶

Pyruvic acid — β-D-Mannose

4,6-*O*-(*S*)-(1-Carboxyethylidene)-β-D-mannopyranose + 4,6-*O*-(*R*)-(1-Carboxyethylidene)-β-D-mannopyranose

Scheme 3.11

Other derivatives are used for particular purposes, such as the formation of immobilised derivatives, and these are discussed in the relevant chapter (see later, Chapter 13).

CHAPTER 4

Analytical methods for identification and determination of macromolecular structure

---

In this chapter the reader is provided with an introduction to the various analytical methods available and the chemistry involved therein. The chapter discusses some of the modern physico-chemical techniques which involve sophisticated equipment but which provide more detailed aspects of overall structure. Key references have been included to give the reader a lead to practical details (see also Various Authors, 1968 onwards).

The normal sequence of events in structural analysis after identification of the macromolecule (by its biological activity for example), isolation and purification is component analysis and sequence determination (to give primary structure) using chemical and biological methods. Secondary, tertiary and quaternary structures (see Chapter 3) are obtained by physico-chemical methods which give information on the size and shape of a molecule.

No one method available for structural analysis will provide sufficient data to allow the structure of a polysaccharide to be defined in terms of its component monosaccharide units, the inter-unit linkages, and the sequence in which the units are linked. This information can only be obtained using a number of different techniques in conjunction with one another. Full structural determination is very important for a full understanding of the activity and function of a polysaccharide and its role in nature, but the necessary practical work is very time consuming, the required equipment is often very expensive, and the results are difficult to interpret on account of the many similar structures which exist. For most carbohydrate-containing macromolecules full structures have not been determined and for many the primary structure is still only partially known.

ISOLATION AND PURIFICATION

The first problem of structural analysis is one which is common to the analysis of other groups of macromolecules, and is that of isolating the material in a pure form. The definition of purity is not as clear-cut as was originally thought because microheterogeneity, the phenomenon of minor variations within a single species of compound, is now well recognized. The ensuing discussion on the separation of the carbohydrate species from various impurities, including inorganic salts and low molecular weight materials, and also macromolecular species such as proteins and lignins, is written from the general viewpoint, but it must be borne in mind that each polysaccharide will have its own peculiarities.

Wherever possible, the first stage involves solubilisation in an aqueous or an aprotic solvent such as ethylene glycol or dimethyl sulphoxide, but care must be taken to ensure that the method used and solvents chosen do not modify or degrade the structure of the macromolecule. This eliminates the use of acids, alkalis, or enzymes. The removal of low molecular weight impurities is readily performed by dialysis (distinction on basis of molecular size), ion exchange chromatography (distinction on basis of molecular charge), or gel filtration (distinction on basis of molecular size). The last two techniques are also used extensively for the separation of the desired material from contaminant macromolecules. Removal of the macromolecule from solution can be achieved by precipitation with solvents such as ethanol or acetone, or by complexation with, for example, metal ions, and, for acidic polysaccharides, quaternary ammonium salts.

Purification stages must proceed until a material of constant composition is obtained. In the past the criteria for estimation of constant composition have relied on physical and chemical measurements such as functional group analysis, specific rotation, and carbohydrate composition after hydrolysis. More recently, ultracentrifugation which gives a measure of sedimentation in a high force field, electrophoresis, which measures the mobility of the polysaccharide in an applied electric field, and the aforementioned chromatographies have all been used to investigate the purity of a preparation. The best definition of homogeneity is one based on assessment by two or more methods which rely on different criteria.

COMPLETE HYDROLYSIS

Once a pure sample of the polysaccharide has been obtained the first step in elucidating its structure is to identify and estimate the component monosaccharides. The molecule is broken down by acid hydrolysis by a mechanism which is the reverse of the Fischer reaction (see Chapter 3, p. 61) and the hydrolysate analysed by chromatographic techniques (see later this chapter). The ion-exchange chromatography of the borate complexes of neutral monosaccharides is the most common method which is used as the basis for fully automatic carbohydrate analysers.

The conditions for hydrolysis must be controlled such that complete hydrolysis is achieved with little or no degradation of the monosaccharide units – use of more than one set of hydrolysis conditions may be necessary. The ease of hydrolysis of different linkages and the stabilities of the various monosaccharides mean that the optimum conditions for each polysaccharide have to be determined. Polysaccharides containing furanose or 5-amino-3,5-dideoxy-D-*glycero*-D-*galacto*-2-nonulopyranonic acid residues, and 2-deoxy-hexoses or -pentoses, are more readily hydrolysed than those containing hexuronic acids or 2-amino-2-deoxyhexoses, with hexose-containing polysaccharides being intermediate. Conditions which have been found appropriate for hexose-containing polysaccharides are 1 M sulphuric acid at 100°C for four hours, with the use of 0.25 M sulphuric acid at 70°C for pentose-containing polysaccharides. Degradation frequently occurs in direct hydrolysis whatever conditions are used. For example, in the case of glycosaminoglycans, 4 M hydrochloric acid at 100°C for nine hours is necessary to liberate all the 2-amino-2-deoxyhexose residues, but under such conditions the majority of the hexuronic acid residues are decomposed.

Partial hydrolysis (under appropriately milder conditions) to give a small number of oligosaccharides can be useful for structural analysis, but care must be taken in interpreting the results since monosaccharides can recombine, under certain hydrolysis conditions, to give oligosaccharides linked in a manner different from that in the original polysaccharide. This process is called reversion (see Chapter 3). Certain functional groups, etc., are lost on hydrolysis and specific methods are required for their estimation. These groups include acetyl, carboxyl, carbonyl, and ether, and methods for their estimation have been reviewed (Aminoff *et al.*, 1970).

METHYLATION ANALYSIS

Once the individual monosaccharide components have been estimated, the manner in which they are linked to each other, and the sequence, has to be determined. If all the free hydroxyl groups in the polysaccharide can be reacted to form derivatives which are stable to acid hydrolysis, the hydroxyl groups produced by hydrolysis of the glycosidic linkages will indicate where linkage points were formerly located on each monosaccharide residue (see Chapter 3 for the formation of methyl ethers).

The fully methylated polysaccharide is hydrolysed to its constituent methylated monosaccharides using sulphuric acid or trifluoroacetic acid. The hydrolysis mixture can be fractionated by partition chromatography on cellulose or silica gel, by adsorption chromatography, or, best of all, by gas-liquid chromatography of volatile derivatives such as the methyl per-*O*-methyl glycosides, partially methylated alditol acetates, or partially methylated *O*-trimethylsilyl ethers. An important extension to gas-liquid chromatography for the further identification of these volatile derivatives is the use of mass spectrometry linked to gas-liquid chromatography, or more recently, the use of mass spectrometry *per se* (Rauvala *et al.*, 1981.) Reviews have been published giving the characteristic data for known, standard, partially methylated compounds (Dutton 1973 and 1974, and Bjornda *et al.*, 1970).

Methylation analysis is not without its problems, the most common being polysaccharides which contain hexuronic acid residues. These can be methylated only with considerable difficulty using the thallium salt of the hexuronic acid and reacting it with methyl iodide and thallium hydroxide. This reaction must be carefully controlled to avoid degradation and demethylation of the polysaccharide. Recently the Hakomori method has been applied to hexuronic acid-containing polysaccharides to achieve complete methylation of hydroxyl and carboxyl groups in one step. Another method by which the difficulty due to uronic acid residues can be overcome is by (indirect) reduction of the acid to an alcohol with sodium borohydride. Methylation analysis on its own will not give structural sequence data, but does identify the monosaccharide components of the polysaccharide and the position of the intermonosaccharide linkages involved (see review by Rauvala *et al.*, 1981).

PARTIAL HYDROLYSIS

If the hydrolysis of the polysaccharide is stopped before the reaction goes to completion, fragments of intermediate molecular weight can be isolated and fractionated using a number of chromatographic techniques such as gel filtration, ion-exchange chromatography, and partition chromatography. Determination of the structure of these simpler oligosaccharides is generally easier than determinations carried out on the parent polysaccharide. If the glycosidic linkages in the polysaccharide are all hydrolysed at the same rate as in, for example, the linear homopolysaccharides, the product of partial hydrolysis will consist, in the case of amylose, of a range of oligosaccharides such as maltotetraose, maltotriose, maltose and D-glucose. In heteropolysaccharides there are a number of types of glycosidic linkages and their respective rates of hydrolysis will differ, giving a degree of selectivity to the reaction. In general terms furanosides are hydrolysed at a greater rate than pyranosides by factors between 10 and 1000 which will result in removal of, for example, arabinofuranosyl residues attached to xylanopyranosyl residues in arabinoxylans. The conditions of the hydrolysis will also effect the specificity of the degradation. In mineral acids, (1→6)-linkages are more stable than (1→4)-linkages, but if this reaction were to be carried out in acidified acetic anhydride (containing approx. 5% sulphuric acid) the (1→6)-linkages are less stable. The use of both these methods of hydrolysis will lead to different fragments which will give overlapping data to provide a better picture of the complete polysaccharide. The concentration of carbohydrate material must be kept below about 0.5% to prevent acid-catalysed polymerisation of the fragments (acid reversion), which leads to artifacts in structural analysis. Some glycoside bonds in polysaccharides can be cleaved specifically by enzymes to give oligosaccharides in a controlled manner. This method will be discussed later in this chapter.

PERIODATE OXIDATION

Oxidation of monosaccharides by glycol cleavage is a widespread method of analysis. The use of lead(IV) acetate has found little application to polysaccharide chemistry owing to the lack of suitable solvents for the carbohydrates in which the reagent does not decompose, although the use of lead(IV) acetate in pyridine has been found to oxidise rigidly-held diaxial diols which do not react readily with periodic acid. In a much more commonly employed method, periodic acid and its salts are used in aqueous solutions of pH 3–5 to avoid acid hydrolysis and the non-selective oxidations which can occur at higher pH values. The reagent reacts with vicinal hydroxyl groups to cleave the linkage between them with the consumption of one mole of periodate per diol. The products of the reaction depend on the linkages between the monosaccharide units. Oxidation of a primary hydroxyl group, adjacent to a secondary hydroxyl group, as in the case of a furanose ring structure leads to the formation of formaldehyde, whilst vicinal triol groups yield formic acid. Reaction of polysaccharides with periodic acid is followed by measurement of the amount of reagent consumed and of the formic acid and, less frequently, of the formaldehyde produced (Fig. 4.1). This allows the distinction between 1,3- and 1,6-disubstituted residues and allows a measurement of the total amount of 1,2- and 1,4-disubstituted residues.

terminal non-reducing unit

IO_6^-, 2 moles → + $H\cdot{}^3CO_2H$

terminal reducing units

2-substituted

IO_4^-, 4 moles → + $H\cdot{}^6CHO$ + $H\cdot{}^5CO_2H$ + $H\cdot{}^4CO_2H$

3-substituted

IO_4^-, 4 moles → + $H\cdot{}^6CHO$ + $H\cdot{}^5CO_2H$ + $H\cdot{}^1CO_2H$

4-substituted

IO_4^-, 3 moles → + $H\cdot{}^6CHO$ + $H\cdot{}^2CO_2H$ + ${}^5H\cdot{}^1CO_2H$

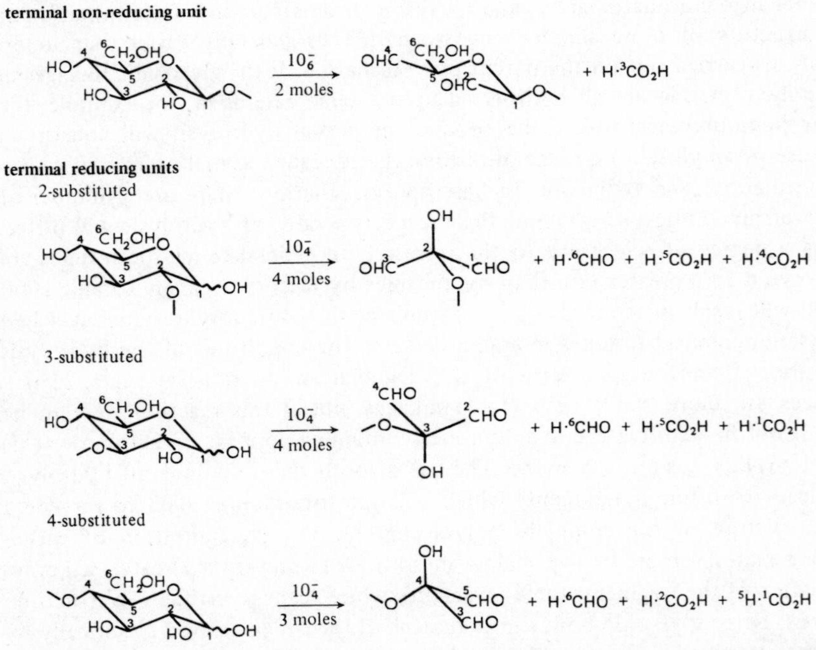

Fig. 4.1 – Periodate oxidation of substituted hexose residues (*continued on next page*).

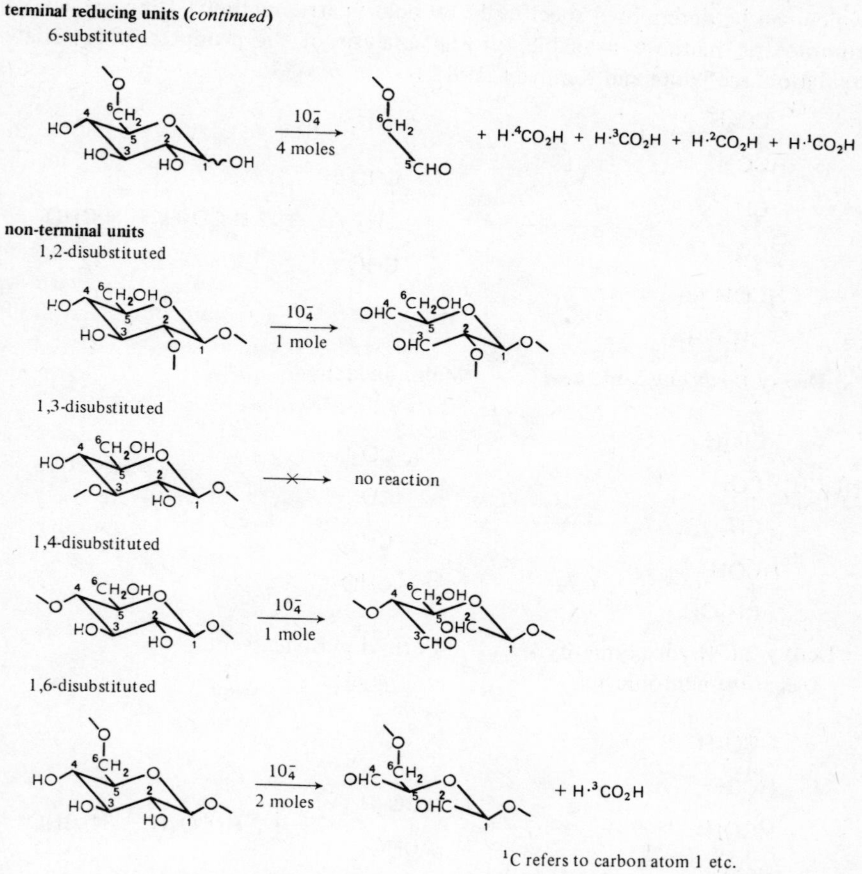

terminal reducing units (*continued*)
6-substituted

$\xrightarrow[\text{4 moles}]{IO_4^-}$ + $H\cdot{}^4CO_2H$ + $H\cdot{}^3CO_2H$ + $H\cdot{}^2CO_2H$ + $H\cdot{}^1CO_2H$

non-terminal units
1,2-disubstituted

$\xrightarrow[\text{1 mole}]{IO_4^-}$

1,3-disubstituted

no reaction

1,4-disubstituted

$\xrightarrow[\text{1 mole}]{IO_4^-}$

1,6-disubstituted

$\xrightarrow[\text{2 moles}]{IO_4^-}$ + $H\cdot{}^3CO_2H$

[1]C refers to carbon atom 1 etc.

Fig. 4.1 – Periodate oxidation of substituted hexose residues.

The dialdehyde-type products of periodate oxidation are unstable in water and it is therefore desirable to reduce them, usually with sodium borohydride, to alcohols before acid hydrolysis (to split the oxidised material into the component units) is carried out. Analysis of these component products is essential because it provides a means of distinguishing between 1,2- and 1,4-disubstituted residues. The products of hydrolysis, such as glycerol, glycol aldehyde, glyceraldehyde, tetritols (such as D-erythritol), and free monosaccharides (resulting from periodate-resistant residues) are usually determined by gas-liquid chromatography as their trimethylsilyl ethers. This method again does not give complete linkage sequences, but gives information which is used in conjunction with other methods. Periodate oxidation of the products obtained by the action of alkalis on carbohydrates (see Chapter 3 and later this chapter) gives products (Fig. 4.2) which can be determined specifically by colorimetric methods. (For a review of colorimetric methods available for the analysis of the products of periodate oxidation, see White and Kennedy, 1981.)

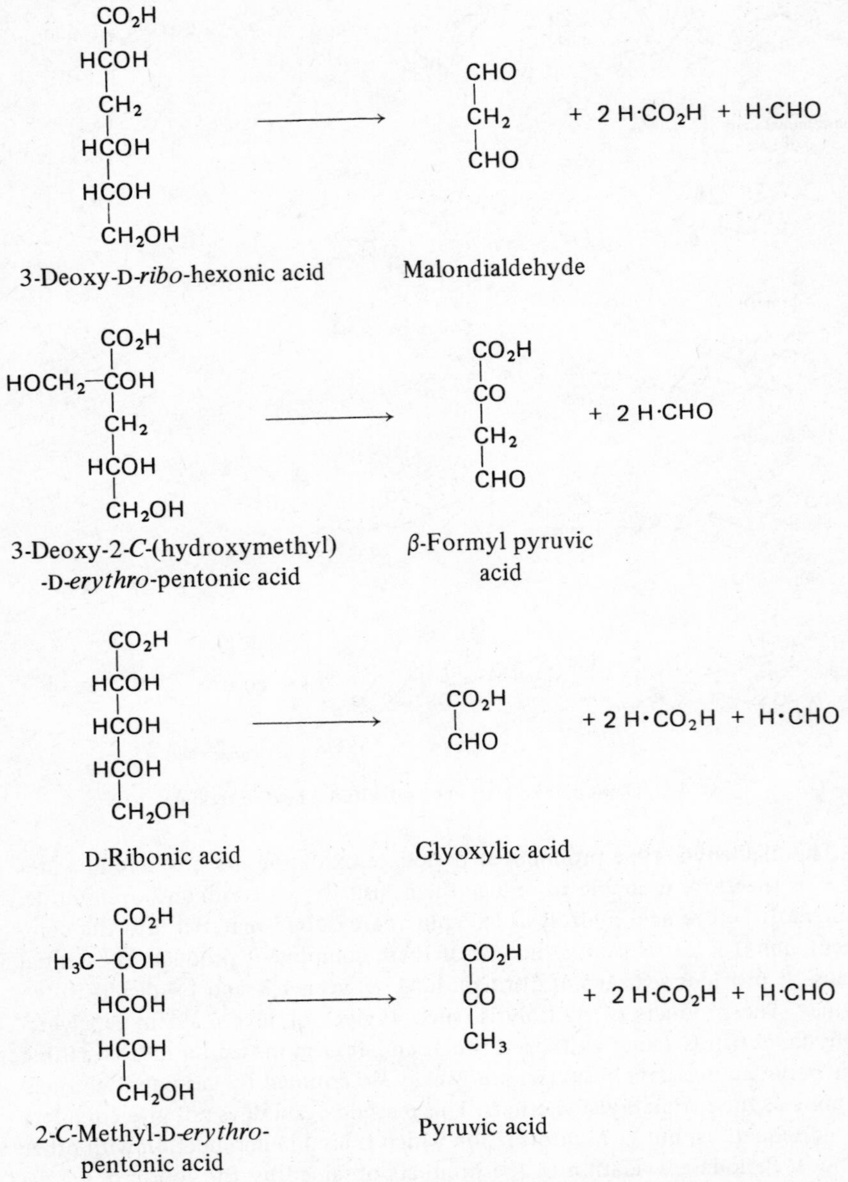

$$\begin{array}{c} CO_2H \\ | \\ HCOH \\ | \\ CH_2 \\ | \\ HCOH \\ | \\ HCOH \\ | \\ CH_2OH \end{array} \longrightarrow \begin{array}{c} CHO \\ | \\ CH_2 \\ | \\ CHO \end{array} + 2\,H{\cdot}CO_2H + H{\cdot}CHO$$

3-Deoxy-D-*ribo*-hexonic acid — Malondialdehyde

$$\begin{array}{c} CO_2H \\ | \\ HOCH_2{-}COH \\ | \\ CH_2 \\ | \\ HCOH \\ | \\ CH_2OH \end{array} \longrightarrow \begin{array}{c} CO_2H \\ | \\ CO \\ | \\ CH_2 \\ | \\ CHO \end{array} + 2\,H{\cdot}CHO$$

3-Deoxy-2-*C*-(hydroxymethyl) -D-*erythro*-pentonic acid — β-Formyl pyruvic acid

$$\begin{array}{c} CO_2H \\ | \\ HCOH \\ | \\ HCOH \\ | \\ HCOH \\ | \\ CH_2OH \end{array} \longrightarrow \begin{array}{c} CO_2H \\ | \\ CHO \end{array} + 2\,H{\cdot}CO_2H + H{\cdot}CHO$$

D-Ribonic acid — Glyoxylic acid

$$\begin{array}{c} CO_2H \\ | \\ H_3C{-}COH \\ | \\ HCOH \\ | \\ HCOH \\ | \\ CH_2OH \end{array} \longrightarrow \begin{array}{c} CO_2H \\ | \\ CO \\ | \\ CH_3 \end{array} + 2\,H{\cdot}CO_2H + H{\cdot}CHO$$

2-*C*-Methyl-D-*erythro*-pentonic acid — Pyruvic acid

Fig. 4.2 – Periodate oxidation of alkali resistant residues.

Smith degradation

An important modification to the periodate oxidation, borohydride reduction and total hydrolysis sequence described above is that known as the Smith degradation, which uses mild hydrolysis of the product from borohydride reduction (usually dilute mineral acid at room temperature) to cause partial degradation with the production of specific glycosides of oligosaccharides characteristic of the original polysaccharide (Perlin, 1959). This relies on the comparative stability of the glycosidic linkage between a sugar residue (the original periodate-resistant residue) and an alditol. For example, Smith degradation of nigeran [an α-D-glucan with alternating (1→3) and (1→4) linkages, see Table 2.6] results in the production of 2-*O*-α-D-glucopyranosyl-erythritol (18, see Scheme 4.1).

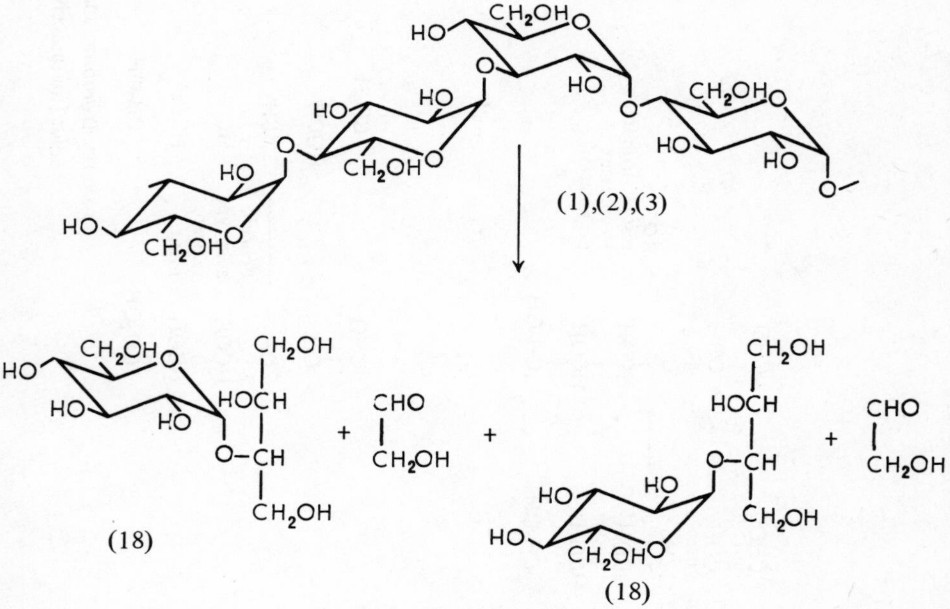

Reagents: (1) periodic acid; (2) sodium borohydride; (3) mild acid.

Scheme 4.1

ALKALINE DEGRADATION

This method of analysis of polysaccharides provides little information about the overall structure, which cannot be obtained by acid hydrolysis, but since the use is often made of alkali in the isolation of a purified sample, the type of reactions which occur should be understood (for a review see Whistler and BeMiller, 1958,

3-substituted residue

$$\begin{array}{c} CHO \\ | \\ HCOH \\ | \\ ROCH \\ | \\ HCOH \\ | \\ HCOH \\ | \\ CH_2OH \end{array} \underset{}{\overset{^{-}OH}{\rightleftharpoons}} \begin{array}{c} CHO^{-} \\ \| \\ COH \\ | \\ ROCH \\ | \\ HCOH \\ | \\ HCOH \\ | \\ CH_2OH \end{array} \longrightarrow \begin{array}{c} CHO \\ | \\ COH \\ \| \\ CH \\ | \\ HCOH \\ | \\ HCOH \\ | \\ CH_2OH \end{array} + RO^{-}$$

R = rest of polysaccharide

RO^{-}: rest of chain free to continue degradation

$$\rightleftharpoons$$

$$\begin{array}{c} CHO \\ | \\ CO \\ | \\ CH_2 \\ | \\ HCOH \\ | \\ HCOH \\ | \\ CH_2OH \end{array} \longrightarrow \begin{array}{c} CO_2H \\ | \\ HOCH \\ | \\ CH_2 \\ | \\ HCOH \\ | \\ HCOH \\ | \\ CH_2OH \end{array} + \begin{array}{c} CO_2H \\ | \\ HCOH \\ | \\ CH_2 \\ | \\ HCOH \\ | \\ HCOH \\ | \\ CH_2OH \end{array}$$

(arrow: benzilic acid type rearrangement)

3-Deoxy-D-*arabino*-and-D-*ribo*-hexonic acids (metasaccharinic acids*)

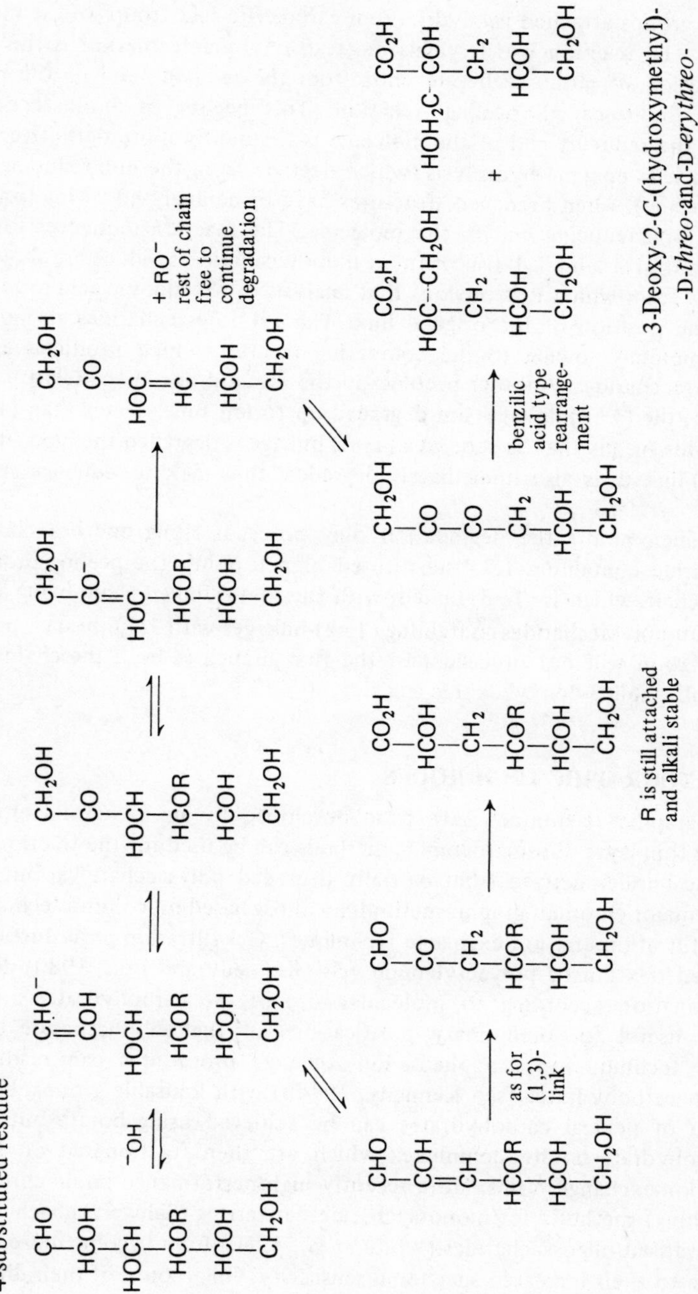

Scheme 4.2

* trivial names which are not accepted for current usage.

and Hough and Richardson, 1979). The most common reactions are the hydrolysis of ester groups attached *via* hydroxyl or carboxylic acid groups of the monosaccharide. The reaction which yields most structural information is the progressive erosion of monosaccharide units from the reducing end of the polysaccharide, the so-called 'peeling' reaction. This peeling of monosaccharide units from the reducing end of the molecule is frequently more definitive than methods such as enzyme hydrolysis (which degrade from the non-reducing end of the molecule), when branched structures have to be analysed owing to there being only one reducing end to the molecule. The reaction sequences for the degradation of 1,3- and 1,4-disubstituted monosaccharide residues are shown in Scheme 4.2 from which it is obvious that analysis of the aldonic acid produced will give the position of the original link. The 1,4-polysaccharides are not degraded completely, owing to the competing reaction which produces alkali-stable polysaccharides. Another problem in the method lies in the relative rates of reaction; the (1→3)-linkages are degraded up to ten times faster than (1→4)-linkages. This means that as soon as a (1→4)-linkage is degraded the next unit, if it is (1→3)-linked, is also immediately degraded, thus making sequence studies difficult.

At branch points the degradation only proceeds along one branch. In a polysaccharide containing 1,3,4-substituted branch points the peeling proceeds along the chain which is (1→3)-linked, with the (1→4)-linked chain being alkali-stabilised. In polysaccharides containing (1→4)-linkages with 1,4,6-branch points, the degradation will not proceed past the first branch as both the chains give rise to alkali-stable 3-deoxyhexonic acids.

CHROMATOGRAPHIC TECHNIQUES

Chromatographic techniques have been developed to aid structural analysis. Paper and thin-layer chromatographic methods can be used for the fractionation of oligosaccharides derived from partially degraded polysaccharides, but nowadays the major chromatographic methods are those based on column techniques, using gel filtration and ion exchange techniques. Gel filtration procedures using cross-linked dextran or polyacrylamide gels (Kennedy and Fox, 1980) depend on fractionation according to molecular size of the carbohydrates and are especially useful for preliminary purifications of the polysaccharide before structural techniques are applied. Ion-exchange procedures can readily be applied to carbohydrates (see Kennedy, 1974b) with ionisable groups, but the separation of neutral carbohydrates can be achieved using borate buffers to give carbohydrate–borate complexes which are then fractionated on borate forms of ion-exchange resins. More recently high performance liquid chromatographic (hplc) methods for monosaccharide derivatives (Schwarzenbach, 1979) or underivatised oligosaccharides (White *et al.*, 1980) have become more appropriate due to their increased speed and sensitivity. On account of their different criteria for separation, column techniques are often used in conjunction with each other as a means of determining the purity and any microheterogeneity of a carbohydrate sample.

From the standpoint of preparative methods, column chromatography is most useful in that large amounts of sample can often be fractionated with complete recovery of the sample in underivatised form. Nevertheless, achievement of good resolution is often only obtained with slow flow rates. For this reason, new techniques of column chromatography have also been developed for structural analysis and for the preparation of fractions for structural analysis. Affinity- or adsorption-chromatography has been used extensively for the purification of non-carbohydrate molecules and is based on selective adsorption on to an insoluble adsorbent which contains groups/molecules which interact specifically with the molecule to be purified, for example, enzyme inhibitors for enzyme purification and antibodies for antigen purification, and the technique is now being extended into the field of carbohydrate fractionation. Non-interacting impurity is removed from the bound carbohydrate, which is subsequently desorbed by disrupting the interaction in a way which does not cause degradation. The methods for desorption include changes of pH, ionic strength, or the use of an inhibitor to the interaction. The use of an immobilised form (see Chapter 13) of concanavalin A, a lectin (formerly a phytohaemagglutinin) which specifically reacts with branched-chain polysaccharides of a particular structural type, has been made for the fractionation of a number of polysaccharides and the principle has now been extended to the use of a whole series of lectins in immobilised form. Column supports coated with polyaromatic compounds have also been found to be of some use in the fractionation of polysaccharides. Recent developments in the technology of packing materials for high performance liquid chromatography mean that faster and more selective separations can be obtained and methods are reported of fractions of small oligosaccharides which take less than one hour to complete (White *et al.*, 1980).

Gas-liquid chromatography has found limited use in the structural analysis of polysaccharides, owing to the inherent requirements of the method, namely volatility and stability, under the conditions used. In practice the method has been limited to the analysis of component monosaccharides, after hydrolysis and, more importantly, to the analysis of partially methylated sugars produced in methylation analysis. Disaccharides can be made sufficiently volatile for gas-liquid chromatography. The method has the advantage of rapid analysis compared with column techniques, but the method will also, at the same time, separate the various anomers and different ring sizes of a particular monosaccharide, making the number of peaks on the chromatograms larger than the number of component monosaccharides. The need for volatile compounds for analysis has led to a number of methods of derivatisation, methyl ethers, alditol acetates, and trimethylsilyl ethers being used commonly, but other methods based on volatile products have also been employed, including the use of isopropylidene derivatives. The use of trimethylsilyl ethers of carbohydrates is preferred, owing to the ease of preparation of the derivatives at room temperature in a few minutes. This method has been extended to include derivatives of acidic monosaccharides and basic monosaccharides. More-recent developments that increase the value of gas-liquid chromatography in the structural analysis include the use of specific detectors and direct coupling of the gas liquid chromatograph to radioactive counters (gas-liquid radiochromatography) and to mass spectrometers.

MASS SPECTROMETRY

Mass spectrometry plays a large role in the structural analysis of polysaccharides, not only in the identification of compounds derived from methylation analysis, but also in the analysis of oligosaccharides directly after preparation of one of the volatile derivatives mentioned earlier (see Lönngren and Svennson, 1974). The molecular weight of small oligosaccharides can be measured and the sequence of monosaccharide units and position of glycosidic bonds have been determined, although some information on the nature of the residues present is also usually needed. The direct mass spectrometric identification of oligosaccharides containing more than four residues with the use of trimethylsilyl derivatives is difficult, but characteristic fragmentation patterns of peracetylated glycoside derivatives of pentasaccharides have been obtained and, more recently, a method has been described for the detection of D-fructose residues in permethylated oligosaccharides and which also gives information on the ratio of pyranose to furanose residues and the positions of the glycosidic linkages.

Methods involving chemical ionisation rather than electron impact are more sensitive than conventional methods of analysis, and such methods have been used in the analysis of oligopeptides, low molecular weight fragments obtained from hydrolysis of glycoproteins. Not only was the aminoacid sequence obtained but the carbohydrate-peptide linkages could be determined by comparing the fragmentation patterns with those obtained for the various monosaccharide-aminoacid derivatives (Morris, 1980).

NUCLEAR MAGNETIC RESONANCE SPECTROSCOPY

A method which will give information on the anomeric protons in polysaccharides, provided the monosaccharide components and substitution positions are known, has been developed using nuclear magnetic resonance (n.m.r.) spectroscopy. The hydroxyl groups of the sugar residues are preferably converted to derivatives such as their methyl or trimethylsilyl ethers to eliminate from the spectrum the peak due to hydroxyl groups. The protons of the anomeric carbon atom occur at lower field than protons on the other carbon atoms, with those in the equatorial position showing larger chemical shifts than those in axial positions. Complete structural analyses of polysaccharides have been obtained with ^{1}H n.m.r. spectroscopy of methylated monosaccharides, and simpler polysaccharides such as glycogens.

^{13}C n.m.r., despite its fairly recent introduction as a tool in carbohydrate analysis, has already proved to be a powerful technique for the structural determination of polysaccharides, providing information on their composition, sequence and comformation. Using the Fourier transform method, it allows spectra of the polysaccharides to be obtained using only their natural abundance ^{13}C atoms; it complements ^{1}H n.m.r. in that it gives better signal separation owing to the wider range of chemical shifts involved. The technique can be used on relatively small amounts of material, making it particularly suited to the study of polysaccharides of biological origin, where the lack of sufficient material has, in the past, frustrated many attempts at structural elucidation (for a review of the methodology and interpretation of spectra see Jennings and Smith, 1978 and Coxon, 1980).

A recent innovation is the use of two-dimensional ^{13}C n.m.r. in which the spectrum is displayed in two dimensions, showing signal strength as a function of chemical shift in one dimension and as a function of coupling constant in the other, the net effect of which is to produce a series of peaks for each carbon resonance which is characteristic of the ^{13}C–^{1}H coupling constants (Hall and Morris, 1980). The sensitivity of ^{13}C n.m.r. can be increased by increasing the contents of ^{13}C atoms in the carbohydrate (see Chapter 5, p. 92).

The use of ^{19}F, and ^{31}P n.m.r. may also be useful in determination of the position of substitution of monosaccharides by another, by using derivatives of the polysaccharides such as [^{19}F] trifluoroacetates.

ELECTROPHORETIC TECHNIQUES

Electrophoresis is not a substitute for chromatography, but provides very useful complementary information because it utilizes different criteria for separation, namely molecular charge, size and shape. The use of high-voltage paper electrophoresis has been applied to the separation not only of monosaccharides but also oligosaccharides. The method is not restricted to carbohydrate derivatives which possess an electric charge of their own, such as acidic monosaccharides, basic monosaccharides, and monosaccharide sulphates and phosphates, but has been extended to include neutral compounds which can form electrically charged complexes with electrolytes such as sodium borate, arsenite, or molybdate. The relative mobilities of the carbohydrates can be varied by changing the complexing agent used, when steric factors often determine the formation of different complexes. Choice of electrolytes will often lead to identification of the carbohydrate and its structure and bonding. Separations of acidic polysaccharides have been obtained directly, using high-voltage paper electrophoresis, but separation of neutral polysaccharides requires pre-conversion to their borate derivatives.

The development of better supporting media, such as cellulose acetate strips and polyacrylamide gels in the form of rods or slabs, for electrophoretic purposes has meant that purer chromatographic materials, which are of homogenous character and possess minimal adsorption properties, are available, thus reducing tailing and resulting in quicker separations on a small scale. Methods are reported for the separation of acidic polysaccharides on cellulose acetate and polyacrylamide gels, and the application of the latter method to molecular weight determination has been discussed (Mathews, 1976).

IMMUNOCHEMICAL REACTIONS

Polysaccharides have been found to be determinants of the immunological specificites of many types of micro-organisms. The specific interaction depends on the interaction of multiple reactive groups in both the polysaccharide antigen and protein antibody, and so a method based on this type of interaction is usually specific for the structure of a polysaccharide of known structure; it can be used to indicate structural similarities in unknown polysaccharides. An example of this specificity was shown in the discovery of the heterogeneity of a beef lung D-galactan. The precipitate formed with the anti-*Pneumococcus* type XIV sera contained proportions of D-galactose and D-glucuronic acid different from those in the original separation (Heidelberger *et al*., 1955).

THE USE OF ENZYMES IN STRUCTURAL ANALYSIS

Hydrolysis by enzymes provides an alternative method for the controlled hydrolysis of polysaccharides. The information obtained is not limited to that obtainable by analysis of the hydrolysis fragments because the specificity of enzyme action, a specificity based on type of monosaccharide and type of linkage, leads to significant data being obtained, by a process of elimination, from enzyme-resistant structures and partially hydrolysed structures. The enzymes which hydrolyse polysaccharides are, for convenience, divided into two groups, *endo*- and *exo*-polysaccharide hydrolases. *endo*-Polysaccharide hydrolases are specific for linkage and monosaccharide residue, and cause random fragmentation of homopolysaccharides to give a homologous series of oligosaccharides. Examples of this type of enzyme include α-amylase which gives a random series of D-glucose oligomers on reaction with amylose. *exo*-Polysaccharide hydrolases are specific for monosaccharide unit and stereochemistry at C-1 but do not differentiate between the residues attached glycosidically at C-1. They cleave polysaccharides by sequential removal of residues from one end of the molecule, usually the nonreducing end, which is the opposite end to that from which alkaline degradation starts. Examples of *exo*-polysaccharide hydrolases include β-amylase, which removes maltose units sequentially from amylose, producing an almost quantitative amount of maltose if the reaction goes to completion.

The first uses of enzyme analysis were in the determination of chain-length and degrees of branching in highly branched polysaccharides such as glycogen and amylopectin. Traditional methods for this analysis required the estimation of nonreducing end groups by chemical methods. The use of enzymes allows smaller quantities of material to be used and increases the speed of the determination. The method used by Lee and Whelan (1966) was based on the use of two enzymes, one of which (pullulanase) specifically hydrolyses the (1→6)-links at the 1,4,6-branch points to give linear chains of (1→4)-linked α-D-glucose units which are degraded by β-amylase to give maltose and D-glucose units, the latter arising from degradation of chains with an odd number of D-glucosyl residues. Analysis of the hydrolysis mixture for D-glucose gives a measure of the chain length because one D-glucose molecule is produced from one chain containing an odd number of D-glucosyl residues and, using the assumption that there is an equal number of odd and even chains, one D-glucose molecule is produced from two unit chains.

More recently, a number of glycoside hydrolases have been produced in sufficiently pure form to allow the development of a method of determination of monosaccharide sequences based on these enzymes. These enzymes will remove specific monosaccharide units linked by specific linkages from the nonreducing end of a polysaccharide. For example, β-D-galactosidase will remove D-galactosyl residues linked β-glycosidically to the polysaccharide. Table 4.1 gives a listing of the major polysaccharide-degrading enzymes, and reviews of the sources and methods of purification of these enzymes are available (Kennedy, 1971–1981). The method adopted can rely on the use of the enzymes sequentially or together, as in the case of enzymic degradation of keratan sulphate with β-D-galactosidase, β-D-2-acetamido-2-deoxyglucosidase, and a sulphatase, a method which showed that some D-galactosyl and 2-acetamido-2-deoxy-D-glucosyl groups at the nonreducing end of the molecule are non-sulphated. The use of sequential enzyme hydrolyses is a well-established technique particularly for the analysis of the carbohydrate residues of macromolecules (see Li and Li, 1976).

A number of glycopeptidases which will certainly prove to be of great value in the analysis of the carbohydrate moiety of glycoproteins have recently been isolated. These enzymes cleave the carbohydrate glycosidic bond adjacent to the aminoacid residue. An example of this type of enzyme is N^4-(2-acetamido-2-deoxy-β-D-glucopyranosyl) hydrogen L-asparaginate amidohydrolase [4-*N*-(2-β-D-glucosaminyl)-L-asparaginase, EC 3.5.1.26], which cleaves the bond between the 2-acetamido-2-deoxy-D-glucose units of glycoproteins containing the sequence D-mannose linked to *N,N'*-diacetylchitobiose linked in turn to an L-asparagine residue (19), that is, the glycopeptide linkage of the structure N^4-(2-acetamido-2-deoxy-β-D-glucopyranosyl) hydrogen L-asparaginate.

-β-D-Man*p*-(1→4)-β-D-Glc*p*NAc-(1→4)-β-D-Glc*p*NAc-(1→4′)-L-Asn

(19)

The hydrolysis of glycosidic linkages by enzymes involves scission of the glycosyl-oxygen bond (see Fig. 4.3), but a number of enzymes known as eliminases or lyases, usually of bacterial origin, react by a different mechanism and cause cleavage of the oxygen-aglycone bond (see Fig. 4.3) in acidic polysaccharides (such as pectins), producing unsaturated hexuronic acid units.

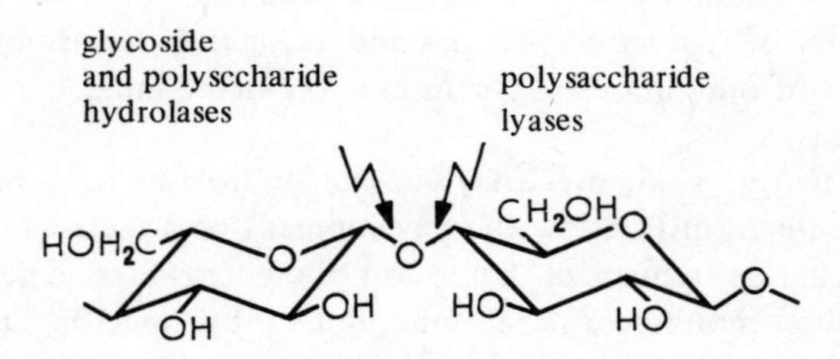

Fig. 4.3 – Position of action of glycoside and polysaccharide hydrolases and polysaccharide lyases.

Interpretation of the results from enzyme analysis must be carried out with caution. The mode of action of the *exo*-polysaccharide hydrolases is such that it is not possible to determine from which branch(es) the terminal residue(s) have been removed: this is in contrast to the results from the alkaline degradation method. Microheterogeneity of chains within the same molecule will also make interpretation uncertain. It is essential that the enzyme used is highly purified, as other glycoside hydrolases present in the enzyme can also lead to ambiguities and incorrect assumptions. It was originally thought, for example, that all D-mannose residues in glycoproteins were α-linked, but the use of α-D-mannosidase, purified to remove all traces of β-D-mannosidase activity, disproved this.

MOLECULAR SIZE AND SHAPE

A complete description of a polysaccharide involves an estimate of its molecular size and shape. Some of the methods described above will give a measure of the size and shape as part of the analysis (for example, gel filtration, and nonreducing

Table 4.1 Enzymes which may be used for the structural analysis of carbohydrate-containing macromolecules[a]

Trivial name	Systematic name	EC number[b]
α-*N*-Acetyl-D-galactosaminidase	2-Acetamido-2-deoxy-α-D-galactoside acetamidodeoxygalactohydrolase	3.2.1.49
β-*N*-Acetyl-D-galactosaminidase	2-Acetamido-2-deoxy-β-D-galactoside acetamidodeoxygalactohydrolase	3.2.1.53
endo-α-*N*-Acetyl-D-galactosaminidase	D-Galactosyl-*N*-acetamidodeoxy-α-D-galactoside D-galactosyl-*N*-acetamidodeoxy-D-galactohydrolase	3.2.1.97
endo-β-*N*-Acetyl-D-glucosaminidase	Mannosyl-glycoprotein 1,4-*N*-acetamidodeoxy-β-D-glycohydrolase	3.2.1.96
α-*N*-Acetyl-D-glucosaminidase	2-Acetamido-2-deoxy-α-D-glucoside acetamidodeoxyglucohydrolase	3.2.1.50
β-*N*-Acetyl-D-glucosaminidase	2-Acetamido-2-deoxy-β-D-glucoside acetamidodeoxyglucohydrolase	3.2.1.30
β-*N*-Acetyl-D-hexosaminidase	2-Acetamido-2-deoxy-β-D-hexoside acetamidodeoxyhexohydrolase	3.2.1.52
exo-β-*N*-Acetylmuramidase	Mucopolysaccharide β-*N*-acetylmuramoylexohydrolase	3.2.1.92
N-Acetylmuramoyl-L-alanine amidase	Mucopeptide amidohydrolase	3.5.1.28
Agarase	Agarose 3-glycanohydrolase	3.2.1.81
Alginate lyase	Poly(1,4-β-D-mannuronide) lyase	4.2.2.3
α-Amylase	1,4-α-D-Glucan glucanohydrolase	3.2.1.1
β-Amylase	1,4-α-D-Glucan maltohydrolase	3.2.1.2
Amylo-1,6-glucosidase	Dextrin 6-α-D-glucosidase	3.2.1.33
β-L-Arabinosidase	β-L-Arabinoside arabinohydrolase	3.2.1.88
α-L-Arabinofuranosidase	α-L-Arabinofuranoside arabinofuranohydrolase	3.2.1.55
Arylsulphatase	Aryl-sulphate sulphohydrolase	3.1.6.1
exo-Cellobiohydrolase	1,4-β-D-Glucan cellobiohydrolase	3.2.1.91
Cellulase	1,4-(1,3;1,4)-β-D-Glucan 4-glucanohydrolase	3.2.1.4
Cerebroside-sulphatase	Cerebroside-3-sulphate 3-sulphohydrolase	3.1.6.8
Chitinase	Poly(1,4-β-(2-acetamido-2-deoxy-D-glucoside)) glycanohydrolase	3.2.1.14
Chondroitin ABC lyase	Chondroitin ABC lyase	4.2.2.4
Chondroitin AC lyase	Chondroitin AC lyase	4.2.2.5
Chondro-4-sulphatase	Δ4,5-β-D-Glucuronosyl-(1,4)-2-acetamido-2-deoxy-D-galactose-4-sulphate 4-sulphohydrolase	3.1.6.9

Table 4.1 (*continued*)

Trivial name	Systematic name	EC number[b]
Chondro-6-sulphatase	Δ4,5-β-D-Glucuronosyl-(1,4)-2-acetamido-2-deoxy-D-galactose-6-sulphate 6-sulphohydrolase	3.1.6.10
Cyclomaltodextrinase	Cyclomaltodextrin dextrin-hydrolase (decyclizing)	3.2.1.54
Dextranase	1,6-α-D-Glucan 6-glucanohydrolase	3.2.1.11
2,6-β-D-Fructan 6-levanbiohydrolase	2,6-β-D-Fructan 6-β-D-fructofuranosyl fructohydrolase	3.2.1.64
β-D-Fructofuranosidase	β-D-Fructofuranoside fructohydrolase	3.2.1.26
exo-β-D-Fructosidase	β-D-Fructan fructohydrolase	3.2.1.80
Fucoidanase	Poly(1,2-α-L-fucoside-4-sulphate) glycanohydrolase	3.2.1.44
α-L-Fucosidase	α-L-Fucoside fucohydrolase	3.2.1.51
β-D-Fucosidase	β-D-Fucoside fucohydrolase	3.2.1.38
α-D-Galactosidase	α-D-Galactoside galactohydrolase	3.2.1.22
β-D-Galactosidase	β-D-Galactoside galactohydrolase	3.2.1.23
Galactosylceramidase	D-Galactosyl-*N*-acylsphingosine galactohydrolase	3.2.1.46
Galactosylgalactosylglucosylceramidase	D-Galactosyl-D-galactosyl-D-glucosylceramide galactohydrolase	3.2.1.47
endo-1,3-α-D-Glucanase	1,3-(1,3;1,4)-α-D-Glucan 3-glucanohydrolase	3.2.1.59
exo-1,3-α-Glucanase	1,3-α-D-Glucan 3-glucohydrolase	3.2.1.84
endo-1,3-β-D-Glucanase	1,3-β-D-Glucan glucanohydrolase	3.2.1.39
endo-1,3(4)-β-D-Glucanase	1,3-(1,3;1,4)-β-D-Glucan 3(4)-glucanohydrolase	3.2.1.6
4-*N*-(2-β-D-Glucosaminyl)-L-asparaginase	4-*N*-(2-Acetamido-2-deoxy-β-D-glucopyranosyl)-L-asparagine amidohydrolase	3.5.1.26
α-D-Glucosidase	α-D-Glucoside glucohydrolase	3.2.1.20
β-D-Glucosidase	β-D-Glucoside glucohydrolase	3.2.1 21
exo-1,4-α-D-Glucosidase	1,4-α-D-Glucan glucohydrolase	3.2.1.3
exo-1,6-α-D-Glucosidase	1,6-α-D-Glucan glucohydrolase	3.2.1.70
exo-1,3-β-D-Glucosidase	1,3-β-D-Glucan glucohydrolase	3.2.1.58
exo-1,4-β-D-Glucosidase	1,4-β-D-Glucan glucohydrolase	3.2.1.74
Glucosylceramidase	D-Glucosyl-*N*-acylsphingosine glucohydrolase	3.2.1.45
β-D-Glucuronidase	β-D-Glucuronide glucuronosohydrolase	3.2.1.31
Glycosylceramidase	Glycosyl-*N*-acylsphingosine glycohydrolase	3.2.1.62
Heparin lyase	Heparin lyase	4.2.2.7

Table 4.1 (*continued*)

Trivial name	Systematic name	EC number[b]
Heparitinsulphate lyase	Heparin-sulphate lyase	4.2.2.8
Hyaluronate lyase	Hyaluronate lyase	4.2.2.1
Hyaluronoglucosaminidase	Hyaluronate 4-glycanohydrolase	3.2.1.35
L-Iduronidase	Mucopolysaccharide α-L-iduronohydrolase	3.2.1.76
Inulinase	2,1-β-D-Fructan fructanohydrolase	3.2.1.7
Isoamylase	Glycogen 6-glucanohydrolase	3.2.1.68
exo-Isomaltohydrolase	1,6-α-D-Glucan isomaltohydrolase	3.2.1.94
exo-Isomaltotriohydrolase	1,6-α-D-Glucan isomaltotriohydrolase	3.2.1.95
Isopullulanase	Pullulan 4-glucanohydrolase	3.2.1.57
Lichenase	1,3-1,4-β-D-Glucan 4-glucanohydrolase	3.2.1.73
Lysozyme	Mucopeptide *N*-acetylmuramoylhydrolase	3.2.1.17
α-D-Mannosidase	α-D-Mannoside mannohydrolase	3.2.1.24
β-D-Mannosidase	β-D-Mannoside mannohydrolase	3.2.1.25
exo-1,2-1,3-α-D-Mannosidase	1,2-1,3-α-D-Mannan mannohydrolase	3.2.1.77
Neuraminidase	Acylneuraminyl hydrolase	3.2.1.18
Oligo-1,6-glucosidase	Dextrin 6-α-D-glucanohydrolase	3.2.1.10
Pectate lyase	Poly(1,4-α-D-galacturonide) lyase	4.2.2.2
Pectinesterase	Pectin pectylhydrolase	3.1.1.11
Pectin lyase	Poly(methoxygalacturonide) lyase	4.2.2.10
Polygalacturonase	Poly(1,4-α-D-galacturonide) glycanohydrolase	3.2.1.15
exo-Polygalacturonase	Poly(1,4-α-D-galacturonide) galacturonohydrolase	3.2.1.67
exo-Poly-α-D-galacturonosidase	Poly(1,4-α-D-galactosiduronate) digalacturonohydrolase	3.2.1.82
Pullulanase	Pullulan 6-glucanohydrolase	3.2.1.41
α-L-Rhamnosidase	α-L-Rhamnoside rhamnohydrolase	3.2.1.40
β-L-Rhamnosidase	β-L-Rhamnoside rhamnohydrolase	3.2.1.43
Sucrose α-D-glucohydrolase	Sucrose α-D-glucohydrolase	3.2.1.48
α,α-Trehalase	α,α-Trehalose glucohydrolase	3.2.1.28
endo-1,3-β-D-Xylanase	1,3-β-D-Xylan xylanohydrolase	3.2.1.32
endo-1,4-β-D-Xylanase	1,4-β-D-Xylan xylanohydrolase	3.2.1.8
D-Xylose isomerase	D-Xylose ketol isomerase	5.3.1.5
exo-1,3-β-D-Xylosidase	1,3-β-D-Xylan xylohydrolase	3.2.1.72
exo-1,4-β-D-Xylosidase	1,4-β-D-Xylan xylchydrolase	3.2.1.37

[a] The nomenclature used in this Table is identical to that used in Recommendations (1978) although that work does not necessarily comply with that recommended by IUPAC/IUB for carbohydrates.

[b] EC number stands for Enzyme Commission number, see Chapter 5, p.98.

end group analysis by methylation, or periodate oxidation), but specific methods are available for characterisation of the polysaccharides to give data on molecular weight, size, and distribution in any given sample.

The use of electron microscopy has been limited by the small size of the molecules which are being dealt with. They are too small to scatter electrons themselves, but the technique of casting a metal shadow on the molecules has led to single molecular patterns being obtained, but the method more frequently gives information only on molecular aggregates and conformational shape.

X-ray diffraction is another method which provides information and polysaccharides which form fibres gives satisfactory diffraction diagrams. These are usually linear molecules but the attachment of single unit side chains, if not too frequent, does not interfere with the formation of crystals and hence with the method. In highly branched polysaccharides crystallinity is only found if the high degree of substitution shows a regular pattern, but in the majority of polysaccharides crystallinity is only partial, resulting in dislocation in the crystal lattice and large areas of amorphous material, which makes the interpretation of results more difficult. This method of analysis has shown, for example, how the repeating units in glycosaminoglycans which consist of a disaccharide unit (see Chapter 9) are arranged in chains.

Colligative property measurements have been used for molecular weight determination. Below a molecular weight of 20 000 the method involving measurement on differences in vapour pressure or boiling points of pure solvent and solutions can be used, but these are limited by the sensitivity of the techniques available for measuring the small differences. Above a molecular weight of 20 000 the only method which can satisfactorily be used is the measurement of osmotic pressure. The limitation on this method is the sensitivity of measurement of the pressure differences for high molecular weight substances, the upper limit of molecular weight being of the order of 500 000. At the opposite end of the molecular weight range the nature of the semipermeable membrane dictates the limitation of the method. The newer techniques of vapour phase osmometry and dynamic osmometry have been used successfully.

Light scattering by dilute solutions of the polysaccharide provides an absolute method for the determination of molecular weights. Solutions of the polysaccharides are illuminated with monochromatic polarised or unpolarised light, and the scattered light intensity is measured as a function of the scattering angle. From these data the shape of the molecules and the molecular weight can be obtained.

Molecular weights and shapes can also be obtained from studies of sedimentation velocities of a solution of the polysaccharides under the influence of a high force field by following the changes in refractive index gradients. The rate of sedimentation obtained from these ultracentrifugation studies provides a measure of molecular weight and, on comparison with the calculated behaviour for molecular models, provides a basis for assessment of molecular shape. Fractionation by ultracentrifugal methods is now very popular, and effective separations of, for example, proteoglycans have been obtained using caesium chloride density gradients, whilst the use of neutral salt solutions at high concentrations is made for large scale preparations.

Using some, or all of the above techniques it has been possible to obtain detailed three dimensional structures for some polysaccharides and carbohydrate-containing macromolecules (for example, glycosaminoglycans). These are discussed in the relevant sections of the forthcoming chapters.

CHAPTER 5

Chemical and biochemical syntheses

This chapter describes the various reactions used to prepare monosaccharides and their polymerisation to give oligo- and poly-saccharides and discusses the reactions carried out by nature in performing synthesis of macromolecules from simple building blocks (biosynthesis).

CHEMICAL SYNTHESIS OF MONOSACCHARIDES

The total synthesis of many sugars has been achieved from non-carbohydrate materials but complex mixtures of products are obtained. In the 1860s the action of dilute alkali on formaldehyde was reported to produce a sweet syrup, which was called formose and has subsequently been shown to contain a mixture of aldoses and ketoses, including D- and L-glucose. The action of dilute alkali on acrolein dibromide does not produce the expected glyceraldehyde but a mixture of sugars named acrose. Initially a racemic mixture of D- and L-glyceraldehyde is produced some of which undergoes conversion into 1,3-dihydroxy-2-propanone (dihydroxyacetone) and then, by an aldol condensation reaction, produces a mixture of ketohexoses (Scheme 5.1). The major problem with these methods is that, at some stage of the reaction sequences, resolution of the racemic mixture (D- and L-forms) is necessary if a pure compound is to be produced. However, Fischer successfully used the second method to prepare α-acrose (a mixture of D- and L-fructose, see Scheme 5.1) from which he prepared many of the previously unknown aldohexoses using some of the reactions discussed below.

The fact that a number of monosaccharides are readily available from natural sources means that they provide suitable starting materials for the less readily available sugars. By using a naturally occurring product only one optical isomer is used, thus eliminating the need for resolution of racemic mixtures of products in most cases. A number of methods have been devised for increasing or decreasing the number of carbon atoms in the molecules (that is, ascent or descent of the homologous series) and for the interconversion of monosaccharides without changing the number of carbon atoms present. These methods have been fully reviewed (Hough and Richardson, 1972 and 1979, and Szarek, 1973) and are described here only briefly.

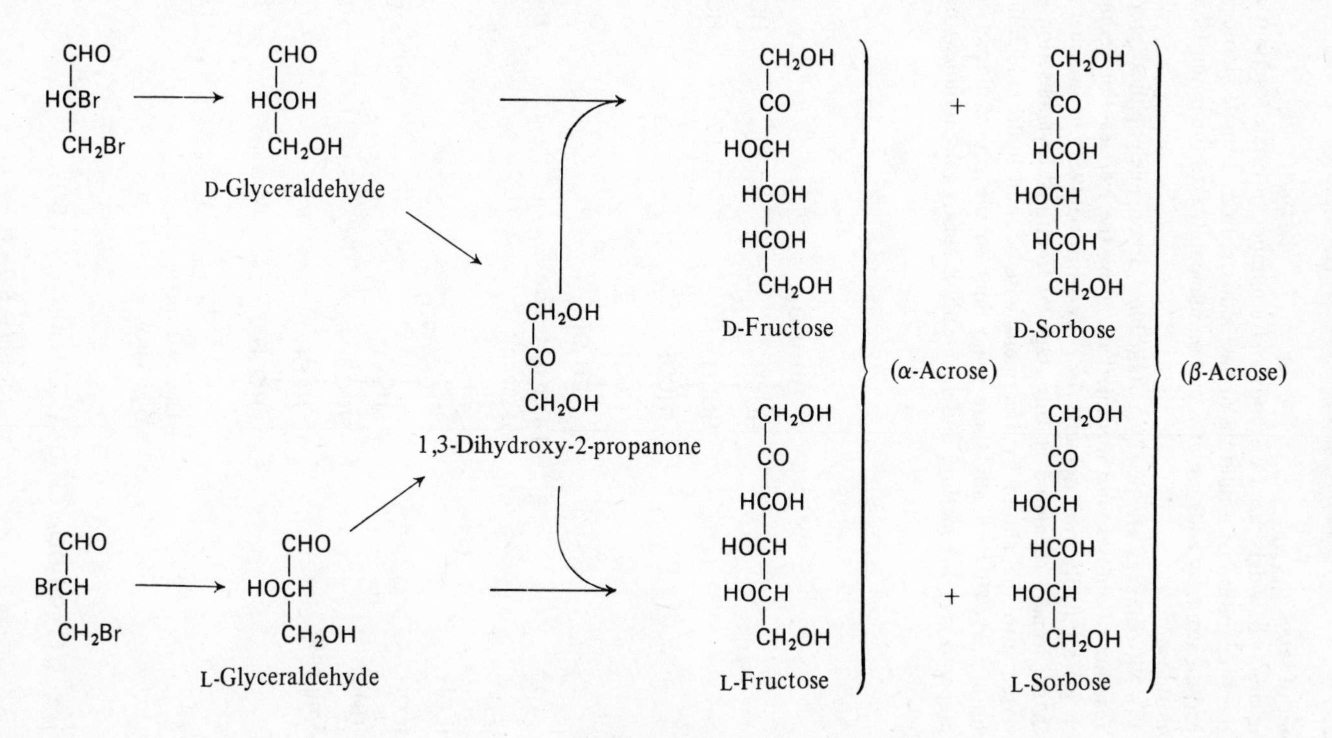

D-Glyceraldehyde
L-Glyceraldehyde
1,3-Dihydroxy-2-propanone
D-Fructose
D-Sorbose
L-Fructose
L-Sorbose
(α-Acrose)
(β-Acrose)

Scheme 5.1

Ascent of series

There are two widely used methods and a number of other reactions available for the preparation of a mixture of two aldoses, from a single aldose with one less carbon atom, by addition of a 'carbon nucleophile' to the potential aldehyde of the aldose.

(a) Nitroalkanes are readily obtainable 'carbon nucleophiles' prepared by the reaction of aldoses with nitromethane using the Nef reaction (Scheme 5.2) which gives initially two isomeric 1-deoxy-1-nitro-alditols in unequal amounts which are usually separated at this stage by fractional crystallisation and then converted into the aldose by treatment with sodium hydroxide followed by sulphuric acid. Theoretically there is no limit to the size of aldose which can be prepared by this method and nine-carbon sugars (aldononoses) have been prepared.

CHO
HOCH
HCOH
HCOH
CH_2OH
D-Arabinose

$\xrightarrow{(1)}$

$CH{=}NO_2$
HOCH
HOCH
HCOH
HCOH
CH_2OH
1-Deoxy-1-nitro-D-mannitol

$\xrightarrow{(2), (3)}$

CHO
HOCH
HOCH
HCOH
HCOH
CH_2OH
D-Mannose

+

$CH{=}NO_2$
HCOH
HOCH
HCOH
HCOH
CH_2OH
1-Deoxy-1-nitro-D-glucitol

$\xrightarrow{(2), (3)}$

CHO
HCOH
HOCH
HCOH
HCOH
CH_2OH
D-Glucose

Reagents: (1) CH_3NO_2, CH_3ONa; (2) NaOH; (3) H_2SO_4

Scheme 5.2

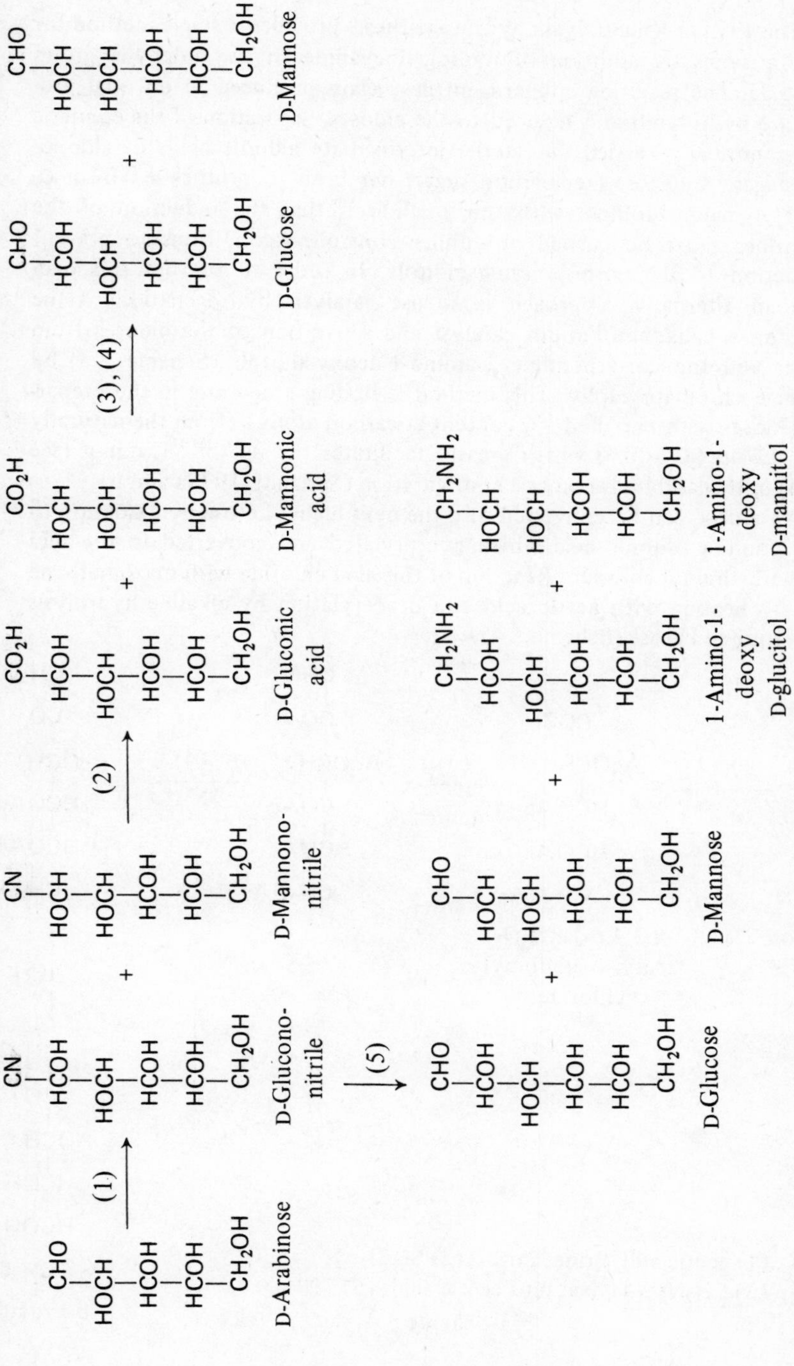

D-Arabinose (CHO, $HOCH$, $HCOH$, $HCOH$, CH_2OH) $\xrightarrow{(1)}$ D-Glucononitrile (CN, $HCOH$, $HOCH$, $HCOH$, $HCOH$, CH_2OH) + D-Mannononitrile (CN, $HOCH$, $HOCH$, $HCOH$, $HCOH$, CH_2OH)

$\xrightarrow{(2)}$ D-Gluconic acid (CO_2H, $HCOH$, $HOCH$, $HCOH$, $HCOH$, CH_2OH) + D-Mannonic acid (CO_2H, $HOCH$, $HOCH$, $HCOH$, $HCOH$, CH_2OH)

$\xrightarrow{(3), (4)}$ D-Glucose (CHO, $HCOH$, $HOCH$, $HCOH$, $HCOH$, CH_2OH) + D-Mannose (CHO, $HOCH$, $HOCH$, $HCOH$, $HCOH$, CH_2OH)

Glucono- and mannono-nitriles $\xrightarrow{(5)}$ D-Glucose (CHO, $HCOH$, $HOCH$, $HCOH$, $HCOH$, CH_2OH) + D-Mannose (CHO, $HOCH$, $HOCH$, $HCOH$, $HCOH$, CH_2OH) + 1-Amino-1-deoxy D-glucitol (CH_2NH_2, $HCOH$, $HOCH$, $HCOH$, $HCOH$, CH_2OH) + 1-Amino-1-deoxy D-mannitol (CH_2NH_2, $HOCH$, $HOCH$, $HCOH$, $HCOH$, CH_2OH)

Reagents: (1) HCN or NaCN; (2) NaOH; (3) heat to form lactone; (4) Na/Hg; (5) Pd/$BaSO_4$

Scheme 5.3

(b) The Fischer-Kiliani cyanohydrin synthesis provides a single method for ascending a series by addition of hydrogen cyanide to the aldehyde group (Scheme 5.3). The resulting epimeric nitriles, again produced in unequal proportions, are hydrolysed and reduced to the aldoses. Separation of the epimeric products is normally carried out at the intermediate aldonic acids or aldonolactones stages. A decose (ten-carbon sugar) has been so synthesised from D-glucose. The major problem with this method is that the reduction of the aldonolactones must be carried out under controlled conditions to prevent over reduction to the corresponding alditols. In order to prevent this over reduction an alternative approach is to use catalytic hydrogenation of the nitriles using a palladium/barium catalyst and separation of the aldoses from each other and the corresponding 1-amino-1-deoxy-alditols (Scheme 5.3) by ion-exchange chromatography. This method is finding a new use in the preparation of aldoses with enriched ^{13}C content at carbon atom 1 (from the naturally occurring 1% to 90–100%) which greatly facilitates the use of ^{13}C n.m.r. (see Chapter 4) in the study of anomeric configuration (Serianni, 1979a and b).

(c) An aldose can be converted into the next higher ketose by oxidation to the corresponding aldonic acid which is acetylated and converted to the acid chloride with thionyl chloride. Reaction of this acid chloride with diazomethane followed by heating with acetic acid and deacetylation by alkaline hydrolysis gives the required ketose (Scheme 5.4).

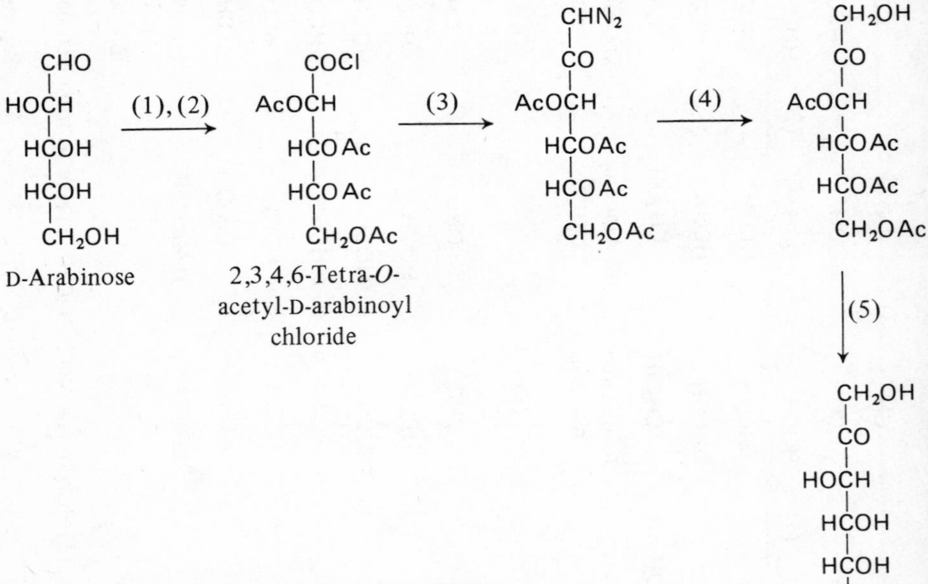

CHO–HOCH–HCOH–HCOH–CH_2OH (D-Arabinose) —(1),(2)→ COCl–AcOCH–HCOAc–HCOAc–CH_2OAc (2,3,4,6-Tetra-*O*-acetyl-D-arabinoyl chloride) —(3)→ CHN_2–CO–AcOCH–HCOAc–HCOAc–CH_2OAc —(4)→ CH_2OH–CO–AcOCH–HCOAc–HCOAc–CH_2OAc —(5)→ CH_2OH–CO–HOCH–HCOH–HCOH–CH_2OH (D-Fructose)

Reagents: (1) acetic anhydride, $ZnCl_2$; (2) $SOCl_2$;
(3) CH_2N_2; (4) heat plus acetic acid; (5) ^-OH

Scheme 5.4

Other methods which have been used to ascend a series include the use of Grignard reagents, malonate esters and the Wittig reaction.

Descent of series

Removal of one of the terminal carbon atoms from an aldose to give the corresponding lower aldose can be achieved by a number of methods. The oldest laboratory method is the Wohl degradation which involves heating the aldose oxime with acetic anyhydride and zinc chloride or sodium acetate so that the oxime undergoes dehydration and acetylation to give an *O*-acetylated nitrile. Hydrogen cyanide is removed by an ammonia-induced reaction and the required aldose with one less carbon atom is then obtained by mild acid hydrolysis (Scheme 5.5). The yield can be increased by using sodium methoxide in chloroform rather than the traditional silver oxide/ammonia combination.

D-Glucose		D-Glucose oxime		2,3,4,5,6-penta-*O*-acetyl-D-Glucononitrile		1,1-Bisacetamido-D-arabinose
CHO		CH = NOH		CN		$CH(NHCOCH_3)_2$
HCOH		HCOH		HCOAc		HOCH
HOCH	(1) →	HOCH	(2) →	AcOCH	(3) →	HCOH
HCOH		HCOH		HCOAc		HCOH
HCOH		HCOH		HCOAc		CH_2OH
CH_2OH		CH_2OH		CH_2OAc		

↓ (4)

CHO
HOCH
HCOH
HCOH
CH_2OH

D-Arabinose

Reagents: (1) H_2NOH; (2) acetic anhydride, $ZnCl_2$; (3) NH_3; (4) H^+

Scheme 5.5

The Ruff degradation method involves the oxidation of the calcium salt of an aldonic acid with hydrogen peroxide in the presence of iron(III) ions (Fenton's reagent) to give a 2-ketoaldonic acid and, by subsequent loss of carbon dioxide, the required aldose (Scheme 5.6). Further degradation lowers the yield although the use of ion-exchange resins helps to eliminate these degradation losses.

$$
\begin{array}{c} CO_2^- \cdot \frac{1}{2}Ca^{2+} \\ | \\ HCOH \\ | \\ HOCH \\ | \\ HCOH \\ | \\ HCOH \\ | \\ CH_2OH \end{array}
\xrightarrow{(1)}
\begin{array}{c} CO_2^- \cdot \frac{1}{2}Ca^{2+} \\ | \\ HCO \\ | \\ HOCH \\ | \\ HCOH \\ | \\ HCOH \\ | \\ CH_2OH \end{array}
\longrightarrow
\begin{array}{c} CHO \\ | \\ HOCH \\ | \\ HCOH \\ | \\ HCOH \\ | \\ CH_2OH \end{array}
+ CO_2
$$

Calcium D-glucuronate — Calcium D-*erythro*-2-pentulosonate — D-Arabinose

Reagents: (1) H_2O_2, Fe^{3+}

Scheme 5.6

$$
\begin{array}{c} CHO \\ | \\ HCOH \\ | \\ HOCH \\ | \\ HCOH \\ | \\ HCOH \\ | \\ CH_2OH \end{array}
\xrightarrow{(1)}
\begin{array}{c} CH(SC_2H_5)_2 \\ | \\ HCOH \\ | \\ HOCH \\ | \\ HCOH \\ | \\ HCOH \\ | \\ CH_2OH \end{array}
\xrightarrow{(2)}
\begin{array}{c} CH(SO_2C_2H_5)_2 \\ | \\ HCOH \\ | \\ HOCH \\ | \\ HCOH \\ | \\ HCOH \\ | \\ CH_2OH \end{array}
\xrightarrow{(3)}
\begin{array}{c} CHO \\ | \\ HOCH \\ | \\ HCOH \\ | \\ HCOH \\ | \\ CH_2OH \end{array}
+ CH(SO_2C_2H_5)_2
$$

D-Glucose — D-Glucose di-*S*-ethyl dithioacetal — D-Glucose di-(ethylsulphone) — D-Arabinose

Reagents: (1) C_2H_5SH, HCl;
(2) $C_2H_5CO_3H$; (3) NH_3

Scheme 5.7

In the disulphone degradation method, the aldose is converted to a dithioacetal (mercaptal) by reaction with an alkanethiol, followed by oxidation with a peracid to give the disulphone which, in dilute ammonia, undergoes cleavage of the C-1–C-2 bond to give the required product (Scheme 5.7). Due to the high yields and purity of the final products the method is to be preferred.

Other methods which can be used to descend a series include the Hofmann degradation of an acid amide with hypochlorite and the use of glycol cleavage agents to degrade a suitably blocked derivative.

Interconversion without changing the number of carbon atoms

The chemistry of conversion of aldoses into ketoses, using mildly basic conditions such as aqueous calcium hydroxide, has already been described (Chapter 3), whilst the reverse process can be carried out *via* reduction of the ketose to alditols, careful oxidation to aldonic acids and subsequent reduction to give a mixture of aldoses. The chemistry of these reactions is discussed in Chapter 3.

The preparation of some rare sugars has been achieved by inversion of configuration at one or more chiral centres (epimerisation) of a readily available monosaccharide. Due to the extensive rearrangements which occur with free sugars during base catalysed epimerisation these reactions are rarely carried out on unprotected monosaccharides. In order to illustrate the many methods which are, or have been, used to invert the configuration of hydroxyl groups the epimerisation of D-glucose at C-5 to give L-idose (Scheme 5.8) is shown. The 5,6-ditosylated D-glucofuranose derivative (20) undergoes nucleophilic displacement on treatment with sodium benzoate in *N,N*-dimethylformamide to give an L-idose derivative (21). As will be seen later in this chapter, an increase in the

CH_2OH HO HO HO OH — D-Glucose —many steps→ $TsOH_2C$ OTs H AcO O O O H_3C CH_3 (20) —(1)→ $BzOH_2C$ H OBz HO O O O H_3C CH_3 (21) → HO O CH_2OH HO HO OH L-Idose

Reagents:
(1) sodium benzoate in *N,N*-dimethylformamide
Ts = 4-toluenesulphonyl (tosyl), Bz = benzoyl

Scheme 5.8

understanding of biological reactions is eliminating the need to rely on purely chemical methods since enzymes have been isolated which can perform many of these interconversions specifically (for example, epimerases EC 5.1.3. etc. and isomerases EC 5.3.1. etc.) on unprotected monosaccharides.

CHEMICAL SYNTHESIS OF OLIGOSACCHARIDES

The synthesis of oligosaccharides etc. from monosaccharides generally takes place *via* glycosidation reactions, such as Fischer glycosidation (see Chapter 3), using fully protected monosaccharides with only the positions required for bond formation free. Hence D-glucose protected at positions 2, 3 and 6 produces the expected oligo- and poly-saccharides containing (1→4)-linkages but with random arrangements of α- and β-linkages. It also contains some head to head termination products (products formed by linking C-1 of 1 residue to C-1 of another residue, as in trehaloses, Chapter 2). Separation of this mixture presents complex problems and is often impossible to achieve.

A stepwise addition of a monosaccharide (or oligosaccharide) to another monosaccharide (or oligosaccharide) can be achieved using the Koenigs-Knorr method (see Chapter 3). The carbohydrate unit which will contain the reducing end of the final product is suitably protected such that only that hydroxyl group required for bond formation remains unsubstituted (22). This is then reacted with the appropriate glycosyl halide (23) and the final product (24) is obtained with a β-D-linkage (Scheme 5.9). To synthesise an oligosaccharide with an α-D-linkage the glycosyl halide must contain a protecting group at C-2 such as nitrate which has no neighbouring group effect (25). Thus, isomaltose (26) can be prepared with only a small amount of gentiobiose (24) by-product *via* Scheme 5.10. An alternative method involves the use of 1,2-anhydro derivatives, such as Brigl's anhydride (27), which, at elevated temperatures, interacts with the acetate protecting group at C-6 (28) producing a mixture of anomers. At normal temperatures β-D-glycosides are produced (Scheme 5.11).

As should be obvious from the above descriptions of synthesis of monosaccharides and glycoside bond formation, the chemical synthesis of oligosaccharides is not a simple procedure due to the multistage reactions required to produce the properly substituted carbohydrates but small oligosaccharides have been produced successfully. The transferase activities of carbohydrate hydrolase type enzymes can also be used to build up oligosaccharide structures.

CHEMICAL SYNTHESIS OF POLYSACCHARIDES AND CARBOHYDRATE-CONTAINING MACROMOLECULES

For more-complex polysaccharides the above methods are not appropriate due to the losses, which occur at each stage, severely limiting the amount of material available for subsequent reactions. Attempts have been made to convert some polysaccharides into others by chemical means. For example, synthesis of the glycosaminoglycan heparin (see Chapter 9) has been attempted by sulphation

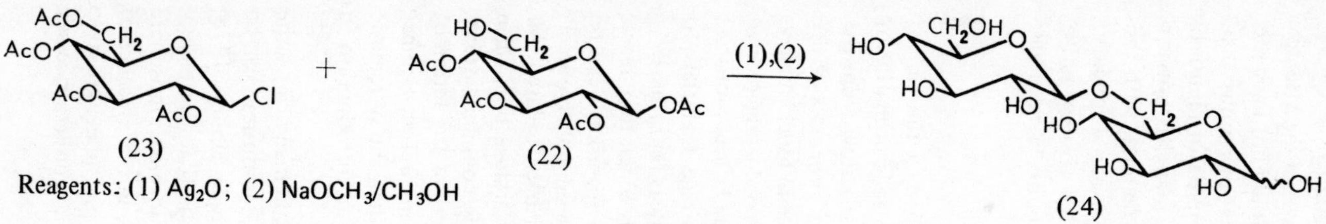

AcO—CH₂, AcO, AcO, AcO, O, Cl
(23)

\+

HO—CH₂, AcO, AcO, AcO, O, OAc
(22)

(1),(2) →

HO, CH₂OH, O, HO, HO, O—CH₂, HO, O, HO, HO, OH
(24)

Reagents: (1) Ag_2O; (2) $NaOCH_3/CH_3OH$

Scheme 5.9

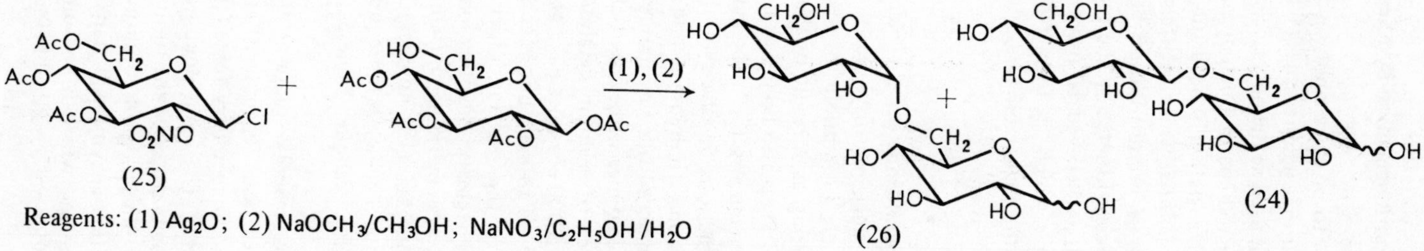

AcO—CH₂, AcO, AcO, O₂NO, O, Cl
(25)

\+

HO—CH₂, AcO, AcO, AcO, O, OAc

(1), (2) →

HO, CH₂OH, O, HO, HO, O—CH₂, HO, O, HO, HO, OH
(26)

\+

HO, CH₂OH, O, HO, HO, O—CH₂, HO, O, HO, HO, OH
(24)

Reagents: (1) Ag_2O; (2) $NaOCH_3/CH_3OH$; $NaNO_3/C_2H_5OH/H_2O$

Scheme 5.10

CH₂OAc, AcO, O, O, OAc
(27)

(1) →

AcO—CH₂, AcO, AcO, HO, O, OR

\+

AcO—CH₂, AcO, AcO, HO, O, OR

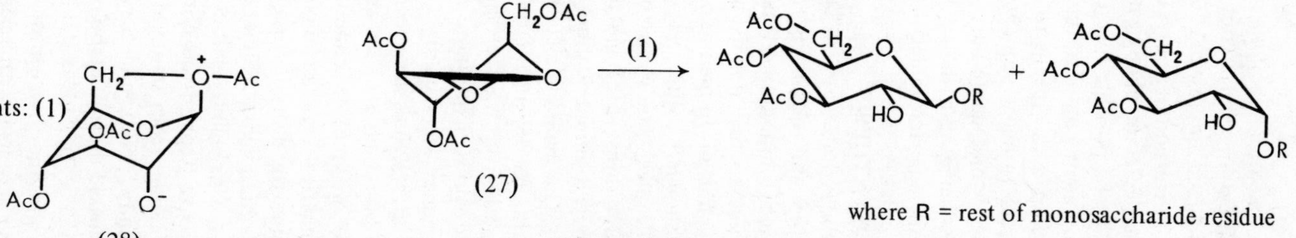

where R = rest of monosaccharide residue

Reagents: (1) CH₂—O⁺—Ac, OAc, O, AcO, O⁻
(28)

Scheme 5.11

and oxidation reactions on chitin to produce a polysaccharide containing sulphamido groups and hexopyranosyluronic acid residues (Whistler and Kosik, 1971). Starch and amylose have also been used as the polysaccharide base in the attempted conversion of a polysaccharide to heparin. Such preparations did not have the same degree of biological activity as naturally occurring heparin; this shows that it may be possible to prepare a material which satisfies the criteria of identical component composition and degree of substitution but that man has not yet developed the means to reproduce synthetically the required material with the correct primary sequence which is believed to produce the exact secondary, tertiary and quaternary structures which, in turn, result in the correct active sites being constructed to give the high biological activity found in naturally occurring macromolecules.

BIOSYNTHESIS

Carbohydrates are of central significance in the balance between the earth's 'living' and 'non-living' carbon, since photosynthesis, which leads primarily to neutral monosaccharides, is largely responsible for reversing the flow from the 'living' to the 'non-living' occurring as a result of the normal processes of life (respiration, fermentation, etc.). They also hold a central place in biochemistry, as shown by Fig. 5.1, being precursors for many naturally occurring macromolecules. There are many reactions involved in the degradation processes (catabolism) of D-glucose to non-carbohydrate and of storage materials to simple monomeric units which are controlled by nature's catalysts (enzymes).

The main characteristic of enzymes is the specificity with which they react. Although dependent on the type of reaction catalysed, this specificity can be gauged from the fact that over 2,000 enzymes have been isolated and characterised whilst many more are believed to exist but have not yet been isolated. All enzymes are named in terms of the reaction which they catalyse but, since 1956, enzymes have been given systematic classifications (see Recommendations, 1978) which gives an enzyme a systematic name (which describes the action of the enzyme), a commission number (EC number) based on a four-number numerical code and a common name which is more practical than the systematic name for everyday usage (see, for example, Table 5.1). For an introduction to enzymology see Palmer (1981).

This chapter will not discuss catabolic reactions; they are, in many cases, the reverse steps of the synthetic reactions (anabolism) and full descriptions can be obtained from standard biology/biochemistry texts (for example, Holum, 1978, Candy, 1980, and Stryer, 1981). The main point to note is that the energy released during catabolism is usually stored in the form of adenosine triphosphate (ATP, 29). This is stable under physiological conditions but, in the presence of a phosphoryl acceptor, the correct enzyme will transfer the phosphate group to the acceptor with release of energy which is consumed by anabolic processes. It is these anabolic processes which are described below.

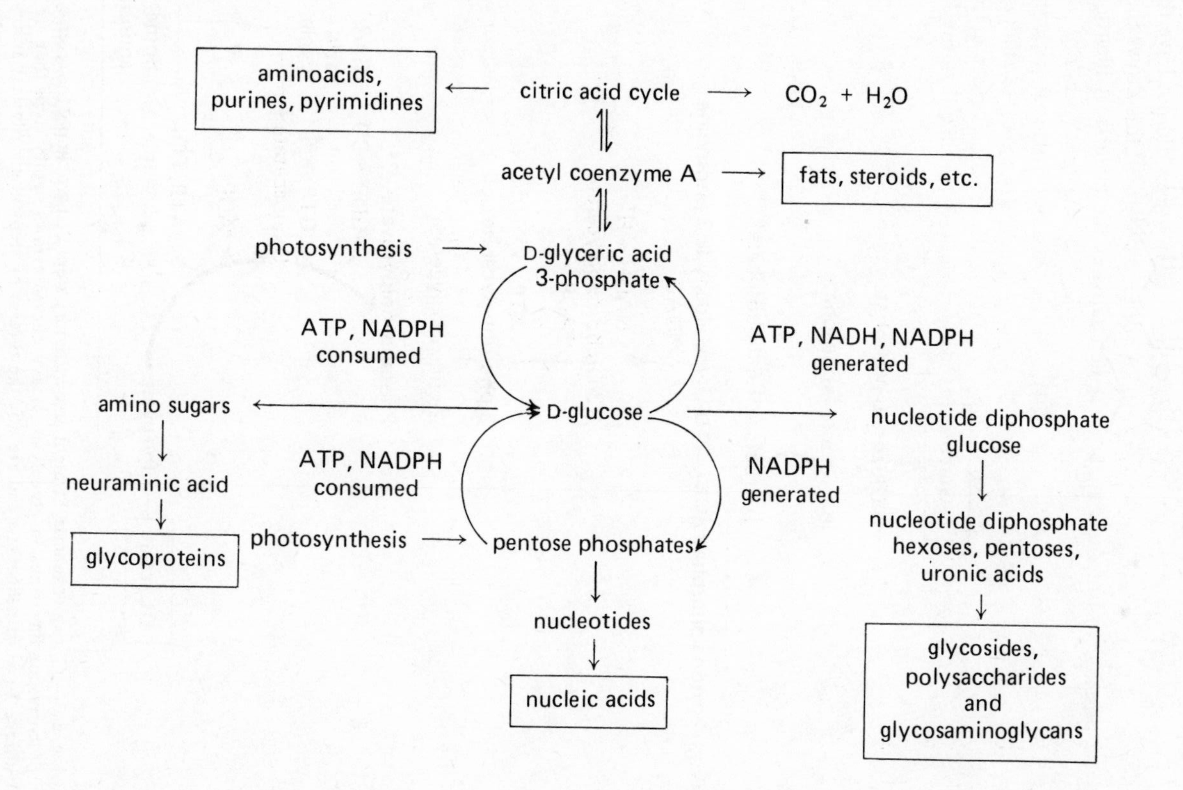

Fig. 5.1 – The central role of D-glucose in biochemistry (see Table 1.1, p. 17 for explanation of abbreviations).

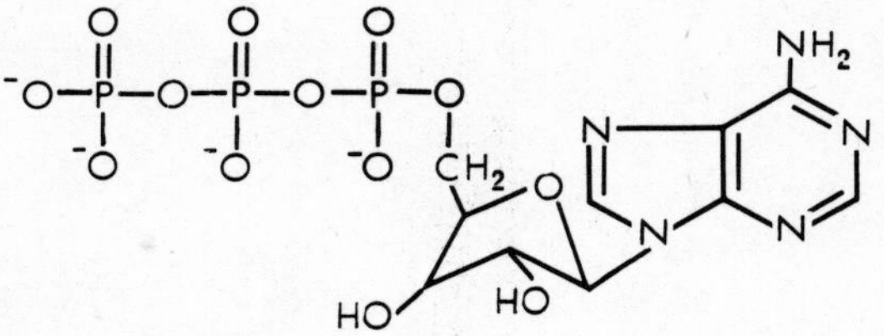

O O O
$^-O-P-O-P-O-P-O$
^-O ^-O ^-O CH_2
NH_2
N N N N O
HO HO

(29)

D-Glucose

D-Glucose 6-phosphate

D-Fructose 6-phosphate

D-Fructose 1,6-diphosphate

Dihydroxyacetone phosphate ⇌ D-Glyceraldehyde 3-phosphate

NAD^+

$NADH + H^+$

1,3-Diphosphoglycerate

ADP

ATP

3-phosphoglycerate

2-phosphoglycerate

Phosphoenolpyruvate

$GDP + CO_2$

GTP

Citric acid cycle

Oxaloacetate

ADP

ATP, CO_2

L-Lactate → Pyruvate

Some aminoacids

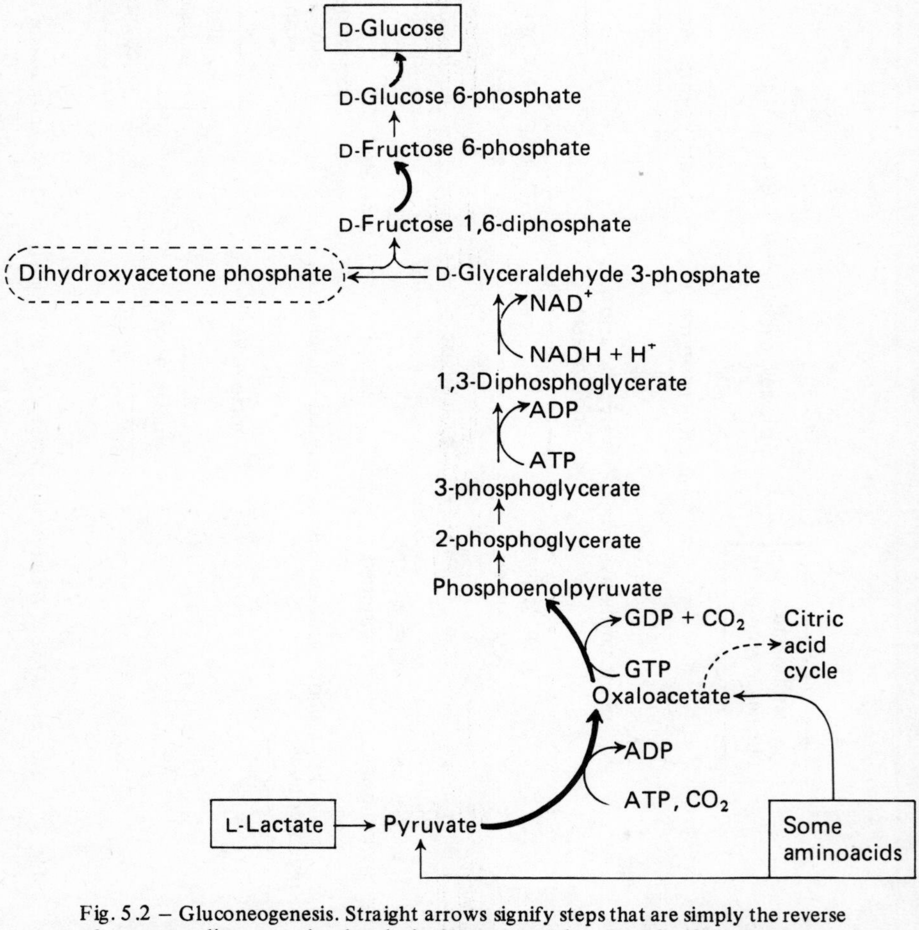

Fig. 5.2 – Gluconeogenesis. Straight arrows signify steps that are simply the reverse of corresponding steps in glycolysis. Heavy curved arrows signify steps that are unique to gluconeogenesis and are not the reverse of steps in glycolysis. Carbon skeletons from nearly all aminoacids can be used in one way or another to make D-glucose (see Table 1.1, p. 17 for explanation of abbreviations).

Anabolism of D-glucose

The biosynthesis of D-glucose from non-carbohydrate materials, known as gluconeogenesis ('glucose-new-genesis'), is outlined in Fig. 5.2. Nearly all amino-acids can be partially degraded to give a suitable precursor for gluconeogenesis but lactate, a reduction product of pyruvate, can also be used in order to maintain the body's metabolic balance. Once D-glucose has been formed, different derivatives from the phosphate esters (for example, ATP) take over the role of providing the driving force for the reactions. Through these derivatives, typified by uridine diphosphate-D-glucose (UDP-D-Glc, 30) formed from α-D-glucopyranose 1-phosphate and uridine triphosphate (UTP, 31) as shown in Scheme 5.12, the majority of interconversions and glycosylations occur. For example, D-glucose can be converted to D-galactose by epimerisation at C-4 catalysed by the enzyme UDP-D-glucose 4-epimerase (EC 5.1.3.2) or to D-glucuronic acid by oxidation catalysed by UDP-D-glucose dehydrogenase (EC 1.1.1.22). Subsequent decarboxylation of D-glucuronic acid by UDP-D-glucuronate decarboxylase (EC 4.1.1.35) produces D-xylose. A list of the various enzymes involved in the biosynthesis of D-glucose and its subsequent conversion into other mono- and poly-saccharides is given in Table 5.1.

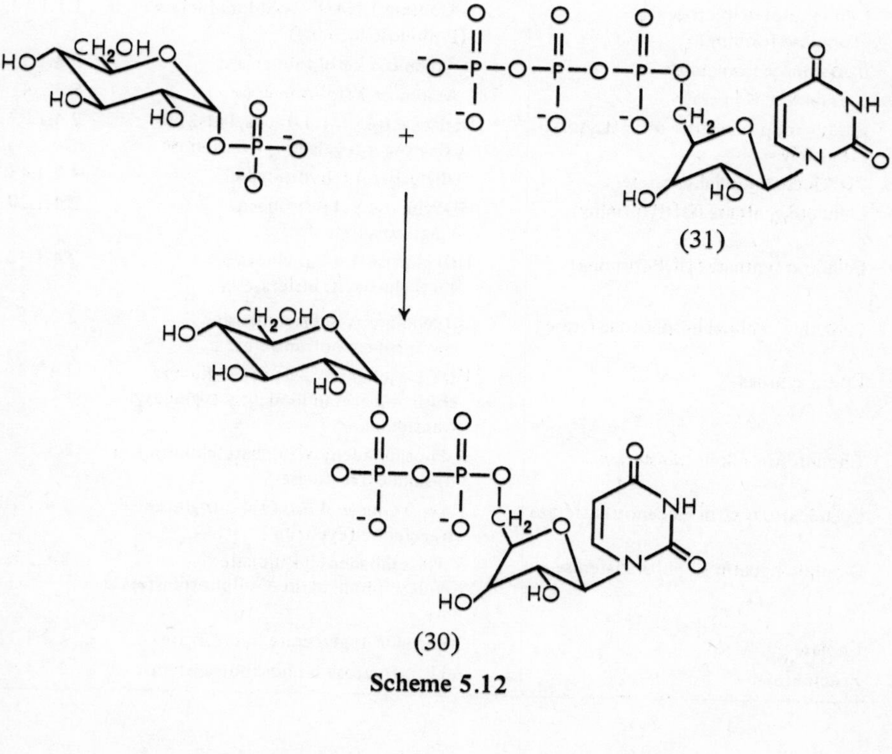

(31)

(30)

Scheme 5.12

Table 5.1 Some of the enzymes involved in the biosynthesis of monosaccharides and other carbohydrate-containing molecules[a]

Trivial name	Systematic name	EC number[b]
Acetylglucosamine phosphomutase	2-Acetamido-2-deoxy-D-glucose-1,6-bisphosphate: 2-acetamido-2-deoxy-D-glucose-1-phosphate phosphotransferase	2.7.5.2
Acylglucosamine 2-epimerase	2-Acylamido-2-deoxy-D-glucose 2-epimerase	5.1.3.8
Acylglucosamine-6-phosphate 2-epimerase	2-Acylamido-2-deoxy-D-glucose-6-phosphate 2-epimerase	5.1.3.9
Acylneuraminate cytidylyltransferase	CTP:*N*-acylneuraminate cytidylyltransferase	2.7.7.43
Aldonolactonase	L-Gulono-γ-lactone lactonohydrolase	3.1.1.18
Alginate synthase	GDPmannuronate:alginate D-mannuronyltransferase	2.4.1.33
Amylosucrase	Sucrose:1,4-α-D-glucan 4-α-D-glucosyltransferase	2.4.1.4
D-Arabinitol dehydrogenase	D-Arabinitol:NAD^+ 4-oxidoreductase	1.1.1.11
L-Arabinitol dehydrogenase	L-Arabinitol:NAD^+ 4-oxidoreductase (L-xylulose-forming)	1.1.1.12
L-Arabinitol dehydrogenase (ribulose-forming)	L-Arabinitol:NAD^+ 2-oxidoreductase (L-ribulose-forming)	1.1.1.13
D-Arabinose isomerase	D-Arabinose ketol-isomerase	5.3.1.3
L-Arabinose isomerase	L-Arabinose ketol-isomerase	5.3.1.4
Blood-group-substance α-D-galactosyltransferase	UDPgalactose:*O*-α-L-fucosyl-(1,2)-D-galactose α-D-galactosyltransferase	2.4.1.37
CDPglucose 4,6-dehydratase	CDPglucose 4,6-hydro-lyase	4.2.1.45
Cellulose synthase (GDP-forming)	GDPglucose:1,4-β-D-glucan 4-β-glucosyltransferase	2.4.1.29
Cellulose synthase (UDP-forming)	UDPglucose:1,4-β-D-glucan 4-β-D-glucosyltransferase	2.4.1.12
Ceramide cholinephosphotransferase	CDPcholine:*N*-acylsphingosine cholinephosphotransferase	2.7.8.3
Chitin synthase	UDP-2-acetamido-2-deoxy-D-glucose: chitin 4-β-acetamidodeoxy-D-glucosyltransferase	2.4.1.16
Chondroitin sulphotransferase	3′-Phosphoadenylylsulphate:chondroitin 4′-sulphotransferase	2.8.2.5
Cyclomaltodextrin glucanotransferase	1,4-α-D-Glucan 4-α-D-(1,4-α-D-glucano)-transferase (cyclizing)	2.4.1.19
Desulphoheparin sulphotransferase	3′-Phosphoadenylylsulphate: *N*-desulphoheparin *N*-sulphotransferase	2.8.2.8
Enolase	2-Phospho-D-glycerate hydro-lyase	4.2.1.11
Fructokinase	ATP:D-fructose 6-phosphotransferase	2.7.1.4

Table 5.1 (*continued*)

Trivial name	Systematic name	EC number[b]
Fructose-bisphosphatase	D-Fructose-1,6-bisphosphate 1-phosphohydrolase	3.1.3.11
Fructose-bisphosphate aldolase	D-Fructose-1,6-bisphosphate D-glyceraldehyde-3-phosphate-lyase	4.1.2.13
Galactokinase	ATP:D-galactose 1-phosphotransferase	2.7.1.6
Galactose-1-phosphate uridylyltransferase	UTP:α-D-galactose-1-phosphate uridylyltransferase	2.7.7.10
Galactosylceramide sulphotransferase	3′-Phosphoadenylylsulphate: galactosyl ceramide 3′-sulphotransferase	2.8.2.11
GDPfucose–glycoprotein fucosyltransferase	GDPfucose:glycoprotein fucosyltransferase	2.4.1.68
GDPmannose 4,6-dehydratase	GDPmannose 4,6-hydro-lyase	4.2.1.47
Glucosamine acetyltransferase	Acetyl-CoA:2-amino-2-deoxy-D-glucose *N*-acetyltransferase	2.3.1.3
Glucosamine-phosphate acetyltransferase	Acetyl-CoA:2-amino-2-deoxy-D-glucose-6-phosphate *N*-acetyltransferase	2.3.1.4
1,4-α-Glucan branching enzyme	1,4-α-D-Glucan:1,4-α-D-glucan 6-α-D-(1,4-α-D-glucano)-transferase	2.4.1.18
1,4-α-D-Glucan 6-α-D-glucosyltransferase	1,4-α-D-Glucan:1,4-α-D-glucan (D-glucose) 6-α-D-glucosyltransferase	2.4.1.24
4-α-D-Glucanotransferase	1,4-α-D-Glucan:1,4-α-D-glucan 4-α-D-glycosyltransferase	2.4.1.25
1,3-β-D-Glucan synthase	UDPglucose:1,3-β-D-glucan 3-β-D-glucosyltransferase	2.4.1.34
Glucokinase	ATP:D-glucose 6-phosphotransferase	2.7.1.2
Glucomannan 4-β-D-mannosyltransferase	GDPmannose:glucomannan 1,4-β-D-mannosyltransferase	2.4.1.32
Glucose-1-phosphate adenylyltransferase	ATP:α-D-glucose-1-phosphate adenylyltransferase	2.7.7.27
Glucose-1-phosphate guanylyltransferase	GTP:α-D-glucose-1-phosphate guanylyltransferase	2.7.7.34
Glucosephosphate isomerase	D-Glucose-6-phosphate ketol-isomerase	5.3.1.9
Glucose-1-phosphate thymidylyltransferase	dTTP:α-D-glucose-1-phosphate thymidylyltransferase	2.7.7.24
Glucose-1-phosphate uridylyltransferase	UTP:α-D-glucose-1-phosphate uridylyltransferase	2.7.7.9
Glucose-6-phosphatase	D-Glucose-6-phosphate phosphohydrolase	3.1.3.9
Glucose-6-phosphate dehydrogenase	D-Glucose-6-phosphate:$NADP^+$ 1-oxidoreductase	1.1.1.49
Glucuronate reductase	L-Gulonate:$NADP^+$ 1-oxidoreductase	1.1.1.19
L-Gulonate dehydrogenase	L-Gulonate:NAD^+ 3-oxidoreductase	1.1.1.45
L-Gulonolactone oxidase	L-Gulono-γ-lactone:oxygen 2-oxidoreductase	1.1.3.8
Glycerate kinase	ATP:D-glycerate 3-phosphotransferase	2.7.1.31

Table 5.1 (*continued*)

Trivial name	Systematic name	EC number[b]
Glycerol kinase	ATP:glycerol 3-phosphotransferase	2.7.1.30
Glycogen (starch) synthase	UDPglucose:glycogen 4-α-D-glucosyltransferase	2.4.1.11
Heparitin sulphotransferase	3′-Phosphoadenylylsulphate: heparitin *N*-sulphotransferase	2.8.2.12
Hexokinase	ATP:D-hexose 6-phosphotransferase	2.7.1.1
myo-Inositol 1-kinase	ATP:*myo*-inositol 1-phosphotransferase	2.7.1.64
myo-Inositol oxygenase	*myo*-Inositol:oxygen oxidoreductase	1.13.99.1
myo-Inositol-1-phosphate synthase	1L-*myo*-Inositol-1-phosphate lyase (isomerizing)	5.5.1.4
Ketohexokinase	ATP:D-fructose 1-phosphotransferase	2.7.1.3
Lactose synthase	UDPgalactose:D-glucose 4-β-D-galactosyltransferase	2.4.1.22
D-Lyxose ketol-isomerase	D-Lyxose ketol-isomerase	5.3.1.15
Mannose isomerase	D-Mannose ketol-isomerase	5.3.1.7
Mannose-1-phosphate guanylyltransferase	GTP:α-D-mannose-1-phosphate guanylyltransferase	2.7.7.13
1,3-β-D-Oligoglucan phosphorylase	1,3-β-D-Oligoglucan:orthophosphate glucosyltransferase	2.4.1.30
Phospho-*N*-acetylmuramoyl-pentapeptide-transferase	UDP-*N*-acetylmuramoyl-L-alanyl-D-γ-glutamyl-L-lysyl-D-alanyl-D-alanine: undecaprenoid-1-ol-phosphate phospho-*N*-acetylmuramoyl-pentapeptide-transferase	2.7.8.13
6-Phosphofructokinase	ATP:D-fructose-6-phosphate 1-phosphotransferase	2.7.1.11
Phosphoglucomutase	α-D-Glucose-1,6-bisphosphate:α-D-glucose-1-phosphate phosphotransferase	2.7.5.1
Phosphoglucomutase (glucose-cofactor)	D-Glucose-1-phosphate:D-glucose 6-phosphotransferase	2.7.5.5
Phosphogluconate dehydratase	6-Phospho-D-gluconate hydro-lyase	4.2.1.12
Phosphogluconate dehydrogenase (decarboxylating)	6-Phospho-D-gluconate:NADP$^+$ 2-oxidoreductase (decarboxylating)	1.1.1.44
6-Phosphogluconolactonase	6-Phospho-D-gluconate-δ-lactone lactonohydrolase	3.1.1.31
Phosphoglycerate kinase	ATP:3-phospho-D-glycerate 1-phosphotransferase	2.7.2.3
Phosphoglycerate phosphomutase	D-Phosphoglycerate 2,3-phosphomutase	5.4.2.1
Phosphoglyceromutase	2,3-Bisphospho-D-glycerate:2-phospho-D-glycerate phosphotransferase	2.7.5.3
Phospho-2-keto-3-deoxy-gluconate aldolase	6-Phospho-2-keto-3-deoxy-D-gluconate D-glyceraldehyde-3-phosphate-lyase	4.1.2.14
Phosphoketolase	D-Xylulose-5-phosphate D-glyceraldehyde-3-phosphate-lyase (phosphate-acetylating)	4.1.2.9

Table 5.1 (*continued*)

Trivial name	Systematic name	EC number[b]
5′-Phosphoribosylamine synthetase	Ribose-5-phosphate:ammonia ligase (ADP-forming)	6.3.4.7
Phosphorylase	1,4-α-D-Glucan:orthophosphate α-D-glucosyltransferase	2.4.1.1
Pyruvate kinase	ATP:pyruvate 2-*O*-phosphotransferase	2.7.1.40
Ribokinase	ATP:D-ribose 5-phosphotransferase	2.7.1.15
Ribosephosphate isomerase	D-Ribose-5-phosphate ketol-isomerase	5.3.1.6
Ribosephosphate pyrophosphokinase	ATP:D-ribose-5-phosphate pyrophosphotransferase	2.7.6.1
Ribulokinase	ATP:L(or D)-ribulose 5-phosphotransferase	2.7.1.16
D-Ribulokinase	ATP:D-ribulose 5-phosphotransferase	2.7.1.47
Ribulosephosphate 3-epimerase	D-Ribulose-5-phosphate 3-epimerase	5.1.3.1
L-Ribulosephosphate 4-epimerase	L-Ribulose-5-phosphate 4-epimerase	5.1.3.4
Sphingomyelin phosphodiesterase	Sphingomyelin cholinephosphate hydrolase	3.1.4.2
Starch (bacterial glycogen) synthase	ADPglucose:1,4-α-D-glucan 4-α-D-glucosyltransferase	2.4.1.21
Sucrose-phosphatase	Sucrose-6′-phosphate phosphohydrolase	3.1.3.24
Sucrose-phosphate synthase	UDPglucose:D-fructose-6-phosphate 2-α-D-glucosyltransferase	2.4.1.14
Sucrose synthase	UDPglucose:D-fructose 2-α-D-glucosyltransferase	2.4.1.13
Sugar-phosphatase	Sugar-phosphate phosphohydrolase	3.1.3.23
Sulphate adenylyltransferase	ATP:sulphate adenylyltransferase	2.7.7.4
dTDPglucose 4,6-dehydratase	dTDPglucose 4,6-hydro-lyase	4.2.1.46
Teichoic-acid synthase	CDPribitol:teichoic-acid phosphoribitoltransferase	2.4.1.55
Transaldolase	Sedoheptulose-7-phosphate: D-glyceraldehyde-3-phosphate dihydroxyacetonetransferase	2.2.1.2
Transketolase	Sedoheptulose-7-phosphate: D-glyceraldehyde-3-phosphate glycolaldehydetransferase	2.2.1.1
Trehalose-phosphatase	Trehalose-6-phosphate phosphohydrolase	3.1.3.12
α,α-Trehalose-phosphate synthase (GDP-forming)	GDPglucose:D-glucose-6-phosphate α-D-glucosyltransferase	2.4.1.36
α,α-Trehalose-phosphate synthase (UDP-forming)	UDPglucose:D-glucose-6-phosphate 1-α-D-glucosyltransferase	2.4.1.15
Triokinase	ATP:D-glyceraldehyde 3-phosphotransferase	2.7.1.28
UDPacetylgalactosamine–protein acetylgalactosaminyltransferase	UDP-2-acetamido-2-deoxy-D-galactose: protein acetamidodeoxygalactosyltransferase	2.4.1.41

Table 5.1 (*continued*)

Trivial name	Systematic name	EC number[b]
UDPacetylglucosamine 4-epimerase	UDP-2-acetamido-2-deoxy-D-glucose 4-epimerase	5.1.3.7
UDP-*N*-acetylglucosamine–glycoprotein *N*-acetylglucosaminyltransferase	UDP-2-acetamido-2-deoxy-D-glucose: glycoprotein 2-acetamido-2-deoxy-D-glucosyltransferase	2.4.1.51
UDPacetylglucosamine pyrophosphorylase	UTP:2-acetamido-2-deoxy-α-D-glucose-1-phosphate uridylyltransferase	2.7.7.23
UDP-*N*-acetylmuramoylalanine synthetase	UDP-*N*-acetylmuramate:L-alanine ligase (ADP-forming)	6.3.2.8
UDP-*N*-acetylmuramoyl-L-alanyl-D-glutamate synthetase	UDP-*N*-acetylmuramoyl-L-alanine:D-glutamate ligase (ADP-forming)	6.3.2.9
UDP-*N*-acetylmuramoyl-L-alanyl-D-glutamyl-L-lysyl-D-alanyl-D-alanine synthetase	UDP-*N*-acetylmuramoyl-L-alanyl-D-glutamyl-L-lysine:D-alanyl-D-alanine ligase (ADP-forming)	6.3.2.10
UDPgalactose–*N*-acylsphingosine galactosyltransferase	UDPgalactose:*N*-acylsphingosine galactosyltransferase	2.4.1.47
UDPgalactose–lipopolysaccharide galactosyltransferase	UDPgalactose:lipopolysaccharide galactosyltransferase	2.4.1.44
UDPgalactose–mucopolysaccharide galactosyltransferase	UDPgalactose:mucopolysaccharide galactosyltransferase	2.4.1.74
UDPgalactose–sphingosine β-D-galactosyltransferase	UDPgalactose:sphingosine β-D-galactosyltransferase	2.4.1.23
UDPglucose–ceramide glucosyltransferase	UDPglucose:*N*-acylsphingosine glucosyltransferase	2.4.1.80
UDPglucose dehydrogenase	UDPglucose:NAD$^+$ 6-oxidoreductase	1.1.1.22
UDPglucose 4-epimerase	UDPglucose 4-epimerase	5.1.3.2
UDPglucose–hexose-1-phosphate uridylyltransferase	UDPglucose:α-D-galactose-1-phosphate uridylyltransferase	2.7.7.12
UDPglucuronate decarboxylase	UDPglucuronate carboxy-lyase	4.1.1.35
UDPglucuronate 4′-epimerase	UDPglucuronate 4′-epimerase	5.1.3.6
UDPglucuronate 5′-epimerase	UDPglucuronate 5′-epimerase	5.1.3.12
UDPglucuronosyltransferase	UDPglucuronate β-D-glucuronosyltransferase (acceptor-unspecific)	2.4.1.17
Xylose isomerase	D-Xylose ketol-isomerase	5.3.1.5
D-Xylulokinase	ATP:D-xylulose 5-phosphotransferase	2.7.1.17
L-Xylulokinase	ATP:L-xylulose 5-phosphotransferase	2.7.1.53
D-Xylulose reductase	Xylitol:NAD$^+$ 2-oxidoreductase (D-xylulose-forming)	1.1.1.9
L-Xylulose reductase	Xylitol:NADP$^+$ 4-oxidoreductase (L-xylulose-forming)	1.1.1.10

[a] The nomenclature used in this table is identical to that used in Recommendations (1978) although that work does not necessarily comply with that recommended by IUPAC/IUB for carbohydrates.

[b] EC number stands for Enzyme Commission number, see this chapter, p. 98.

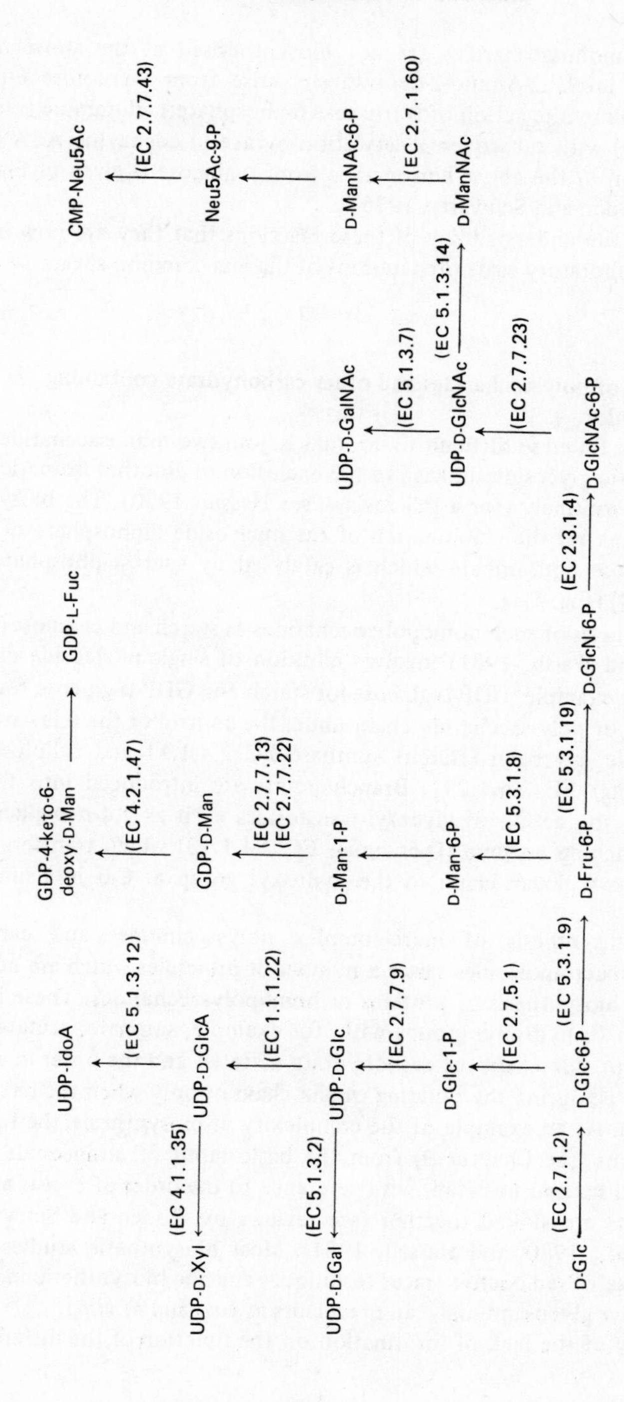

Fig. 5.3 – Biosynthetic conversion of D-glucose into some of the other biologically important monosaccharides.

Certain monosaccharides are not biosynthesised at the above nucleoside diphosphate level. 2-Amino-2-deoxysugars arise from D-fructose 6-phosphate and glutamine by the action of D-fructose 6-phosphate:L-glutamine transamidase (EC 5.3.1.19) with subsequent acetylation by acetyl coenzyme A. A schematic representation of the above conversions from D-glucose is given in Fig. 5.3 (see review by Rodén and Schwartz, 1975).

Such is the understanding of these reactions that they are now being used for *in vitro* laboratory scale preparations of the less common sugars.

Biosynthesis of polysaccharides and other carbohydrate-containing macromolecules

What man has found so difficult to do, that is, join two monosaccharides together with a specific glycosidic linkage to the exclusion of all other isomeric products, nature does routinely (for a full review, see Hassid, 1970). The biosynthesis of sucrose occurs *via* the combination of the nucleoside diphosphate of D-glucose and D-fructose 6-phosphate which is catalysed by sucrose-phosphate synthase (EC 2.4.1.14) (Fig. 5.4).

Biosynthesis of such homopolysaccharides as starch and cellulose (see review by Preiss and Walsh, 1981) involves addition of single nucleoside diphosphate residues (for example, UDP-D-glucose for starch and GDP-D-glucose for cellulose) to an oligo- or poly-saccharide chain under the control of the relevant synthases [for example, glycogen (starch) synthase EC 2.4.1.11 and cellulose synthase (GDP-forming) EC 2.4.1.29]. Branch points are introduced into the growing polymer by the action of glycosyl-transferases such as 1,4-α-D-glucan (amylopectin) branching enzyme (Q-enzyme, EC 2.4.1.18) which transfers a segment of a (1→4)-α-D-glucan chain to the hydroxyl group at C-6 in a similar glucan chain.

The biosynthesis of more-complex polysaccharides and carbohydrate-containing macromolecules raises a number of principles which are not encountered in the biosynthesis of proteins or homopolysaccharides. These include the substitution of hydroxyl groups with, for example, sulphate, acetate and phosphate etc., or side chains of carbohydrate material and the order in which these occur (that is, during the building of the chain or only when the chain has been terminated). As an example of the complexity of biosynthesis, the formation of proteoglycans (see Chapter 9) from the basic input of aminoacids and carbohydrate is discussed in detail, with reference to the order of events and how the various parts are linked together (see reviews by Rodén and Schwartz, 1975, Rodén *et al.*, 1980, and Hassell, 1981). Most biosynthetic studies have been based on use of radioactive tracer techniques and the biosynthetic incorporation of radioactive glycosaminoglycan precursors *in vivo* and *in vitro*.

In view of the lack of information on the function of the different parts of

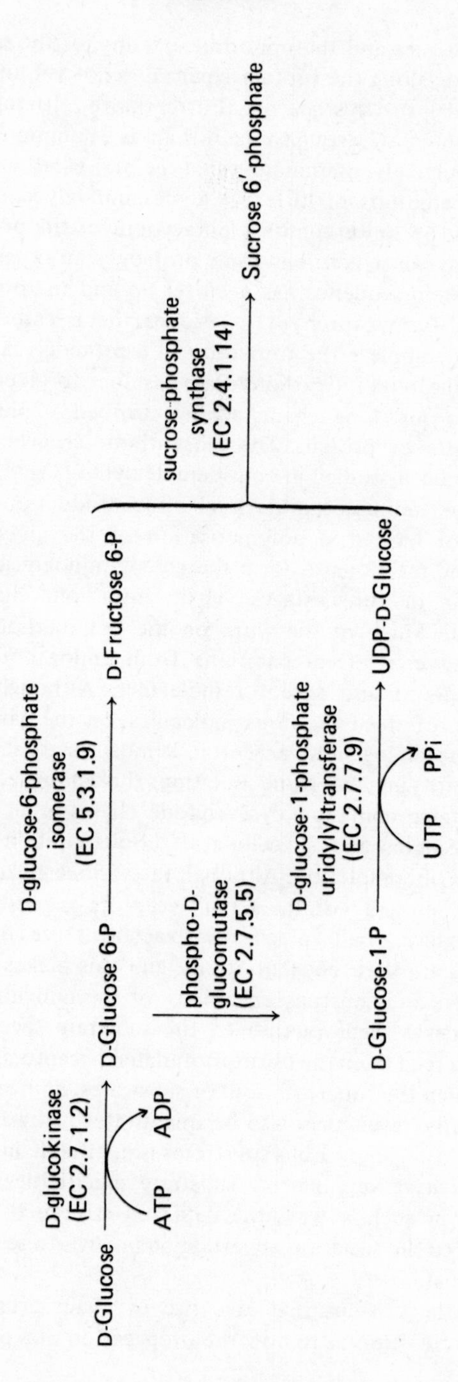

Fig. 5.4 – Biosynthesis of sucrose.

proteoglycan molecules and the importance, if any, of the spacing of the glycosaminoglycan chains along the protein chain, it is not yet known whether a code could possibly exist for this spacing. Furthermore, although different protein chains exist, it will be interesting to see if there is a common aminoacid sequence code for a particular glycosaminoglycan type and whether there is any coding for the different amounts of different glycosaminoglycans in proteoglycans – the situation may be quite complex on account of the presence of more than one glycosaminoglycan structure in some proteoglycan types.

Once the protein sequence has been set up and the protein chain has been produced in intact form, carbohydrate residues (that is, carbohydrate block units) may be added to complete the formation of a proteoglycan molecule. Whereas the addition of the relevant carbohydrate residues to glycosaminoglycan oligosaccharides can occur, long chains are not formed – presumably due to the absence of the primer protein. The biosynthetic aspects of the glycosaminoglycan chain have been studied in considerable detail.

Following the monosaccharide nucleotide production, which provides the basic material for biological polymerisation of the glycosaminoglycan chain (see Fig. 5.5), the reactions to form the glycosaminoglycan chains can be considered to occur in three stages: chain initiation, chain propagation and chain termination. Much of the work on the polymerisation process has been carried out using enzyme preparations from endogenous material but with exogenous and donor and acceptor molecules. Although objections may be raised to the use of the exogenous molecules, on the basis that they may not represent the true endogenous acceptor, it must be realized that there is great difficulty in identifying, let alone isolating, the endogneous acceptor. Cellular biosynthetic systems contain polysaccharide chains in all stages of growth and therefore homogenates and subcellular fractions contain acceptors of various sizes and degrees of completion. Although the various glycosyl transfer reactions should ideally be studied with the actual acceptors or substances that are present *in vivo*, this does have disadvantages; the exact structures of the acceptors are, as yet, unknown as are their concentrations, and this makes determination of the kinetic parameters and substrate specificity of the individual enzymes difficult. In many instances a clear picture of the substrate specificity of the various reactions has emerged from the use of formulated acceptors of defined structures. Furthermore, when the concentrations of substrates, and/or amounts of products are known, enzyme levels may also be quantitated. However, acceptance of the results obtained using exogenous substrates is justifiable in terms of the fact that the enzymes do have very narrow substrate specificities. Once a reaction has been identified in such a way, the data and structural descriptions obtained must be applied to the situation appertaining *in vivo*, to see if the rate of enzyme action is even greater.

The biosynthetic steps that give rise to chain propagation are less well known, and the question as to how the propagation of a particular glycosamino-

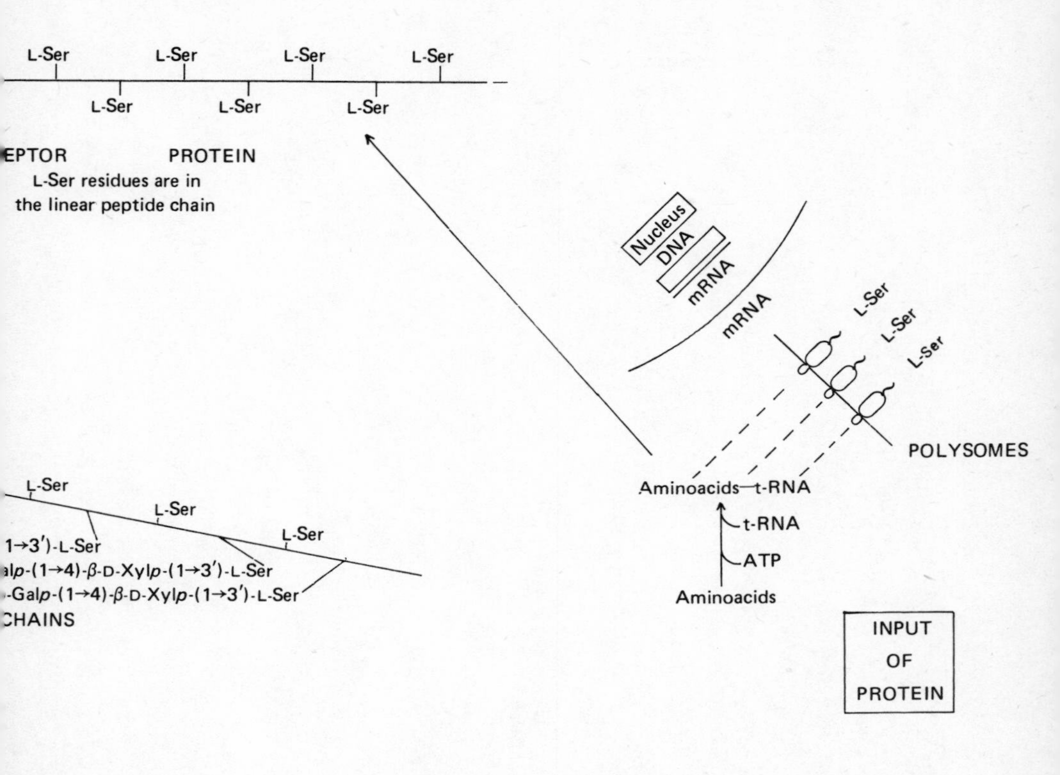

KEY

a	*D-xylopyranosyltransferase*
b	*D-galactopyranosyltransferase (type 1)*
c	*D-galactopyranosyltransferase (type 2)*
d	*D-glucopyranosyluronic acid transferase (type 1)*
e	*D-glucopyranosyluronic acid transferase (type 2)*
f	*D-2-acetamido-2-deoxygalactopyranosyltransferase*
g	*O-sulphatotransferase*
AcCoA	*acetyl coenzyme A*
ADP	*adenosine 5′-diphosphate*
APS	*adenosine 5′-P-(dihydrogen phosphatosulphate)*
ATP	*adenosine 5′-triphosphate*
NAD^+	*nicotinamide adenine dinucleotide (oxidized)*
PAPS	*adenosine 3′-phosphate-5′-P-(dihydrogen phosphatosulphate)*
PPi	*inorganic diphosphate*
UDP	*uridine 5′-pyrophosphate*
UTP	*uridine 5′-triphosphate*

glycan is selected, arising from the common glycopeptide linkage structure for a number of the glycosaminoglycans, remains unanswered. Early studies indicated the formation of a low-sulphated polysaccharide chain on incubation of a cartilagenous preparation with UDP-D-glucuronic acid and UDP-2-acetamido-2-deoxy-D-galactose, and some information came from studies on hyaluronic acid production in *Streptoccoci*. At that time the question remained as to whether the chain was propagated by addition of single monosaccharide block units or pre-produced disaccharide block units. Enzymes which incorporate radioactivity from UDP-D-[^{14}C]-glucuronic acid and UDP-2-acetamido-2-deoxy-D-[^{3}H]-galactose into growing chondroitin 4/6-sulphate chains have been recognized. It is now clear that the polymer results from the concerted action, on the glycopeptide linkage acceptor, of two glycosyltransferases: a 2-acetamido-2-deoxy-D-galactopyranosyltransferase and a D-glucopyranosyluronic acid transferase. These enzymes alternately add the two respective monosaccharide component residues of the repeating disaccharide unit directly to the chain without participation of a free disaccharide unit.

Originally it was considered that the biosynthesis of L-iduronic acid units was analogous to that of D-glucuronic acid units, but there is no evidence of either UDP-L-iduronic acid of an L-idopyranosyluronic acid transferase, and indeed such an enzyme appears to be absent. Important information on the biosynthesis of L-iduronic acid units comes from work with a microsomal fraction of a heparin-producing mastocytoma, incubation of which with UDP-D-[^{14}C]-glucuronic acid and unlabelled UDP-2-acetamido-2-deoxy-D-glucose resulted in incorporation of radioactivity into endogenous polysaccharide. When adenosine 3′-phosphate-5′-*P*-(dihydrogen phosphatosulphate) (a precursor of sulphate groups) was included in the incubate, the product polysaccharide contained L-[^{14}C]-idopyranosyluronic acid as well as D-[^{14}C]-glucopyranosyluronic acid residues. Pulse-chase experiments revealed that D-[^{14}C]-glucuronic acid was incorporated into the polymer during the pulse period (in the absence of adenosine 3′-phosphate-5′-*P*-(dihydrogen phosphatosulphate)), and in the bound form it was subsequently converted to bound L-[^{14}C]-iduronic acid during the chase period (in the presence of adenosine 3′-phosphate-5′-*P*-(dihydrogen phosphatosulphate)). Such experiments lead to the conclusion that, in glycosaminoglycan chains, the L-idopyranosyluronic acid residues are formed by epimerisation of D-glucopyranosyluronic acid residues at the polymer (heparin, dermatan sulphate) level by a D-glucopyranosyluronic acid 5-epimerase.

The actual mechanism of epimerisation involves loss of the hydrogen atom from C-5 of the carbohydrate ring. It must be noted that the epimerisation involves only inversion of configuration of C-5 in the D-glucopyranosyluronic acid residue – this gives an L-configuration and the C-5 epimer of D-glucuronic acid is L-iduronic acid. No alteration of the anomeric bond in the glycosidic linkage is involved – the change from β-D- to α-L- is purely one resulting from nomenclature standardisation.

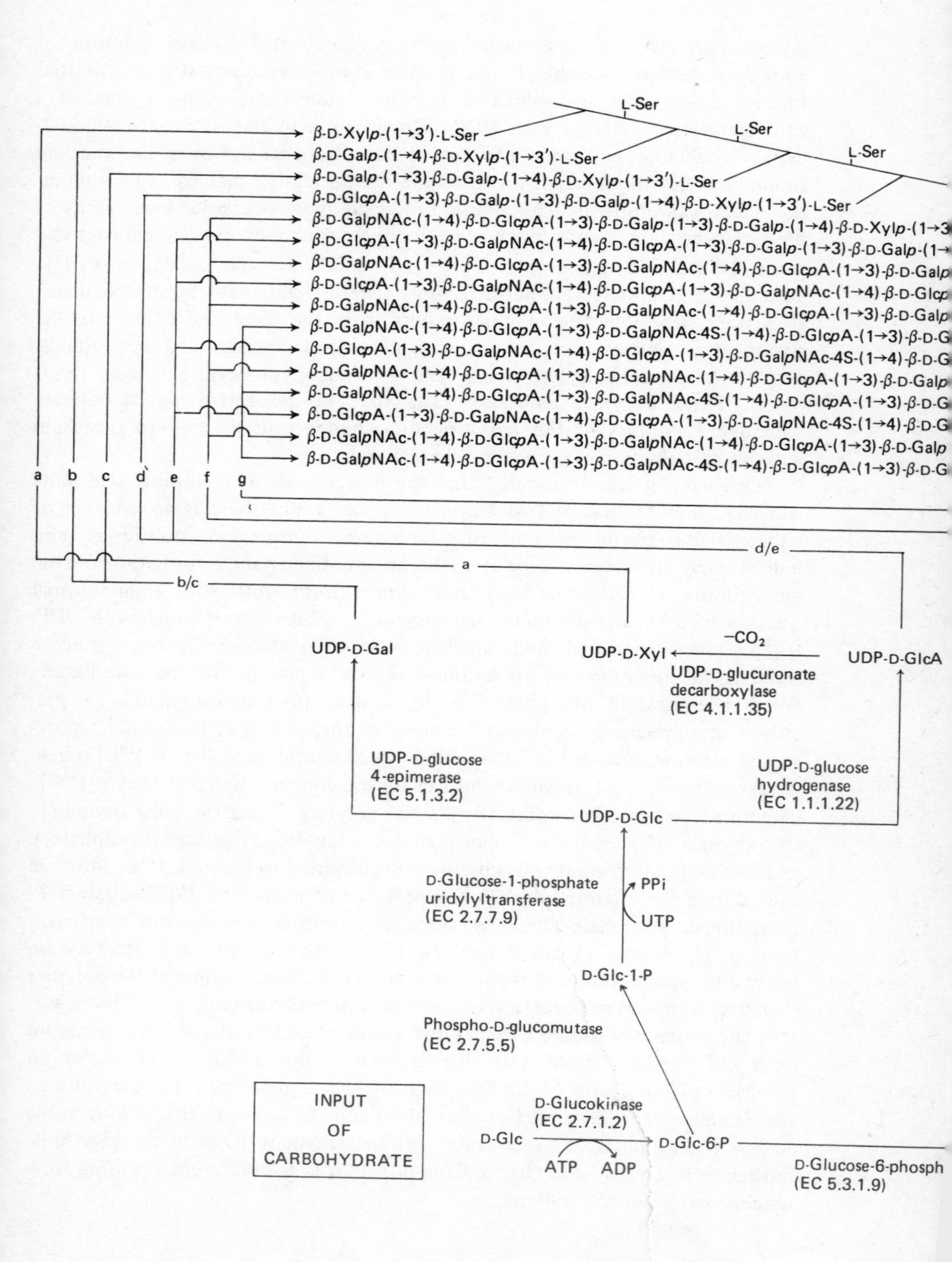
L-Ser
L-Ser
L-Ser
β-D-Xylp-(1→3′)-L-Ser
β-D-Galp-(1→4)-β-D-Xylp-(1→3′)-L-Ser
β-D-Galp-(1→3)-β-D-Galp-(1→4)-β-D-Xylp-(1→3′)-L-Ser
β-D-GlcpA-(1→3)-β-D-Galp-(1→3)-β-D-Galp-(1→4)-β-D-Xylp-(1→3′)-L-Ser
β-D-GalpNAc-(1→4)-β-D-GlcpA-(1→3)-β-D-Galp-(1→3)-β-D-Galp-(1→4)-β-D-Xylp-(1→3
β-D-GlcpA-(1→3)-β-D-GalpNAc-(1→4)-β-D-GlcpA-(1→3)-β-D-Galp-(1→3)-β-D-Galp-(1→
β-D-GalpNAc-(1→4)-β-D-GlcpA-(1→3)-β-D-GalpNAc-(1→4)-β-D-GlcpA-(1→3)-β-D-Galp
β-D-GlcpA-(1→3)-β-D-GalpNAc-(1→4)-β-D-GlcpA-(1→3)-β-D-GalpNAc-(1→4)-β-D-Glcp
β-D-GalpNAc-(1→4)-β-D-GlcpA-(1→3)-β-D-GalpNAc-(1→4)-β-D-GlcpA-(1→3)-β-D-Galp
β-D-GalpNAc-(1→4)-β-D-GlcpA-(1→3)-β-D-GalpNAc-4S-(1→4)-β-D-GlcpA-(1→3)-β-D-G
β-D-GlcpA-(1→3)-β-D-GalpNAc-(1→4)-β-D-GlcpA-(1→3)-β-D-GalpNAc-4S-(1→4)-β-D-G
β-D-GalpNAc-(1→4)-β-D-GlcpA-(1→3)-β-D-GalpNAc-(1→4)-β-D-GlcpA-(1→3)-β-D-Galp
β-D-GalpNAc-(1→4)-β-D-GlcpA-(1→3)-β-D-GalpNAc-4S-(1→4)-β-D-GlcpA-(1→3)-β-D-G
β-D-GlcpA-(1→3)-β-D-GalpNAc-(1→4)-β-D-GlcpA-(1→3)-β-D-GalpNAc-4S-(1→4)-β-D-G
β-D-GalpNAc-(1→4)-β-D-GlcpA-(1→3)-β-D-GalpNAc-(1→4)-β-D-GlcpA-(1→3)-β-D-Galp
β-D-GalpNAc-(1→4)-β-D-GlcpA-(1→3)-β-D-GalpNAc-4S-(1→4)-β-D-GlcpA-(1→3)-β-D-G
a
b
c
d
e
f
g
d/e
a
b/c
UDP-D-Gal
UDP-D-Xyl
−CO2
UDP-D-glucuronate
decarboxylase
(EC 4.1.1.35)
UDP-D-GlcA
UDP-D-glucose
4-epimerase
(EC 5.1.3.2)
UDP-D-glucose
hydrogenase
(EC 1.1.1.22)
UDP-D-Glc
D-Glucose-1-phosphate
uridylyltransferase
(EC 2.7.7.9)
PPi
UTP
D-Glc-1-P
Phospho-D-glucomutase
(EC 2.7.5.5)
INPUT
OF
CARBOHYDRATE
D-Glucokinase
(EC 2.7.1.2)
D-Glc
D-Glc-6-P
ATP
ADP
D-Glucose-6-phosph
(EC 5.3.1.9)

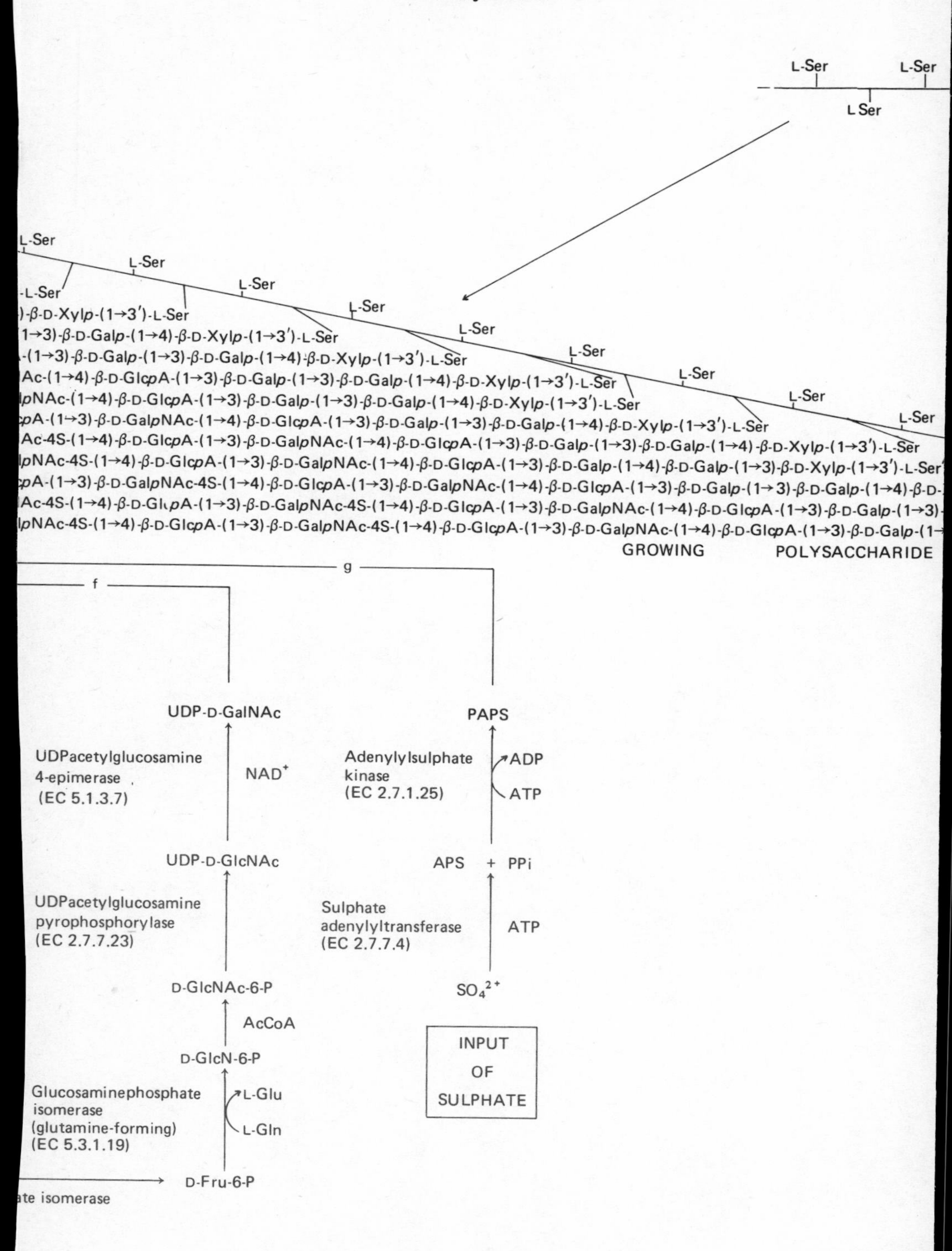

ig. 5.5 – Representative collation of the stages of the biosynthetic pathway to chondroitin 4-sulphate proteoglyca

The incorporation of radioactivity from [^{35}S] sulphate ion into sulphate-containing glycosaminoglycans by *in vitro* and *in vivo* systems has long been known. The fixation of radioactive sulphate depends upon a specific biosynthetic process associated with glycosaminoglycan formation, and not upon an exchange with the sulphate groups already present in the glycosaminoglycan already formed.

The sulphate groups must be transferred to the monosaccharide units at some stage after the attachment of these units to the growing proteoglycan molecule. This transfer must be affected by a sulphatotransferase and the evidence is that the transfer occurs directly from the adenosine 3′-phosphate-5′-*P*-(dihydrogen phosphatosulphate) to the carbohydrate.

The characteristics of the sulphatotransferase have been investigated. Evidence has been obtained from alterations in the relative amounts of chondroitin 4- and 6-sulphate- and keratan sulphate-containing proteoglycans in polysaccharide-rich tissues such as cartilage and intervertebral disc. This, together with other evidence for alterations in the rate at which [^{35}S] sulphate is incorporated into the various polysaccharides, demonstrates that different sulphatotransferases are responsible for the sulphation of the different carbon atoms.

The data available on the biosynthesis of chondroitin 4-sulphate proteoglycan have been sufficient to permit the presentation, in scheme form of many of the stages in the transformation from aminoacid right through to complete proteoglycan. This scheme has now been amplified and modified (Fig. 5.5) and it may be readily deduced from it that a large number of transferases, oxidases, deaminases, etc., are involved in the total biosynthesis of the molecule. The situation is a little more complex where more than one glycosaminoglycan chain type is attached to the protein backbone.

The biosynthesis of proteoglycans is described in detail as an example of the complexity of biosynthesis. The process is similarly very complex for other polysaccharides and carbohydrate-containing macromolecules, even if the compound appears to have a simple repeating structure.

The process of synthesising carbohydrate chains in certain glycoproteins is by preassembly of the chain on a lipid carrier with subsequent transfer of the complete chain to the protein backbone (see review by Lennarz, 1975). Monosaccharides (or monosaccharide 1-phosphates) are transferred to a polyprenol phosphate to give glycosyl phosphoryl polyprenols (for example, β-D-glucosyl dolichol phosphate (32) and β-D-glucosyl phosphoryl undecaprenol (33). These glycosyl phosphoryl polyprenols, and possibly also sugar nucleotides, transfer further carbohydrate residues to the initial glycosyl polyprenol phosphate to give linear chains. These linear chains can be transferred to other linear chains to give branched structures until chain termination occurs, when the branched chain is transferred from the lipid to a protein acceptor (Fig. 5.6).

Once man has a fuller understanding of the enzyme reactions involved and the necessary enzymes become readily available it can be anticipated that many of the chemical methods for synthesis will become redundant as *in vitro* enzyme

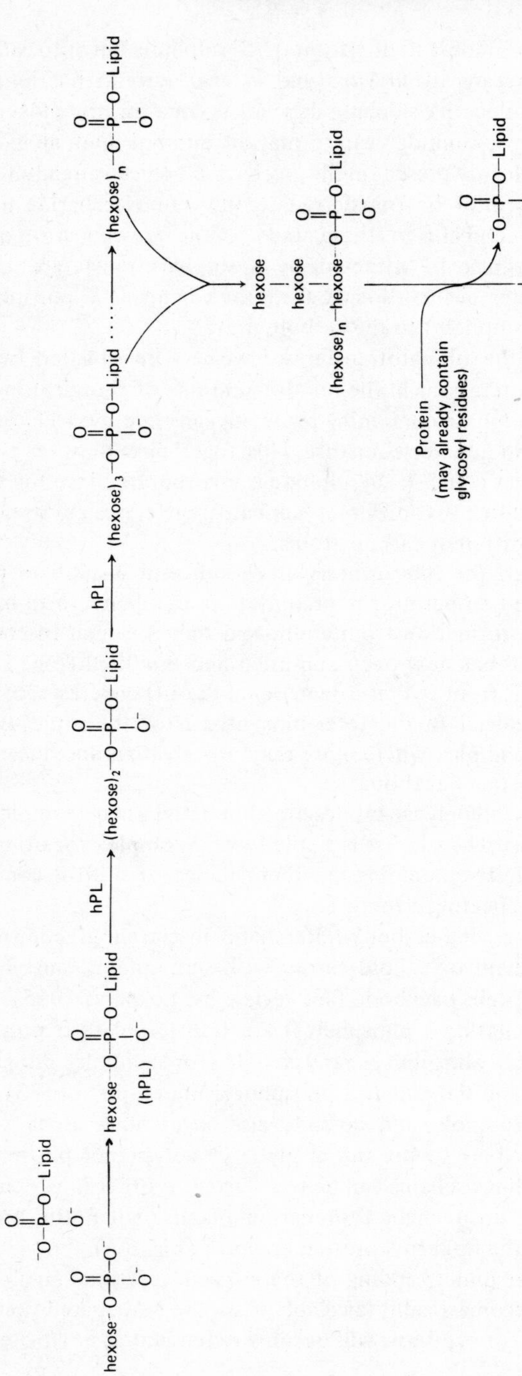

Fig. 5.6 – Schematic representation of the involvement of lipid-linked intermediates in glycoprotein biosynthesis.

controlled synthesis become commonplace. To date some tri- and tetra-saccharides have been synthesised enzymically from the activated monosaccharides (UDP-D-glucose, etc.) which had been chemically synthesised.

$$\text{Glc}-O-\overset{O}{\underset{O^-}{\overset{\|}{\underset{|}{P}}}}-O-CH_2-CH_2-\overset{CH_3}{\overset{|}{CH}}-CH_2-(CH_2-CH=\overset{CH_2}{\overset{|}{C}}CH_2)_{18}-CH_2-CH=C(CH_3)_2$$

(32)

$$\text{Glc}-O-\overset{O}{\underset{O^-}{\overset{\|}{\underset{|}{P}}}}-O-(CH_2-CH=\overset{CH_3}{\overset{|}{C}}-CH_2)_{11}-H$$

(33)

CHAPTER 6

Monosaccharides

The nomenclature and chemistry of neutral monosaccharides have been described in detail in the previous chapters (Chapters 2, 3 and 4). This chapter is devoted to monosaccharides containing either functional groups other than those found in the neutral monosaccharides (for example, carboxyl, amino, halogeno, etc.) or other (hetero) atoms within the ring structure (for example, nitrogen, carbon and sulphur). The chemistry of many of these compounds is the subject of an ongoing series of reviews (Various Authors, 1968 onwards).

ACIDIC SUGARS

There are three important types of acidic sugar; these are classified according to which, otherwise neutral, terminal group has been oxidized (see Chapter 3, page 50). Hence aldonic acids are obtained by oxidation of the carbonyl group, uronic acids by oxidation of the terminal alcohol group and aldaric acids by oxidation of both the carbonyl and terminal alcohol groups. Elimination of water between the carboxyl group and a hydroxyl group leads to the formation of lactones (for example, see Chapter 3, page 51 and structures (10) and (11)). Aldaric acids have no great biological significance but aldonic acids are used in the identification and characterisation of monosaccharides (see Chapter 4, page 76). For a review of acidic sugars see Theander (1980).

By far the most important type of acidic sugars are the uronic acids, which occur almost exclusively as hexuronic acids. In biosynthetic sequences, hexuronic acids are intermediates in the conversion of hexoses into pentoses *via* a process of oxidation and decarboxylation. D-Glucuronic acid and D-galacturonic acid are found in plant gums and bacterial cell-walls, D-mannuronic acid and L-guluronic acid are found in algae (see Chapter 8) whilst D-glucuronic acid and L-iduronic acid are components of proteoglycans (see Chapter 9). The lability of the carboxyl group makes the identification of these residues in polysaccharides difficult due to decarboxylation occurring during hydrolysis, resulting in incorrect interpretation of the primary structures.

The aldonic acid from D-glucose, D-gluconic acid, in the forms of D-gluconic acid 6-phosphate and D-glucono-1,5-lactone 6-phosphate (34), is an important intermediate in carbohydrate metabolism. Another biologically important aldonic acid is vitamin C (L-ascorbic acid, the vitamin responsible for the control of scurvy), which exists as the 1,4-lactone of an unsaturated hexonic acid having an enediol structure at C-2 and C-3 represented by the Fischer structure (35) and the ring structure (36). The biosynthetic reactions of vitamin C, which include the catalysis of the final step in the transfer of hydrogen from substrate to oxygen to form water and the conversion of L-proline to L-hydroxyproline in the newly-forming collagen of tissue growth, are a consequence of the reversible redox reaction (Scheme 6.1).

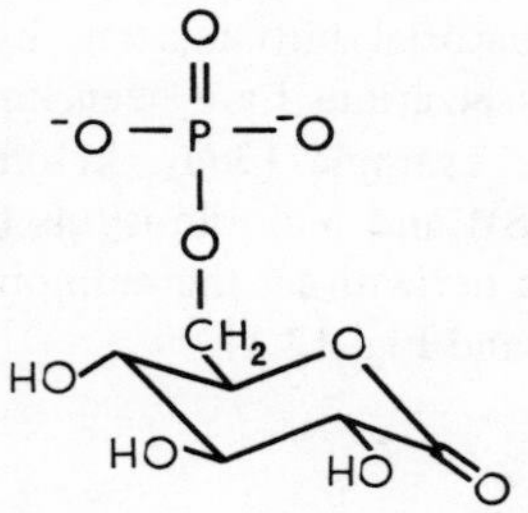

D-Glucono-1,5-lactone 6-phosphate
(34)

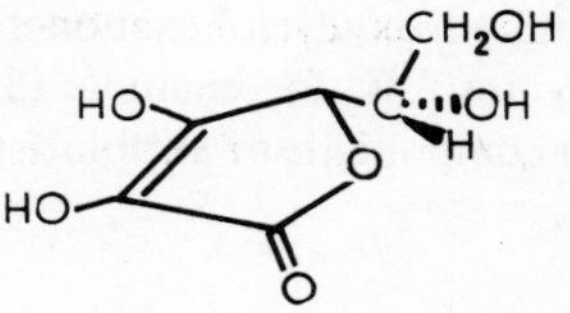

(36)

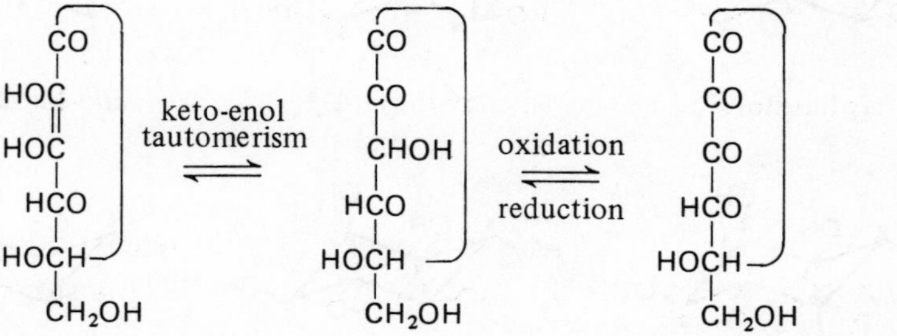

keto-enol tautomerism

oxidation

reduction

L-Ascorbic acid
(35)

Dehydro-L-ascorbic acid

Scheme 6.1

SUGAR ALCOHOLS

There are two groups of sugar alcohols, the first of which, the alditols, are obtained by reduction of neutral monosaccharides and have been described previously (see Chapter 3). A number of alditols occur widely in nature, particularly the hexitols D-mannitol and D-glucitol which occur particularly in seaweeds and mountain ash berries respectively. D-Glycerol and D-ribitol occur in microbial polysaccharides (see Chapter 8).

The second group of sugar alcohols are cyclic derivatives, the most common of which are derivatives of cyclohexane. These 1,2,3,4,5,6 cyclohexanehexols (cyclitols, see Recommendations, 1973), known as inositols (inosit, after the Greek ις, ινος, meaning sinew, fibre or muscle) exist as nine stereoisomers, all but two of which are optically inactive. The structures of all the stereoisomers, which usually exist in chair forms with the greater number of hydroxyl groups in equatorial positions, are given in Fig. 6.1. Chemically, inositols undergo many of the reactions described for neutral monosaccharides and which involve hydroxyl groups. They will, for example, form isopropylidene derivatives usually involving hydroxyl groups which are both equatorial but adjacent hydroxyl groups which are both in axial (that is, diaxial) positions have been known to react. Deoxyinositols (cyclohexanepentols, for example (37)), ketoinositols (pentahydroxycyclohexanones, for example (38)) and aminoinositols (amino-deoxyinositols, for example (39)) are important derivatives, the aminoinositols being components of antibiotics (see Chapter 12 and Fig. 12.1).

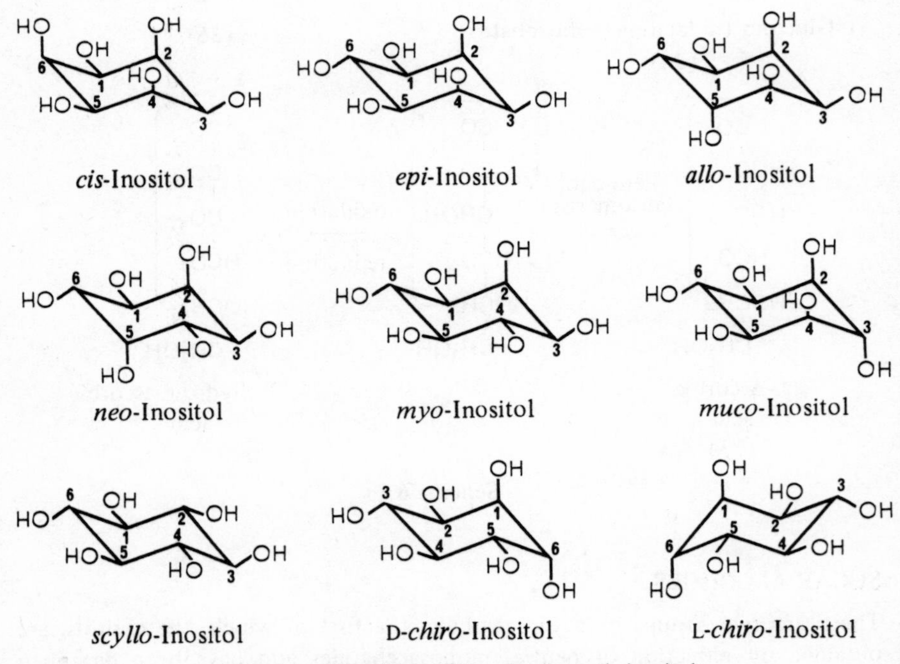

Fig. 6.1 – Structures of the nine isomeric inositols
(1,2,3,4,5,6-cyclohexanehexols, see Recommendations, 1978)

1D-1,2,4/3,5-Cyclohexanepentol
(Viburnitol)
(37)

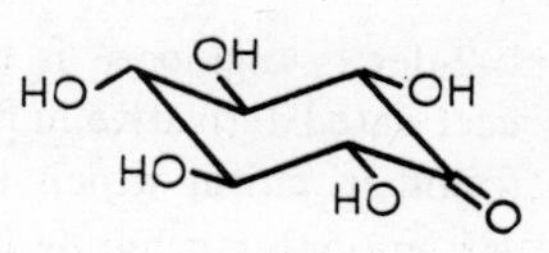

2,4,6/3,5-Pentahydroxycyclohexanone
(38)

1D-1-Amino-1-deoxy-*neo*-inositol
(39)

myo-Inositol (Fig. 6.1) is the most widely distributed inositol and occurs free and combined in tissues of nearly all living species. In animals and micro-organisms it usually occurs combined as phospholipids (see Chapter 10) whilst in plants it also occurs as the hexaphosphate, phytic acid (40). The only other naturally occurring inositols are D-*chiro*- (as D-pinitol, 41), L-*chiro*- (as L-quebrachitol, 42) and *scyllo*-inositol (Fig. 6.1). A comprehensive text on the chemistry, biochemistry and biology of cyclitols is available (Wells and Eisenberg, 1978).

myo-Inositol 1,2,3,4,5,6-hexa-(dihydrogen phosphate)
(Phytic acid)
(40)

(41)

(42)

AMINOSUGARS

2-Amino-2-deoxy-D-glucose is the most abundant aminosugar occurring widely as its *N*-acetylated derivative in polysaccharides, glycoproteins and proteoglycans, and is found in chitin which is a (1→4)-linked 2-acetamido-2-deoxy-β-D-glucopyranose homopolysaccharide (although *N*-acetylation is sometimes incomplete) analogous to cellulose and found in the exoskeletons of crustacea. 2-Acetamido-2-deoxy-D-galactose, -D-mannose and -D-talose also occur in glycoproteins but to a lesser extent. Most aminosugars rarely occur free or as simple derivatives but usually as components in oligo- and poly-saccharides. An exception to this is 5-amino-5-deoxy-D-glucopyranose (nojirimycin (43)) in which the amino group is involved in the ring giving a piperidinose ring structure.

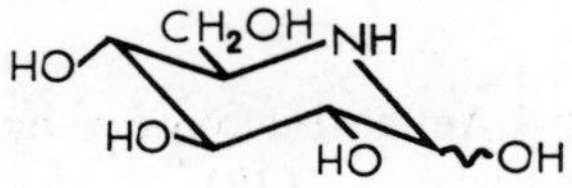

HO, CH_2OH, NH, HO, HO, OH

(43)

Many of the less common, naturally occurring aminosugars such as 3-amino-, 4-amino- and various diamino-sugars and aminouronic acids are found in antibiotics and microbial polysaccharides and are described in the relevant chapters (Chapters 12 and 13 respectively).

A number of higher monosaccharides exist as aminosugars and among these are 2-acetamido-3-*O*-(1-carboxyethyl)-2-deoxy-D-glucose (2-acetamido-2-deoxymuramic acid (44)), which occurs in bacterial cell-walls (see Chapter 8), 5-acetamido-3,5-dideoxy-D-*glycero*-D-*galacto*-2-nonulopyranonic acid (*N*-acetylneuraminic acid (45)), 5-glycolylamido-3,5-dideoxy-D-*glycero*-D-*galacto*-2-nonulopyranonic acid (*N*-glycolylneuraminic acid (46)) and other acetylated and glycolylated derivatives which occur in many glycoproteins (see Chapter 9).

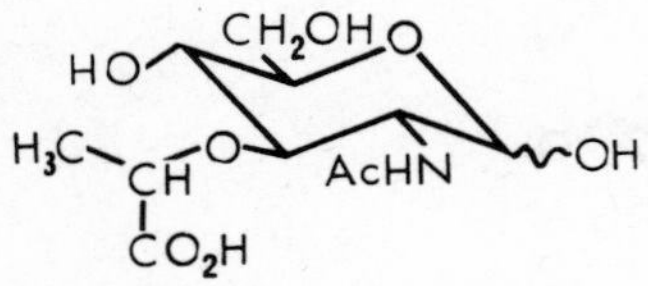

HO, CH_2OH, O, H_3C, CH, O, AcHN, OH, CO_2H

(44)

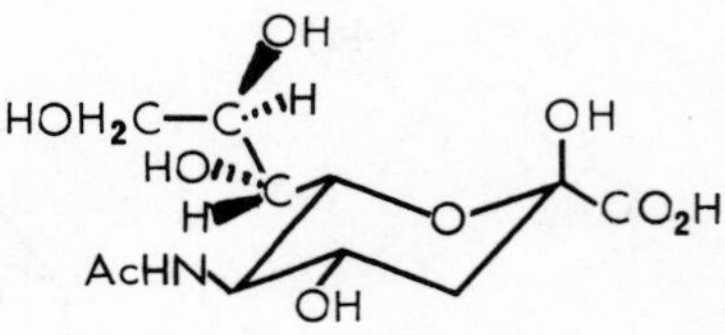

OH, HOH_2C—C—H, HO—C, H, OH, O, CO_2H, AcHN, OH

(45)

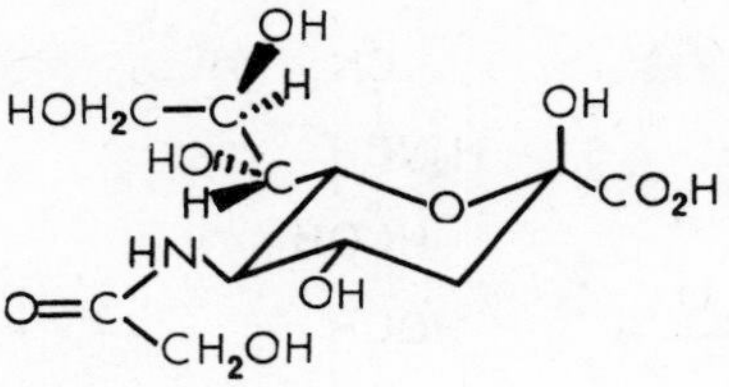

(46)

Aminosugars can be prepared from sulphonate esters (see Chapter 3) by nucleophilic displacement of the ester with ammonia, hydrazine or azide anions. The use of hydrazine is preferred to ammonia because the first-formed amine is more stable and less likely to be displaced. It can also be used to produce diaminosugars (Scheme 6.2). Alternative approaches to the preparation of aminosugars include: the addition of hydrogen cyanide to an aldose in the presence of ammonia and subsequent catalytic hydrogenation over a platinum catalyst to give an epimeric pair of 2-amino-2-deoxyaldoses containing an additional carbon atom (Scheme 6.3); the ring opening of an aldose epoxide with liquid ammonia or azide anions followed by reduction (Scheme 6.4); and the reaction of a dialdehyde with nitromethane followed by reduction (Scheme 6.5) to give a mixture of four isomers due to the formation of 3 new chiral centres with the nitro group of the intermediate product always occupying an equatorial position.

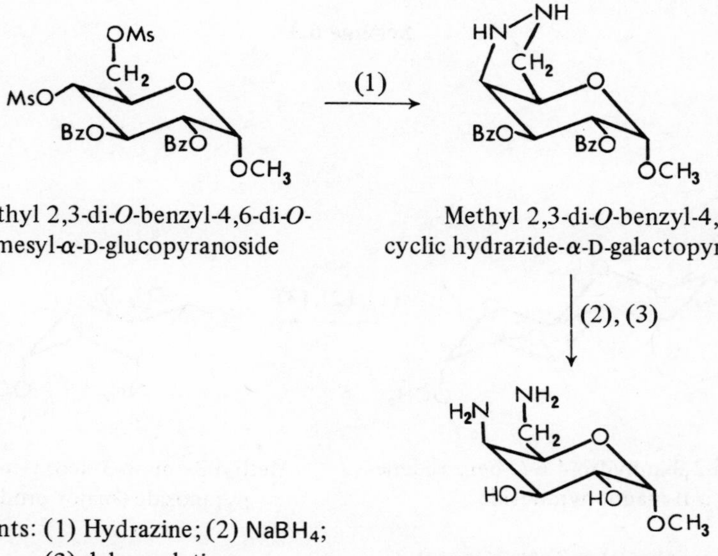

Methyl 2,3-di-*O*-benzyl-4,6-di-*O*-mesyl-α-D-glucopyranoside

(1)

Methyl 2,3-di-*O*-benzyl-4,6-cyclic hydrazide-α-D-galactopyranoside

(2), (3)

Methyl 4,6-diamino-4,6-dideoxy-α-D-galactopyranoside

Reagents: (1) Hydrazine; (2) $NaBH_4$; (3) debenzylation

Scheme 6.2

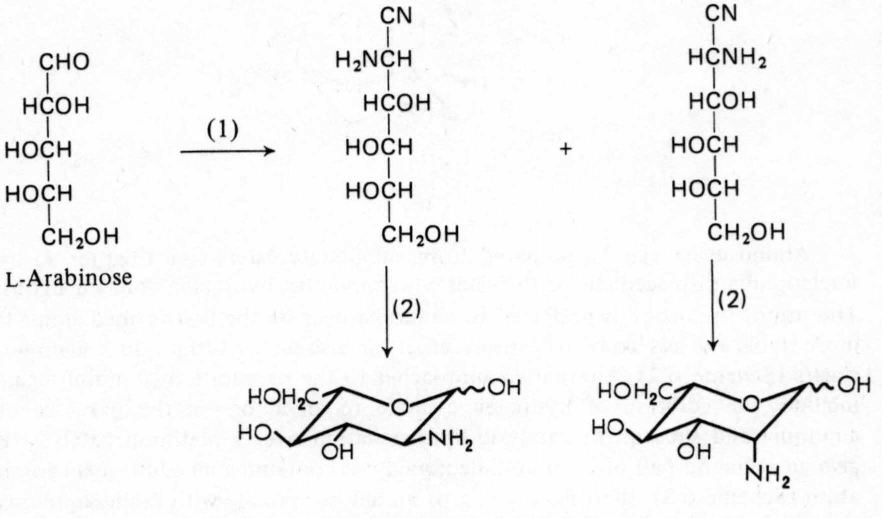

CHO
HCOH
HOCH
HOCH
CH_2OH
L-Arabinose

(1) →

CN
H_2NCH
HCOH
HOCH
HOCH
CH_2OH

\+

CN
$HCNH_2$
HCOH
HOCH
HOCH
CH_2OH

(2) ↓ (2) ↓

HOH_2C O OH HO NH$_2$ OH
HOH_2C O OH HO OH NH$_2$

2-Amino-2-deoxy-L-glucose 2-Amino-2-deoxy-L-mannose

Reagents: (1) HCN, NH_3, CH_3OH; (2) H_2/Pt

Scheme 6.3

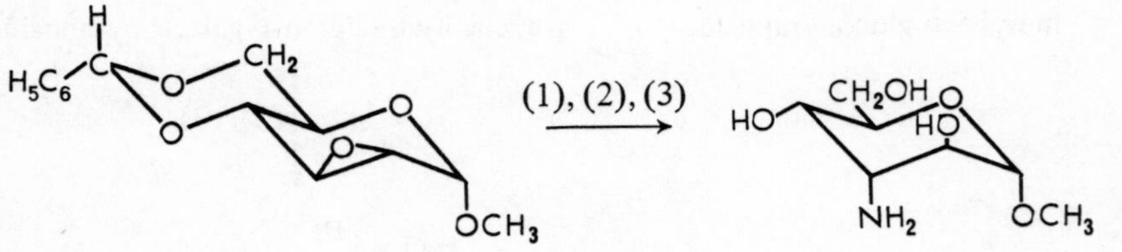

H
H_5C_6 C O CH_2 O O O OCH_3

(1), (2), (3) →

HO CH_2OH O HO NH_2 OCH_3

Methyl 2,3-anhydro-4,6-*O*-benzylidene-α-D-mannopyranoside

Methyl 3-amino-3-deoxy-α-D-altropyranoside (major product)

Reagents: (1) NH_3; (2) $NaBH_4$; (3) deprotect

Scheme 6.4

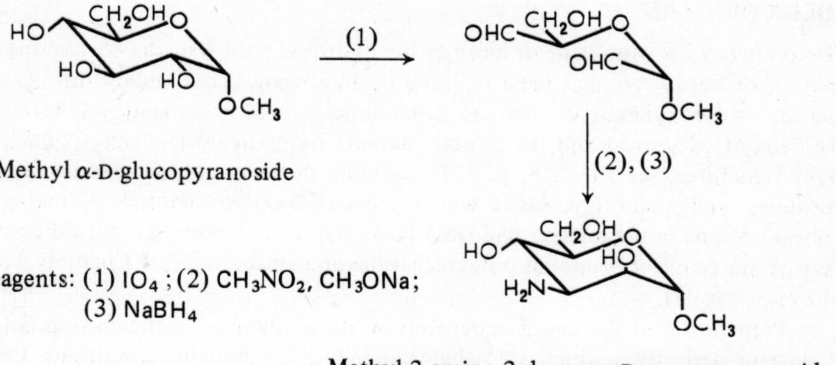

Methyl α-D-glucopyranoside

Reagents: (1) IO_4 ; (2) CH_3NO_2, CH_3ONa; (3) $NaBH_4$

Methyl 3-amino-3-deoxy-α-D-mannopyranoside (also gluco-, galacto- and talo-isomers)

Scheme 6.5

Interconversion of a readily available aminosugar into another aminosugar can be achieved by reaction of the *O*-methanesulphonyl (mesyl) derivative with sodium acetate in aqueous 2-methoxyethanol (Scheme 6.6). A recent review of aminosugars is available (Horton and Wander, 1980).

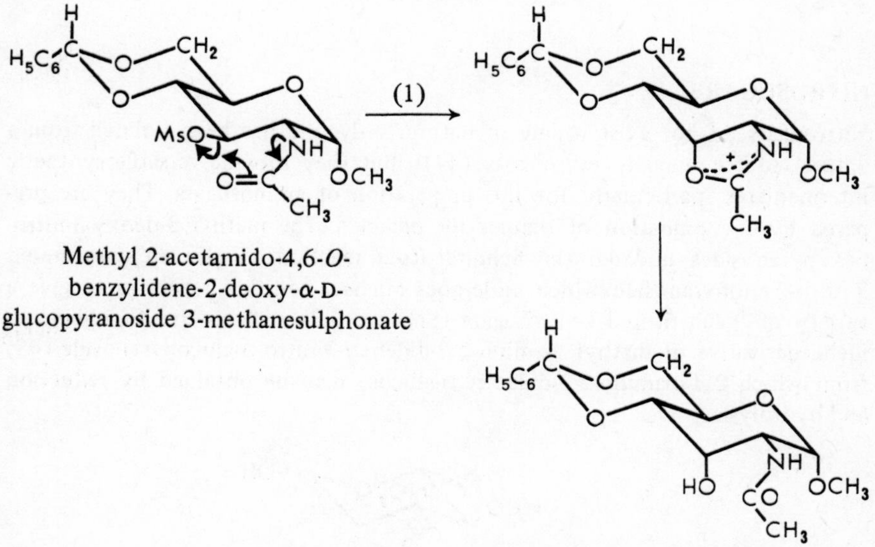

Methyl 2-acetamido-4,6-*O*-benzylidene-2-deoxy-α-D-glucopyranoside 3-methanesulphonate

Methyl 2-acetamido-4,6-*O*-benzylidene-2-deoxy-α-D-allopyranoside

Reagents: (1) CH_3COONa

Scheme 6.6

DEOXYSUGARS

Deoxysugars, in which one or more of the hydroxyl groups of the corresponding aldose or ketose etc. has been replaced by hydrogen, occur widely throughout nature. 6-Deoxyhexoses, such as L-rhamnose (6-deoxy-L-mannose), L-fucose (6-deoxy-L-galactose) and, to a lesser extent, D-quinovose (6-deoxy-D-glucose) (for structures, see Fig. 2.8, p. 33) are found in many polysaccharides, glycoproteins and plant glycosides whilst 2-deoxy-D-*erythro*-pentose (2-deoxy-D-ribose) occurs in the nucleic acid DNA (see Chapter 11). More complex dideoxysugars are found in bacterial polysaccharides and antibiotics (see Chapters 8 and 12 respectively).

Preparation of deoxysugars depends on the availability of the corresponding halogeno derivatives which are dehalogenated under reducing conditions. Iodo- and bromo-derivatives are converted to deoxysugars by removal of the halogen atom through the action of hydrazine and Raney nickel in the presence of sodium acetate. Simple 6-deoxysugars are prepared by reduction of hexose 6-(4-toluene sulphonates) with lithium aluminium hydride. Reactions of 6-deoxysugars are similar to those for neutral monosaccharides except for those which involve the primary hydroxyl group at C-6 in neutral monosaccharides whilst other deoxysugars react as neutral monosaccharides unless the missing hydroxyl group prevents such.

NITROSUGARS

Nitrosugars do not exist widely in nature (only one has been isolated from a natural source, namely evernitrose (47)), but they provide versatile synthetic intermediates, particularly for the preparation of aminosugars. They are prepared by the oxidation of oximes by peracids. The methyl 3-deoxy-3-nitrohexopyranosides undergo the Schmidt-Rutz reaction to give an unsaturated 3-nitro-2-enopyranoside which undergoes Michael addition reactions to give a variety of 2-substituted-3-nitrosugars (Scheme 6.7) such as the 4,6-*O*-benzylidene derivative of methyl 2-amino-2,3-dideoxy-3-nitro-D-glucopyranoside (48) from which 2,3-diamino-2,3-dideoxy-D-glucose may be obtained by reduction and hydrolysis.

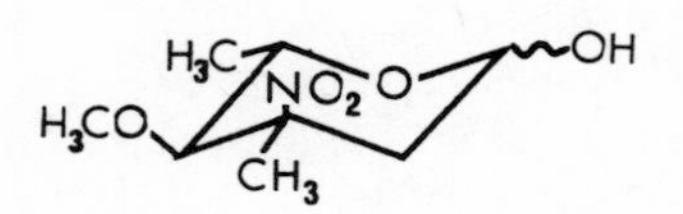

2,3,6-Trideoxy-3-*C*-methyl-4-*O*-methyl-3-nitro-L-*ribo*-hexose (evernitrose)
(47)

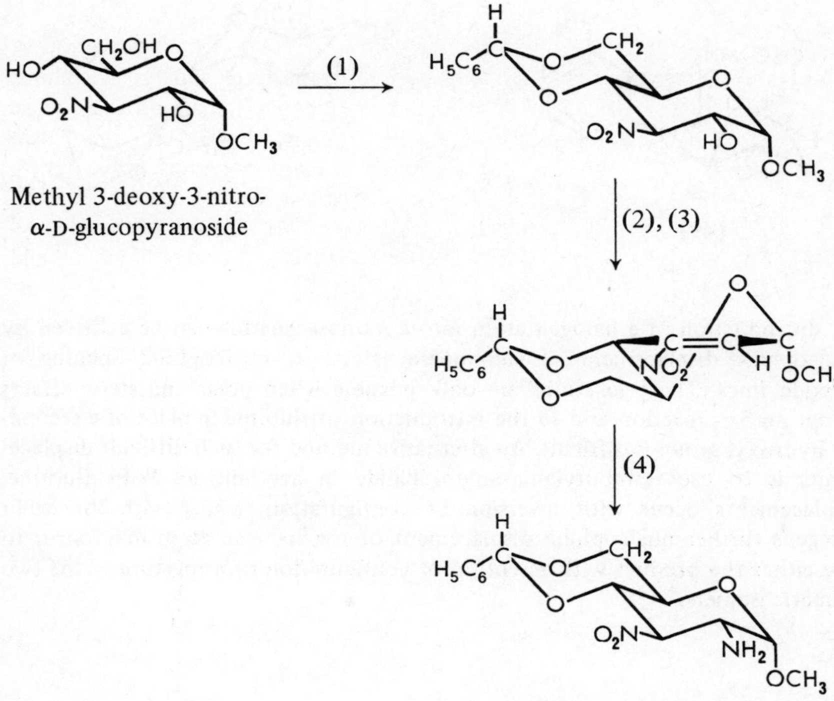

Methyl 3-deoxy-3-nitro-α-D-glucopyranoside

Methyl 2-amino-4,6-*O*-benzylidene-2,3-dideoxy-3-nitro-α-D-glucopyranoside (48)

Reagents: (1) C_6H_5CHO, $ZnCl_2$; (2) Acetic anhydride, BF_3, Et_2O; (3) $NaHCO_3$; (4) NH_3

Scheme 6.7

HALOGENOSUGARS

Only one naturally occurring halogenosugar has been isolated from natural sources, namely the antibiotic nucleocidin (49). However, much interest exists in halogenosugars because in other areas of natural products chemistry the introduction of a halogen atom has often led to the generation of compounds with interesting or different or unique biological activities. Interest in carbohydrate chemistry has, to date, centred on modified antibiotics with improved activity and on synthetic high intensity sweetening agents. A trichloro derivative of sucrose has been prepared (50) which is 2×10^3 times sweeter than sucrose itself, whilst other chlorosucroses have contraceptive properties.

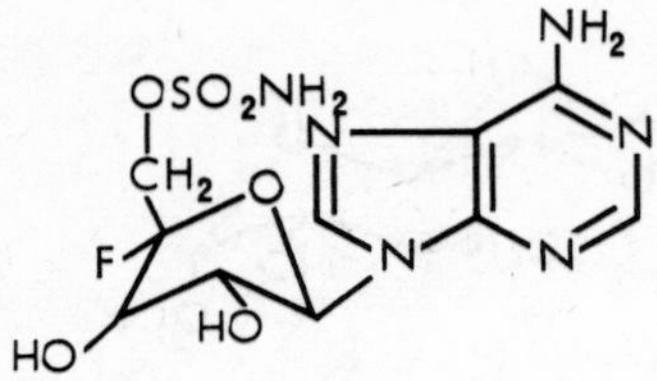

(49)

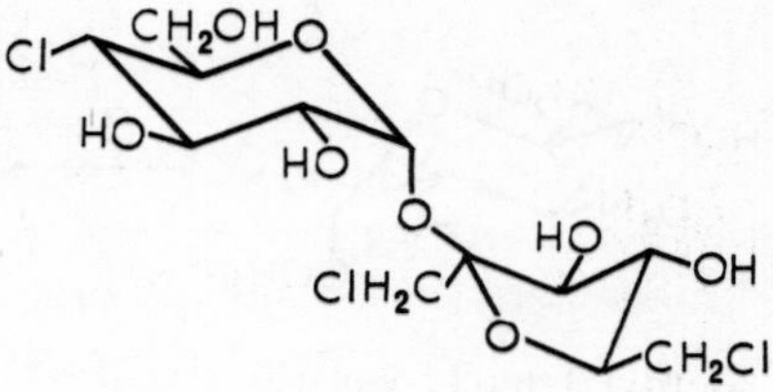

(50)

Introduction of a halogen atom into a monosaccharide can be achieved by nucleophilic displacement of sulphonate esters, or nucleophilic opening of epoxide rings. These reactions are only possible when polar and steric effects favour an S_{N_2} reaction and so the introduction of fluorine in place of a secondary hydroxyl group is difficult. An alternative method for such difficult displacements is to use tetrabutylammonium halide in acetonitrile. With fluorine, displacements occur with inversion of configuration, whilst with the other halogens further nucleophilic displacement of the halogen atom may occur to give either the product with retention of configuration or a mixture of the two epimeric isomers.

THIOSUGARS

The replacement of an oxygen atom in a monosaccharide by sulphur gives rise to thiosugars (Horton and Wander, 1980b), of which the 1-thioglycosides are naturally occurring in the seeds of many plants, as typified by sinigrin (51), isolated from the seeds of black mustard. Over 50 such compounds are known to occur naturally and, because of their flavouring properties, are referred to as the mustard-oil glycosides.

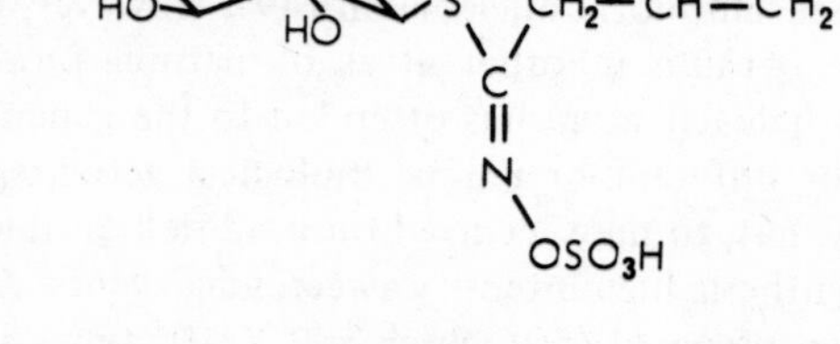

(51)

The greatly enhanced nucleophilicity of sulphur compared to oxygen allows the introduction of sulphur into monosaccharides by nucleophilic displacements of halides and sulphate esters and by ring opening of epoxides. The greater nucleophilicity of sulphur also results in the sulphur atom tending to favour its inclusion in the ring such that 4-thiohexoses, for example 4-thio-D-glucose, exist only in the furanose ring form (52) with the sulphur atom present in the ring, although 6-deoxy-4-thio-D-gulose has only been reported in the pyranose ring form (53) with the sulphur atom not in the ring. 5-Thiohexoses such as 5-thio-D-glucose (54) exist in the pyranose ring form with the sulphur atom present in the ring. Thiosugars frequently act as inhibitors to enzymes for which the corresponding non-sulphur monosaccharide is a substrate. Thus, for example, 5-thio-D-glucose 1-phosphate inhibits the isomerisation of D-glucose 1-phosphate into the 6-phosphate.

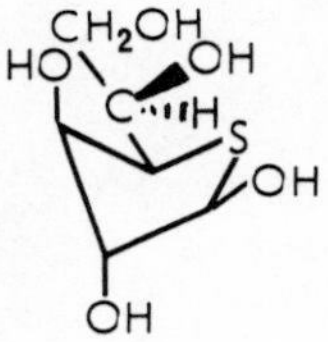

(52) 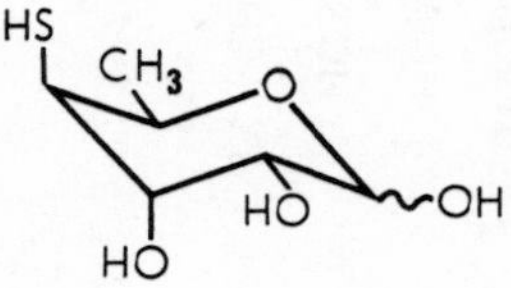 (53) 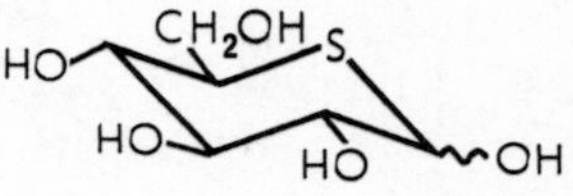 (54)

BRANCHED-CHAIN MONOSACCHARIDES

The term 'branched-chain sugar', used frequently to describe these compounds, is misleading as it can imply a branched-carbohydrate chain polysaccharide as well as a branched-carbon chain monosaccharide and the term 'branched-chain monosaccharide' is to be preferred. Naturally occurring branched-chain monosaccharides are found in antibiotics (see Chapter 12 for structures) and also in several plants including parsley (which contains apiose (55)), and witch-hazel (which contains hamamelose (56) as its digalloyl ester). Other substituents which are known to be attached to a carbon atom to give branched-chain monosaccharides include methyl, ethyl, 2-methyl-1-propenyl [$(CH_3)_2C{=}CH-$] and aldehydo groups (see review by Williams and Wander, 1981).

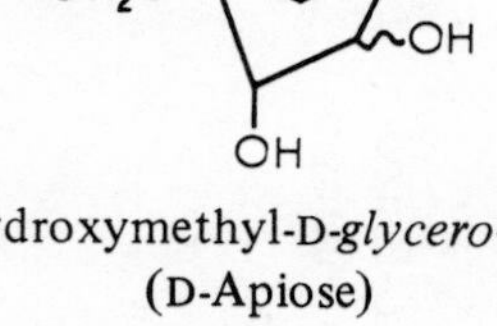

3-*C*-Hydroxymethyl-D-*glycero*-tetrose
(D-Apiose)
(55)

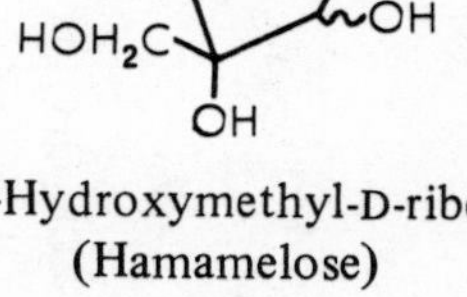

2-*C*-Hydroxymethyl-D-ribose
(Hamamelose)
(56)

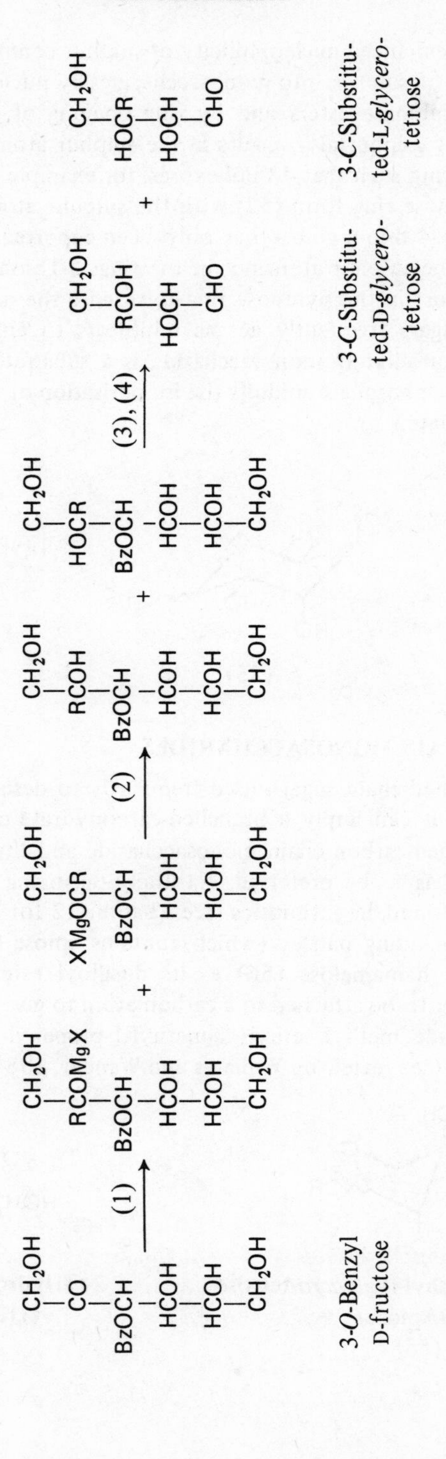

Reagents: (1) RMgX, Et_2O; (2) H^+; (3) $Pb(OAc)_4$; (4) deprotect

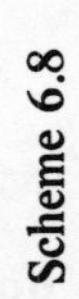

Scheme 6.8

Synthesis of branched-chain monosaccharides is *via* the appropriate keto derivative, either by direct action of a Grignard reagent, diazomethane, nitromethane, Wittig reagents or cyanide (Scheme 6.8) or by treatment with diazomethane in a stereospecific reaction with the nucleophile approaching from an exo-position to give a spiroepoxide (57), hydrolysis of which gives a branched-chain monosaccharide containing a hydroxymethyl group (Scheme 6.9).

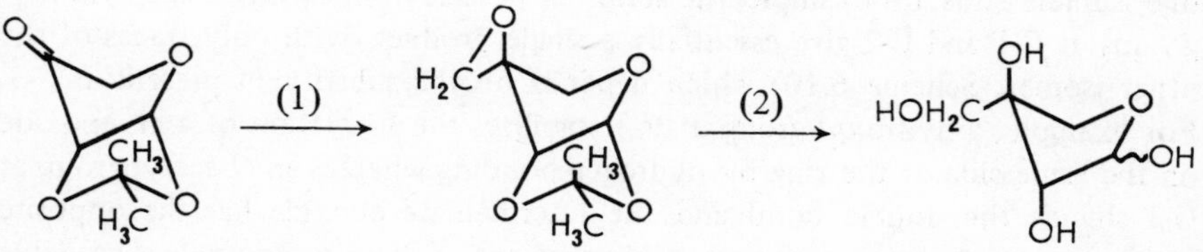

1,2-*O*-Isopropylidene-D-*glycero*-3-tetrulose —(1)→ (57) —(2)→ D-Apiose

Reagents: (1) CH_3N_2; (2) H^+

Scheme 6.9

Systematic naming (Recommendations, 1980d) of branched-chain monosaccharides uses a scheme which regards the compound as being a derivative of a parent unbranched monosaccharide which contains the greatest possible number of carbon atoms, with preference for aldose rather than ketose. Thus (58) is a derivative of an octose not a 2-octulose.

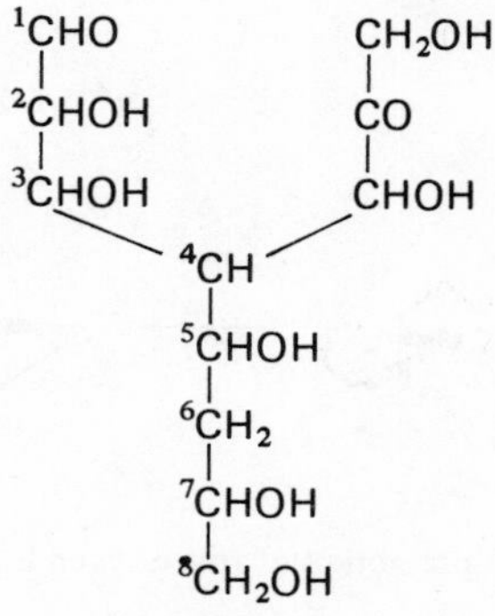

^{1}CHO, ^{2}CHOH, ^{3}CHOH, ^{4}CH (branch: CHOH–CO–CH_2OH), ^{5}CHOH, $^{6}CH_2$, ^{7}CHOH, $^{8}CH_2OH$

(58)

UNSATURATED SUGARS

Naturally occurring unsaturated sugars include vitamin C (36) and antibiotics which contain double bonds or vicinal dideoxy structures which are derived from unsaturated systems. Synthetic unsaturated sugars have recently become valuable intermediates for the preparation of rare sugars. For a review see Ferrier (1980).

1,5-Anhydro-2-deoxy-hex-1-enitols (see Fig. 6.2), trivially referred to as glycals (see Recommendations, 1980e for nomenclature of unsaturated monosaccharides), form useful reagents in which electrophilic attack at C-2 is favoured by the mesomeric interaction of the ring oxygen with the double bond. Whilst it is possible in theory to produce four possible isomers by such reactions, the stereoselectivity of many reagents reduces the number to two or sometimes only one isomer. Thus, for example, the action of peracids, which introduce hydroxyl groups at C-1 and C-2 give essentially a single product (with only traces of the other isomer, Scheme 6.10) which depends on the substituent present at C-3. For example, a hydroxyl group at C-3 stabilises the formation of a cis-epoxide on the same side of the ring by hydrogen bonding whereas an *O*-acetyl group at C-3 shields the double bond and the intermediate epoxide has the opposite configuration. A review of the addition reactions of these compounds is available (Ferrier, 1976). Pyran- and furan-2-enoses and other less-common enoses have been prepared, using the Tipson-Cohen procedure of heating the di-(4-toluenesulphonate) ester (59) with zinc dust and sodium iodide in *N,N*-dimethylformamide (Scheme 6.11), but there have been few reports to date on addition reactions of these compounds.

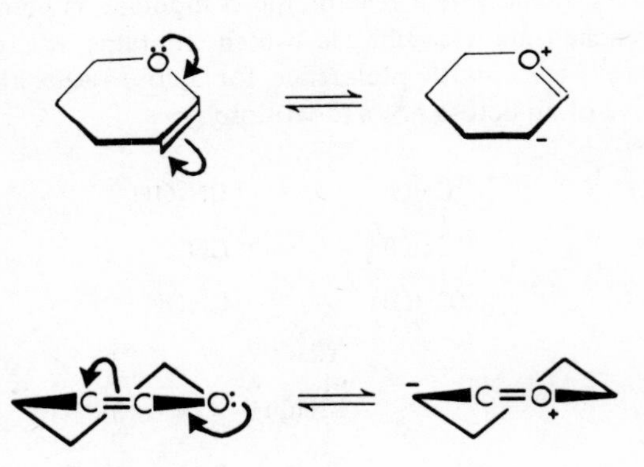

Fig. 6.2 – Mesomeric interaction of ring oxygen in 1,5-anhydro-2-deoxy-hex-1-enitols (glycals).

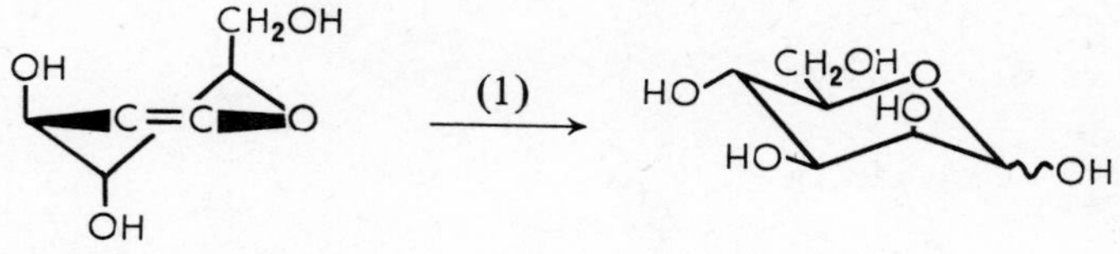

OH, CH$_2$OH, C=C, O, OH — (1) → HO, CH$_2$OH, O, HO, HO, OH

1,5-Anhydro-2-deoxy-D-*arabino*-hex-1-enitol

D-Mannopyranose

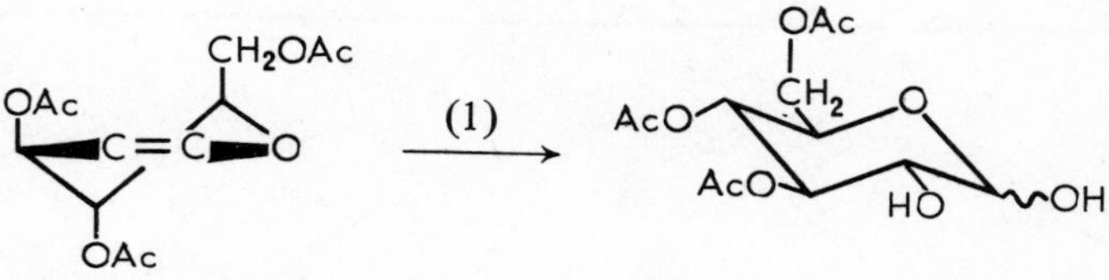

OAc, CH$_2$OAc, C=C, O, OAc — (1) → OAc, CH$_2$, O, AcO, AcO, HO, OH

3,4,6-Tri-*O*-acetyl-1,5-anhydro-2-deoxy-D-*arabino*-hex-1-enitol

3,4,6-Tri-*O*-acetyl-D-glucopyranose

Reagents: (1) Peracid

Scheme 6.10

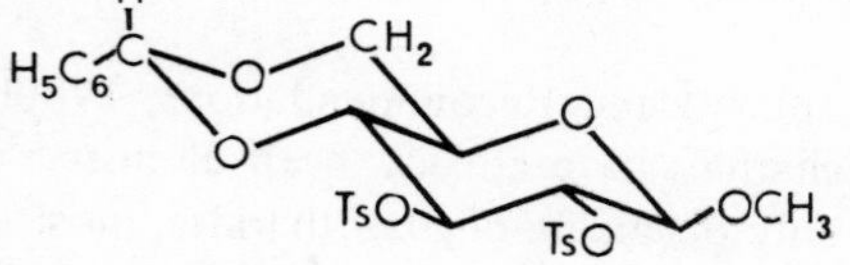

H, H_5C_6, C, O, CH$_2$, O, O, TsO, TsO, OCH$_3$

(59)

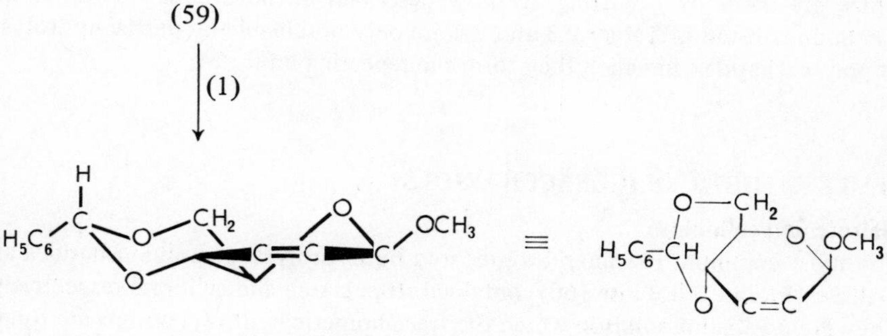

(1)

H, H_5C_6, C, O, CH$_2$, O, C=C, O, C, OCH$_3$ ≡ O—CH$_2$, H_5C_6—CH, O, OCH$_3$, O, C=C

Methyl 4,6-*O*-benzylidine-2,3-dideoxy-β-D-*erythro*-hex-2-enopyranoside

Reagents: (1) NaI, Zn, DMF

Scheme 6.11

CHAPTER 7

Oligosaccharides

Naturally occurring disaccharides and trisaccharides have attracted considerable attention among scientists in the last few years. Biochemists, food scientists, nutritionists, medical scientists, physicians and consumers are greatly interested in the role of these oligosaccharides in nutrition and health. For the chemists and technologists, interest arose partly from the desire to extend their understanding of the fundamental chemistry of these compounds with the view of utilising them as chemical raw materials and industrial 'feedstocks', and partly from their presence in such active materials as antibiotics. For a review of these developments see Lee (1980).

The definition, nomenclature (Chapter 2 and Recommendations, 1980b) and synthesis (Chapter 5) of oligosaccharides have already been discussed in general terms. A number of the commonly occurring oligosaccharides, most of which are naturally occurring, are now described in more detail. Many of the compounds listed in Tables 2.3 and 2.4 are only obtainable by partial hydrolysis of polysaccharides, in which they form the repeating unit.

SEMI-SYNTHETIC OLIGOSACCHARIDES

Maltose and cellobiose

The most common disaccharides prepared by hydrolysis of polysaccharides are maltose (8) and cellobiose (60), obtained from starch and cellulose respectively. They are crystalline solids in which the free anomeric hydroxyl groups are found predominantly in the equatorial position. Thus the structure of β-maltose is 4-*O*-α-D-glucopyranosyl-β-D-glucopyranose whilst β-cellobiose is 4-*O*-β-D-glucopyranosyl-β-D-glucopyranose. They react in a manner similar to reducing monosaccharides and form most of the derivatives which monosaccharides can, having eight free hydroxyl groups.

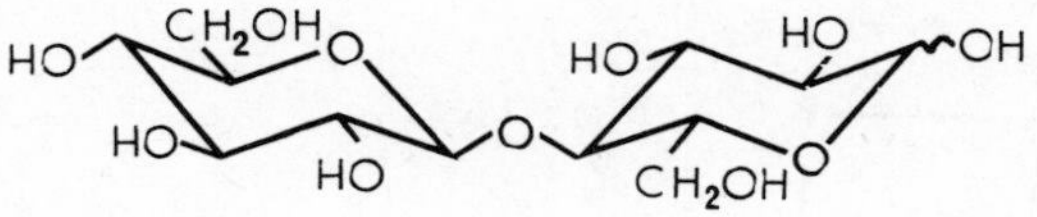

(60)

Cyclomalto-oligosaccharides

A number of cyclomalto-oligosaccharides have been prepared by the action of an enzyme from *Bacillus macerans* on starch, and include cyclomaltohexaose (61), cyclomaltoheptaose and cyclomalto-octaose (which are described in older literature as cycloamyloses, cyclodextrins or as α-, β- and γ-Schardinger dextrins respectively, after F. Schardinger, who characterised these materials as cyclic compounds). They are all crystalline solids and represent the highest molecular weight oligosaccharides which have been obtained in crystalline form. An interesting feature of the cyclic structure (61) is that all the primary hydroxyl groups (attached to C-6 of the carbohydrate residues) lie on one side of the ring whilst all the secondary hydroxyl groups (attached to C-2 and C-3) lie on the other side. These cyclic compounds serve as models for starch because each turn of the polymeric helix structure could be represented by a single closed loop containing 6, 7 or 8 monomer residues.

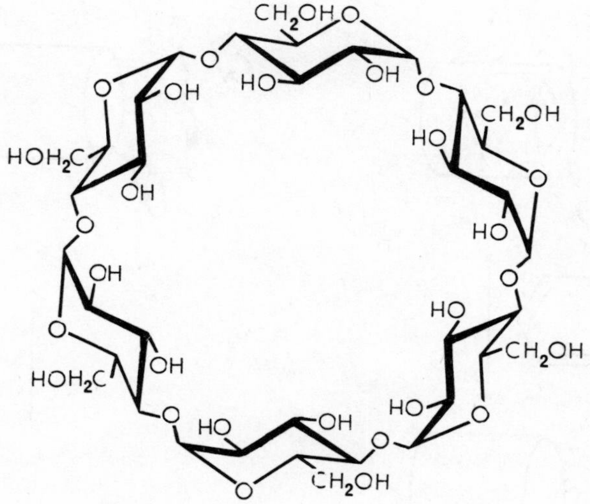

(61)

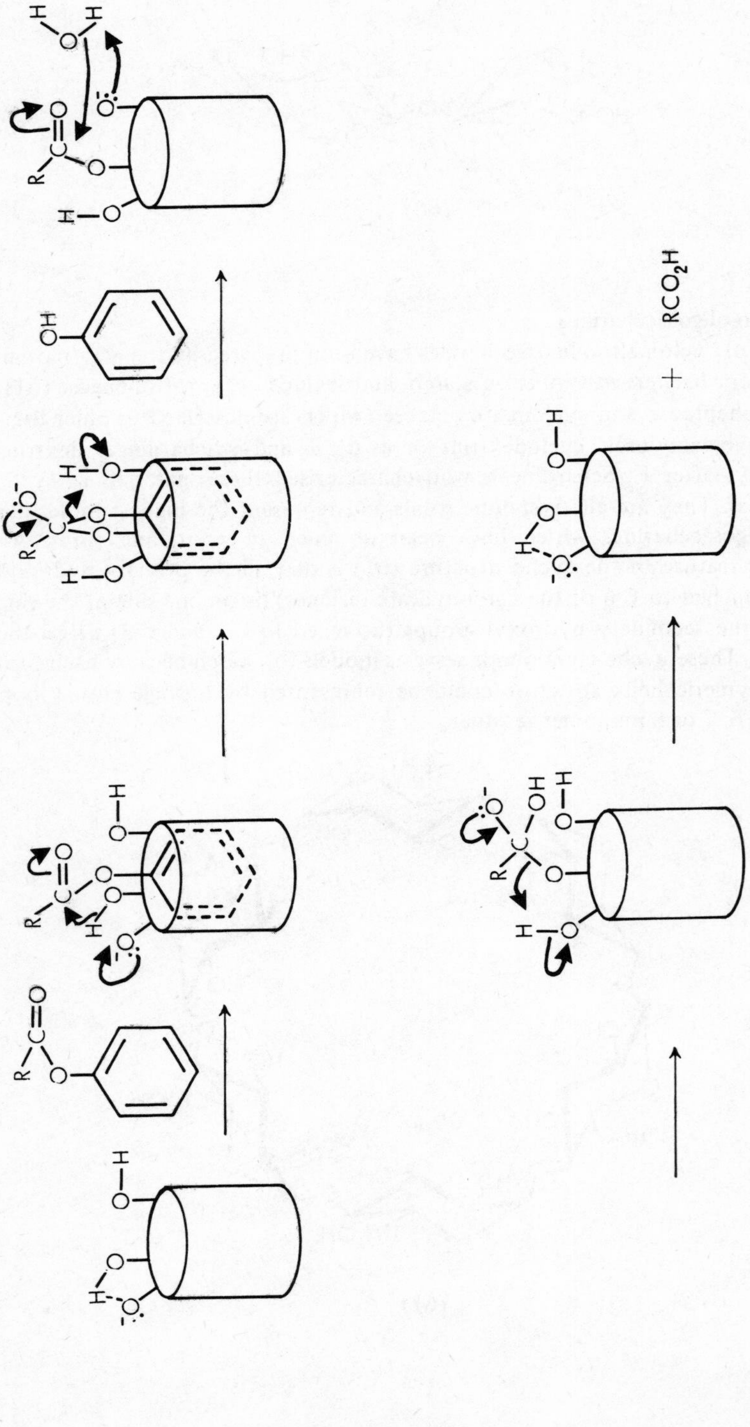

Fig. 7.1 – Schematic representation of the hydrolysis of phenyl esters catalysed by cyclomalto-oligosaccharides.

One of the more unusual properties of these compounds is their ability to bind small molecules within their cyclic structure to form so-called inclusion compounds. These crystalline solids are of interest for scientific research because they are soluble in aqueous solution and can be used to study the hydrophobic interactions which are so important in biological systems. Inclusion of a material within the cyclic structure has been shown to protect this material from the effects of light and atmosphere and to increase the water solubility of the material. Hence cyclomalto-oligosaccharides are used in the production of pharmaceuticals, pesticides, foodstuffs and toiletries as protecting agents, emulsion stabilisers and, since the dry inclusion complexes can be easily handled and stored, as bulking agents (for a review see Saenger, 1980).

The types of materials which form inclusion complexes are many, ranging from methanol and iodine to aliphatic and aromatic acids, esters and many other aromatic compounds, in which the aromatic ring lies within the cyclic structure and substituents interact with the carbohydrate hydroxyl groups at the edge of the ring *via* hydrogen bonding. An extension to this interaction is the ability of these cyclomalto-oligosaccharides to have apparent enzyme action, since they can hydrolyse a number of compounds with which they form inclusion complexes. The most studied example is the hydrolysis of phenyl esters (Fig. 7.1). This apparent enzyme action has led to these cyclomalto-oligosaccharides being used as enzyme models by substituting the free hydroxyl groups with various groups to simulate the active centre of a number of enzymes. A full account of the chemistry of cyclomalto-oligosaccharides and their use as enzyme models can be found in Bender and Komiyama (1978).

NATURALLY OCCURRING OLIGOSACCHARIDES

Sucrose

Sucrose (62) is one of the most common carbohydrates, being widely distributed throughout the plant world. It is the main soluble carbohydrate reserve and energy source and an important dietary material for humans. The principle sources for the commercial production of sucrose are sugar cane, sugar beet and the sap of maple trees.

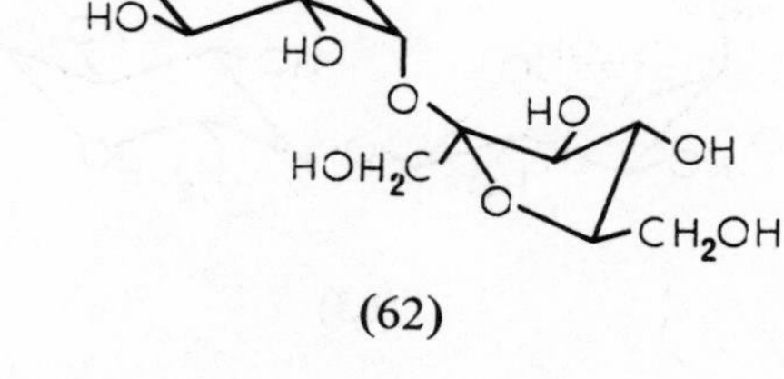

CH_2OH HO HO HO O O HO OH HOH_2C O CH_2OH

(62)

The structure of sucrose, namely α-D-glucopyranosyl-(1↔2)-β-D-fructofuranoside (or β-D-fructofuranosyl-(2↔1)-α-D-glucopyranoside) does not contain an aldehyde function since the glycosidic bond joins the two anomeric carbon atoms and therefore sucrose is not a reducing sugar. Hydrolysis of sucrose, by mildly acidic conditions (heating in pH 5 buffer will suffice) or enzymes, produces a mixture of equal amounts of D-glucose and D-fructose, the process being referred to as 'inversion' and the mixture as 'invert sugar' on account of the change in optical rotation from dextro to laevo due to the high laevorotation of the D-fructose. An enzyme capable of hydrolysing sucrose, β-D-fructofuranosidase and given the trivial name invertase (β-D-fructofuranoside fructohydrolase, EC 3.2.1.26) hydrolyses β-D- and not α-D-fructofuranosides whilst α-D-glucosidase (α-D-glucoside glucohydrolase EC 3.2.1.20) and not β-D-glucosidase (β-D-glucoside glucohydrolase EC 3.2.1.21) also hydrolyses sucrose. These together prove that the configuration of the glycosidic linkages is α for the D-glucose and β for the D-fructose residues.

Lactose

Mammalian milk contains between 2.0 and 8.5% of lactose depending on the mammal either as the free disaccharide (63) or combined with other residues, such as 2-acetamido-2-deoxy-D-glucose, L-fucose, D-galactose and 5-acetamido-3,5-dideoxy-D-*glycero*-D-*galacto*-2-nonulopyranonic acid (hence an old trivial name of lactamic acid) as higher oligosaccharides (see Table 7.1). The concentration of these higher oligosaccharides is reported to be between 0.3 and 0.6%. Lactose has been reported to occur in plants but it is mainly found in milk, hence an old, trivial name 'milk-sugar'. It is not fully explained yet, but in the mammary gland part of the D-glucose present is converted to D-galactose which then combines with more D-glucose to form lactose. The infant to whom the milk is fed must immediately reconvert the D-galactose back to D-glucose during digestion. If not, there can be a number of ill-effects. Why, then, is D-glucose initially converted to D-galactose? The key may lie in an understanding of the higher oligosaccharides and their immunological reactions. It is known, for instance, that lacto-*N*-neotetraose inhibits the cross-reaction between the pneumococcal polysaccharide Type 14 antigen and its antibodies. The presence, therefore, of D-galactose may be to provide the correct structural components for the mother to transmit immunological protection to her infant.

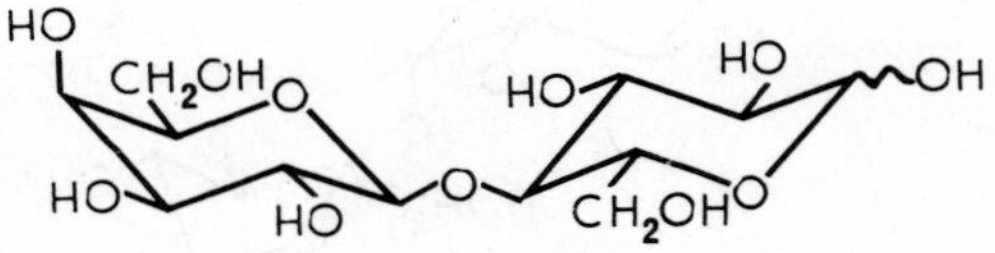

(63)

Table 7.1 Structures of some human milk oligosaccharides

Trivial name	Structure
Disialyllacto-*N*-tetraose	α-Neu*p*5Ac-(2→3)-β-D-Gal*p*-(1→3)-β-D-Glc*p*NAc-(1→3)-β-D-Gal*p*-(1→4)-D-Glc 6 ↑ 1 α-Neu*p*5Ac
2′-Fucosyllactose	α-L-Fuc*p*-(1→2)-β-D-Gal*p*-(1→4)-D-Glc
3-Fucosyllactose	β-D-Gal*p*-(1→4)-D-Glc 3 ↑ 1 α-L-Fuc*p*
Galactosyllactose	[β-D-Gal*p*-(1→3)]$_n$-β-D-Gal*p*-(1→4)-D-Glc (where n = 2 – 5)
Lacto-difucosyltetraose	α-L-Fuc*p*-(1→2)-β-D-Gal*p*-(1→4)-D-Glc 3 ↑ 1 α-L-Fuc*p*
Lacto-*N*-tetraose	β-D-Gal*p*-(1→4)-β-D-Glc*p*NAc-(1→3)-β-D-Gal*p*-(1→4)-D-Glc
Lacto-*N*-neotetraose	β-D-Gal*p*-(1→3)-β-D-Glc*p*NAc-(1→3)-β-D-Gal*p*-(1→4)-D-Glc
Lacto-*N*-fucopentaose I	α-L-Fuc*p*-(1→2)-β-D-Gal*p*-(1→3)-β-D-Glc*p*NAc-(1→3)-β-D-Gal*p*-(1→4)-D-Glc
Lacto-*N*-fucopentaose II	β-D-Gal*p*-(1→3)-β-D-Glc*p*NAc-(1→3)-β-D-Gal*p*-(1→4)-D-Glc 4 ↑ 1 α-L-Fuc*p*
Lacto-*N*-fucopentaose III	β-D-Gal*p*-(1→4)-β-D-Glc*p*NAc-(1→3)-β-D-Gal*p*-(1→4)-D-Glc 3 ↑ 1 α-L-Fuc*p*
Lacto-*N*-difucohexaose I	α-L-Fuc*p*-(1→2)-β-D-Gal*p*-(1→3)-β-D-Glc*p*NAc-(1→3)-β-D-Gal*p*-(1→4)-D-Glc 4 ↑ 1 α-L-Fuc*p*
Lacto-*N*-difucohexaose II	β-D-Gal*p*-(1→3)-β-D-Glc*p*NAc-(1→3)-β-D-Gal*p*-(1→4)-D-Glc 4 3 ↑ ↑ 1 1 α-L-Fuc*p* α-L-Fuc*p*
Lacto-*N*-hexaose	β-D-Gal*p*-(1→4)-β-D-Glc*p*NAc-(1→6)-β-D-Gal*p*-(1→4)-D-Glc 3 ↑ 1 β-D-Gal*p*-(1→3)-β-D-Glc*p*NAc
Lacto-*N*-neohexaose	β-D-Gal*p*-(1→4)-β-D-Glc*p*NAc-(1→6)-β-D-Gal*p*-(1→4)-D-Glc 3 ↑ 1 β-D-Gal*p*-(1→4)-β-D-Glc*p*NAc
Lactosyl-(1→2)-α-D-glucose	β-D-Gal*p*-(1→4)-α-D-Glc*p*-(1→2)-D-Glc
LS tetrasaccharide a	α-Neu*p*5Ac-(2→3)-β-D-Gal*p*-(1→3)-β-D-Glc*p*NAc-(1→3)-β-D-Gal*p*-(1→4)-D-Glc
LS tetrasaccharide b	β-D-Gal*p*-(1→3)-β-D-Glc*p*NAc-(1→3)-β-D-Gal*p*-(1→4)-D-Glc 6 ↑ 1 α-Neu*p*5Ac

Table 7.1 (*continued*)

Trivial name	Structure
LS tetrasaccharide c	α-Neu*p*5Ac-(2→6)-β-D-Gal*p*-(1→4)-β-D-Glc*p*NAc-(1→3)-β-D-Gal*p*-(1→4)-D-Glc
3′-Sialyllactose (neuraminolactose)	α-Neu*p*5Ac-(2→3)-β-D-Gal*p*-(1→4)-D-Glc
6′-Sialyllactose	α-Neu*p*5Ac-(2→6)-β-D-Gal*p*-(1→4)-D-Glc
S-5	Mixture of *N*-acetylneuraminosyl derivatives of lacto-*N*-hexaose (6 isomers) and of lacto-*N*-neohexaose (6 isomers)
S-6	Mixture of L-fucosyl derivatives of S-5 (36 isomers)
N-2	Mixture of L-fucosyl derivatives of lacto-*N*-hexaose (6 isomers) and of lacto-*N*-neohexaose (6 isomers)
N-3	Mixture of di-L-fucosyl derivatives of lacto-*N*-hexaose (15 isomers) and of lacto-*N*-neohexaose (15 isomers)

Lactose is prepared industrially from whey, a by-product of cheese manufacture. Traditionally, the process involved evaporation of the whey until the lactose crystallised from a cold solution as the monohydrate of α-lactose, or from a hot (95°) solution as the anhydrous β-lactose. However, the modern method of ultrafiltration is being applied, on an increasing scale, to purify the lactose, prior to its removal from solution, by removal of the whey proteins and fats. Lactose, as a reducing disaccharide, undergoes the reactions of normal reducing sugars and can be fermented but only by yeasts which have been adapted to lactose. With normal yeasts no fermentation occurs.

Trehaloses

Trehalose is the general name given to a family of three D-glucosyl D-glucosides, nonreducing disaccharides which form the reserve carbohydrate of insects and a number of invertebrates, and are the storage material of fungi, yeasts, algae and members of the Pteridophyta family. In mushrooms, for example, it can account for as much as 15% of the dry weight. Three linkage types are possible, namely α,α; α,β and β,β and all three anomeric configurations of the disaccharide with two pyranose rings are known but in general only α,α-trehalose (64) is the common isomer, although α,β-trehalose (65) has been isolated from honey.

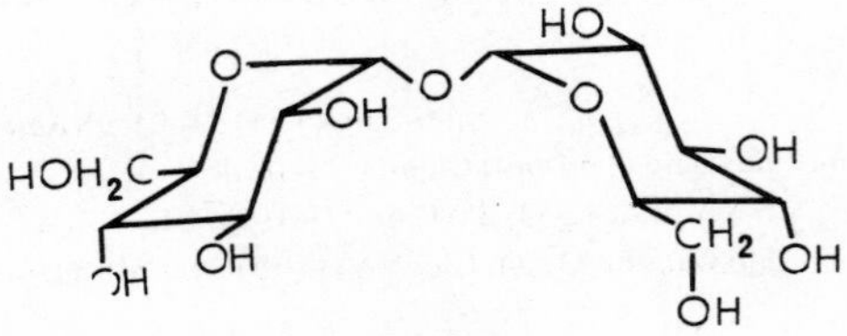

HO
O
O
O
OH
OH
HOH_2C
CH_2
OH
OH
OH
OH

(64)

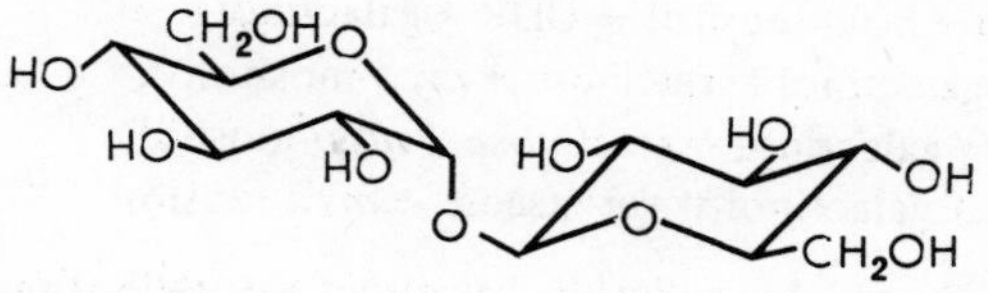

(65)

α,α-Trehalose has the structure α-D-Glc*p*-(1↔1)-α-D-Glc*p* and derives its name from the trehala manna (from which α,α-trehalose was obtained) which forms the cocoons of a beetle of the *Larinus* family. Its trivial name of mycose or mushroom sugar reflects the other major source of the disaccharide. Its biosynthesis involves two enzymes, α,α-trehalose phosphate synthase (UDP-forming) (UDP-D-glucose:D-glucose-6-phosphate-1-α-glucosyltransferase, EC 2.4.1.15) and trehalose phosphatase (trehalose-6-phosphate phosphohydrolase, EC 3.1.3.12) which catalyse the following reactions respectively:

UDP-D-Glc + D-Glc 6-phosphate → α,α-trehalose 6-phosphate + UDP

α,α-trehalose 6-phosphate → α,α-trehalose + orthophosphate

Chemical synthesis of trehaloses tend to favour production of the α,β-isomer but heating of a mixture of 2,3,4,6-tetra-*O*-acetyl-D-glucose and 3,4,6-tri-*O*-acetyl-1,2-anhydro-D-glucose at 100°C produces a mixture of α,α- and α,β-isomers. α,α-Trehalose is relatively inert compared to many carbohydrates due to its nonreducing nature and stable glycosidic linkage (it is much more stable to mild acid than is sucrose), which is possibly why it has transport functions in some organisms. The only reactions which take place with trehaloses are those associated with the primary and secondary hydroxyl groups.

Raffinose

Raffinose is a naturally occurring plant trisaccharide which is as widely distributed as sucrose but only in low concentrations (for example, 0.05% in sugar beet and 1.9% in soybeans) and is related to sucrose as can be seen from its structure of α-D-Gal*p*-(1→6)-α-D-Glc*p*-(1↔2)-β-D-Fru*f* (66) which can be regarded as an *O*-galactosylsucrose. As such, raffinose is the first member of a series of oligosaccharides in which one or more D-galactosyl residues are attached to sucrose. Other members include stachyose (a tetrasaccharide) and verbascose (a pentasaccharide) whose structures are given in Table 2.4.

These oligosaccharides are synthesised from sucrose by an unusual transfer reaction in which an α-D-galactopyranoside of *myo*-inositol, known by the trivial name galactinol (67), is the immediate D-galactosyl donor. The reaction sequences are catalysed by enzymes such as galactinol-raffinose galactosyltransferase (1-*O*-α-D-galactosyl-*myo*-inositol:raffinose galactosyltransferase, EC 2.4.1.67):

UDP-D-Gal + *myo*-inositol → UDP + galactinol,
Sucrose + galactinol ⇌ raffinose + *myo*-inositol,
Raffinose + galactinol ⇌ stachyose + *myo*-inositol,
Stachyose + galactinol ⇌ verbascose + *myo*-inositol.

Raffinose is prepared by isolation by direct crystallisation from the final molasses of the sucrose production from beet. It can also be prepared by extraction from cotton seed meal with subsequent precipitation of the slightly soluble raffinose with calcium or barium hydroxide.

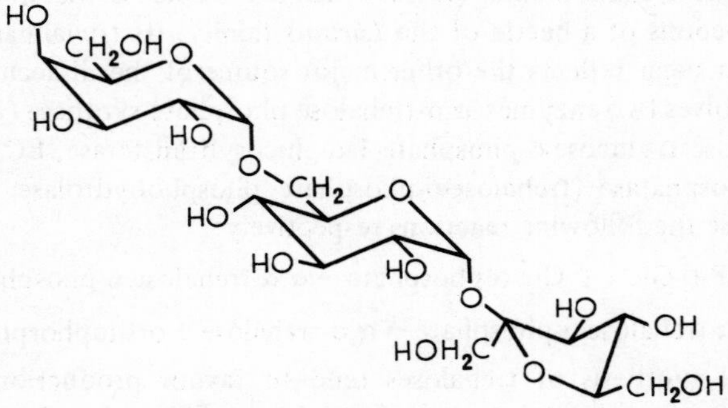

(66)

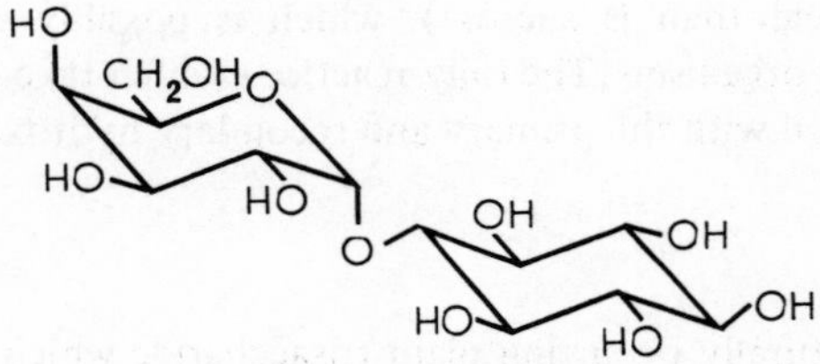

(67)

Melezitose

Melezitose, with the structure α-D-Glc*p*-(1→3)-β-D-Fru*f*-(2↔1)-α-D-Glc*p* (68), is another nonreducing trisaccharide related to sucrose, but in this case the additional monosaccharide unit is attached to the D-fructofuranosyl residue. It occurs in the exudate from insect-produced wounds of many plants including limes, poplars, Douglas fir, Virginia pine, larch (la mélèze – French meaning larch, hence the name), etc. Whether the melezitose is of plant or insect origin is not entirely clear, particularly since the discovery of a glycosyl transferase enzyme in the bodies of the insects, which is capable of synthesising melezitose from sucrose. The presence of melezitose in honeys from the honeydew of many trees is due to interaction of enzymes from the insects with sucrose from the plant-sap and not to the bees producing the trisaccharide. These melezitose-rich honeys, in which crystallisation can occur in the honey comb, provide the best source of the compound, since the crystallised melezitose is easily separated by dilution of the honey with alcohol followed by centrifugation.

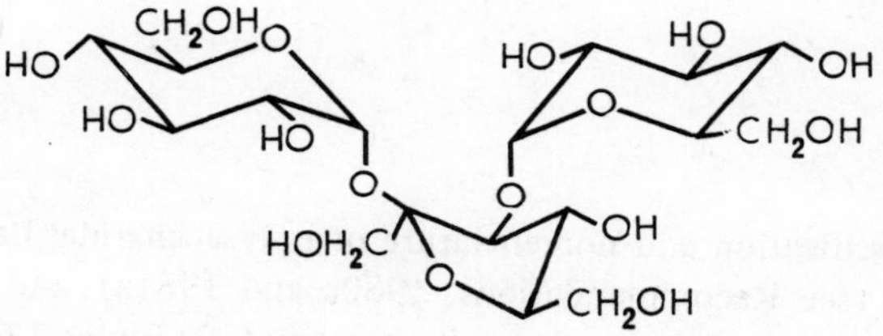

(68)

There are a number of other naturally occurring oligosaccharides containing amino-, deoxy- and other rarer monosaccharide residues which are known as oligosaccharide antibiotics. These are discussed in Chapter 12.

CHAPTER 8

Polysaccharides

The basic classification and nomenclature of polysaccharides has been discussed in Chapter 2 (see Recommendations, 1980c and 1981a), and the sources and structures of some common polysaccharides quoted (Tables 2.6 and 2.7). In the current chapter the structures of specific polysaccharides are described in detail, the polysaccharides being grouped, according to their origin, into plant, algal, microbial, fungal and animal polysaccharides. Where possible, the primary structure or simple repeating units are shown with reference to secondary and higher level structures where these are known.

PLANT POLYSACCHARIDES

Homopolysaccharides

Starch

The principle food-reserve polysaccharides in the plant kingdom are starches. They form the major source of carbohydrates in the human diet and are therefore of great economic importance, being isolated on an industrial scale from many sources (for a review of the biology of starch see Preiss and Walsh, 1981).

Starch has been found in some protozoa, bacteria and algae, but by far the major sources are plants, the carbohydrate occurring in the seeds, fruits, leaves, tubers, and bulbs in various amounts from a few percent to over 75% as in the case of cereal grains. Starch occurs in granular form, the shape of the granules being characteristic of the source of the starch. Isolation of these granules from the plant tissues can be achieved without degradation because they are insoluble in cold water, whereas many of the contaminants are soluble. The granules swell reversibly in cold water, a phenomenon used in the industrial extraction of starch to loosen the granules in the matrix, but as the temperature is raised the process becomes irreversible and eventually the granule bursts to form a starch paste. Not all starch granules in a sample burst at the same temperature, but the range of temperature of gelatinisation is characteristic of a particular starch.

The starch granule can be separated into two distinctly different components, a phenomenon first discovered in the early 1940s, although the heterogeneity of starch had been hinted at earlier. The two components, amylose and amylopectin, vary in relative amount among the different sources from less than 2% of amylose in waxy maize to about 80% of amylose in amylomaize (both corn starches), but the majority of starches contain between 15% and 35% amylose. It has been proposed that the ratio of amylopectin to amylose is a function of the ratio of starch synthetases present in the plant, one synthetase being responsible for the production of amylose whilst the other, a complex, branch-forming synthetase, is responsible for amylopectin production. The determination of this characteristic amylose:amylopectin ratio for a given starch sample is based on the binding capacity of the starch for iodine.

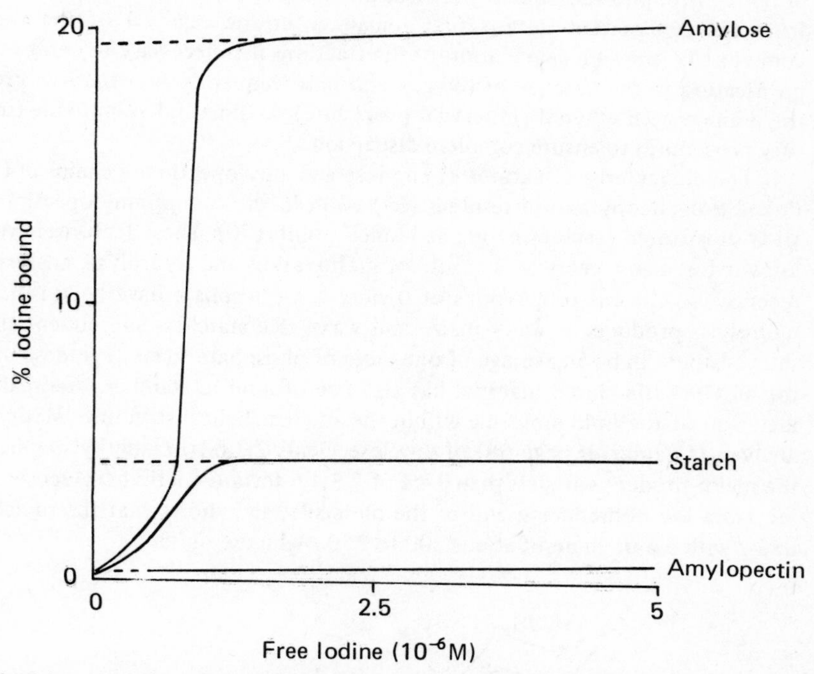

Fig. 8.1 – Absorption of iodine by a starch and its components. The amylopectin content of this starch can be deduced from this graph.

It can be seen from Fig. 8.1 that amylose binds iodine giving a blue colouration, up to a limiting value which is determined partly by the experimental conditions, but in the case illustrated is about 20% of iodine by weight. Amylopectin, on the other hand, binds little iodine and, in so doing, gives a red colouration whilst starch, as a mixture, is of intermediate binding ability. Extrapolation of the linear portions of the binding curve to zero free-iodine concentration, as shown, will give the iodine binding capacity of the starch, etc. The amylose content is obtained by expressing the starch binding capacity as a percentage of the amylose binding capacity. Potentiometric measurements have traditionally been used for such determinations, but it is now being claimed that the use of spectrophotometric techniques at two wavelengths is a much more accurate method.

The most widely used method of fractionating starch into its components is to add polar organic substances to an aqueous dispersion of the starch granules which cause the amylose to form an insoluble complex. The amylose can be purified by reprecipitation usually with a different organic substance. The most suitable precipitants are found to be thymol for the initial precipitation and butan-1-ol for the subsequent purification. The amylopectin fraction is removed from the supernatant liquors that remain after the removal of the amylose complex. To prevent degradation of the fractions it is necessary to carry out the purification in the absence of oxygen and it is frequently necessary to pretreat the granule with ethanol (industrial procedure) or dimethyl sulphoxide (laboratory procedure) to ensure complete dispersion.

The characteristic features of amylose and amylopectin are chains of (1→4)-linked α-D-glucopyranosyl residues (69) with, in the case of amylopectin, 1,4,6-tri-*O*-substituted residues acting as branch points (70). These features have been known for many years as a result of methylation and hydrolysis analysis but, recently, significant proportions of D-glucose 6-phosphate have been isolated as hydrolysis products of waxy maize and waxy rice starches. Subsequent analysis showed there to be an average of one mole of phosphate to six D-glucose units in the amylopectin. Lipid material has also been found in starches, due mainly to inclusion of the lipid molecule within the amylose helical structure. Methylation analysis (Chapter 4, page 68) of amylose yields 2,3,6-tri-*O*-methyl-D-glucose as the major product with less than 0.4% of 2,3,4,6-tetra-*O*-methyl-D-glucose, resulting from the nonreducing end of the molecule, and shows that the molecule is linear with a unit chain of about 200 to 350 D-glucose units.

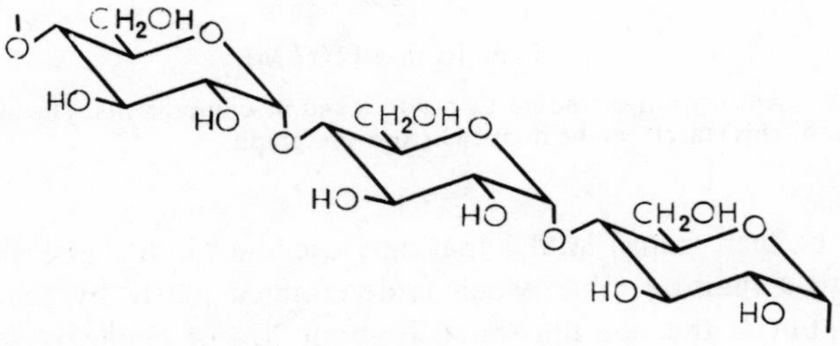

CH_2OH O HO HO O CH_2OH O HO HO O CH_2OH O HO HO

(69)

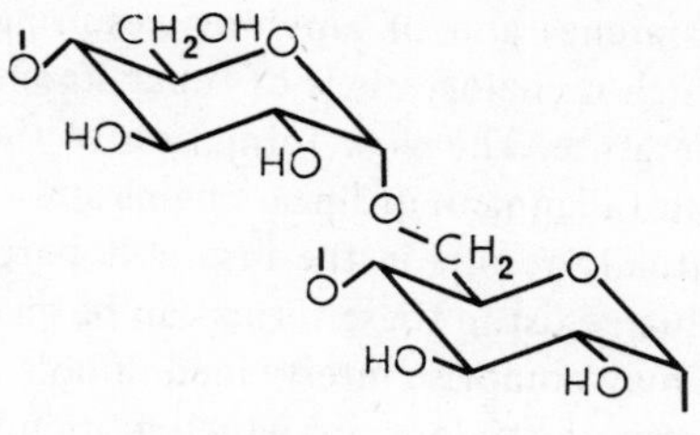

(70)

Osmotically determined molecular weights are in agreement with this chain length, but analysis of an unbranched structure is difficult. Not only is it difficult to measure the small amount of end-groups in relation to the chain-forming units, but degradation has a considerable effect; the rupture of one bond can halve the measured unit chain length. Physical methods of chain length determination, provided that independent methods for proof of homogeneity are used, have shown values for the size of the amylose molecules to be greater than values obtained by chemical methods. Light scattering and ultracentrifugation analyses show that chain lengths of up to 6000 units occur frequently. Enzyme analysis of amylose with β-amylase originally showed the molecule to be linear, producing maltose as its sole degradation product, but contamination of the enzyme with another α-amylolytic enzyme would remove any other α-D-glucosyl groups. A study of the action of pullulanase and other amylolytic enzymes on various amyloses has demonstrated that there are some 1,4,6-tri-*O*-substituted residues which act as branch points in the molecule. This, plus the hydrodynamic behaviour of amylose fractions, has led to the acceptance of there being a limited degree of branching in amylose.

Many of the properties of amylose can be explained in terms of its ability to adopt different molecular conformations in solution. In neutral aqueous solutions the normal conformation is that of a random coil. If there are complexing agents in the solution, amylose will form a helical structure consisting of about six D-glucosyl residues per helical turn. This is the conformation that both gives rise to the characteristic blue colouration of amylose–iodine complexes, and is responsible for the formation of complexes with fats and polar organic solvents, the complexing agent occupying a position at the centre of the helix.

The various forms of retrograded amyloses are due to variations in conformation of the amylose in solution. Retrogradation of amylose is the autodeposition of the polysaccharide in an insoluble form from solution, a phenomenon which rarely occurs with amylopectin. X-ray patterns of the retrograded amyloses has shown that the size and type of amylose and the concentration, temperature and pH of solution all contribute to the structure but two distinct X-ray patterns can be observed. The two crystalline forms of amylose are type A, which is characteristic of cereal starches and of amyloses resulting from retrogradation above 50° and type B which is characteristic of tuber starches and of retrograded amyloses at room temperature. The slow formation of these forms of amylose will allow the formation and alignment of linear chains and through the formation of hydrogen bonds eventually results in the insoluble particles. No clear picture of the conformations which exist in these forms can be given at present, but it is thought that the B-form may comprise intertwined double helices.

A third crystalline form of amylose, for which more information is available, is the so-called V-form which is obtained as a result of retrogradation in the presence of complexing agents. X-ray patterns of this form indicate a flexible helical arrangement with six or seven D-glucose units per turn, depending on the size of the complexing agent. The nature of the B-form has been related to the V-form through a mechanism involving the extension of the helices and changes in hydrogen bonding. Comparisons of X-ray diffraction patterns of the V-form have been made with those from the series of cyclomalto-oligosaccharides which are prepared by the action of an amylase from *Bacillus macerans* cyclomalto-dextrin glucanotransferase (EC 2.4.1.19) on starch (see Chapter 7). Cyclomalto-hexaose gives an X-ray pattern similar to that of the V-form and is considered to be analogous to one turn of the helix.

Amylopectin, on methylation analysis, yields 2,3,6-tri-*O*-methyl-D-glucose as its main product but the amounts of the 2,3,4,6-tetra-*O*-methyl ether of about 4% shows that the unit chain length is smaller than in amylose. Isolation of 2,3-di-*O*-methyl-D-glucose as an additional product indicates the presence of 1,4,6-linked branch points. Measurement of the unit chain length was traditionally carried out by estimating the formic acid liberated by periodate oxidation of the nonreducing end units, but is now more accurately determined by enzymic methods including the use of β-amylase (EC 3.2.1.2) and D-glucose oxidase (EC 1.1.3.4).

Values for the unit chain length are usually within the range 17–26 units. The arrangement of these unit chains to give a branched amylopectin could be in a variety of ways. Fig. 8.2 shows three possible arrangements, the laminated structure (a) proposed by Haworth, the herringbone or comb structure (b) proposed by Staudiner and Husemann, and the branched tree-like structure (c) proposed by Meyer and Bernfeld. There are three types of chain present in these structures; A-chains are side chains linked only through their reducing ends to the rest of the molecules, B-chains are those to which A-chains are attached, and the C-chain which carries the reducing group (there can only be one C-chain per molecule). Enzymic studies based on the degradation with β-amylase and debranching enzymes favour a multiple branched structure, the ratio of A:B chains not being consistent with the expected values for the laminated or herring-bone structures. Recent studies have led to a proposal of a modification to the Meyer tree-like structure, although much of the experimental data would satisfy the Meyer structure (Manners and Matheson, 1981). The cluster model proposed

(Fig. 8.3) satisfies the requirement that a proportion of the B chains must carry more than one A chain to account for differences in the A:B-chain ratio between the original material and partly debranched material. The various A and B chains may actually exist in double-helix conformations but are shown here as straight lines for clarity.

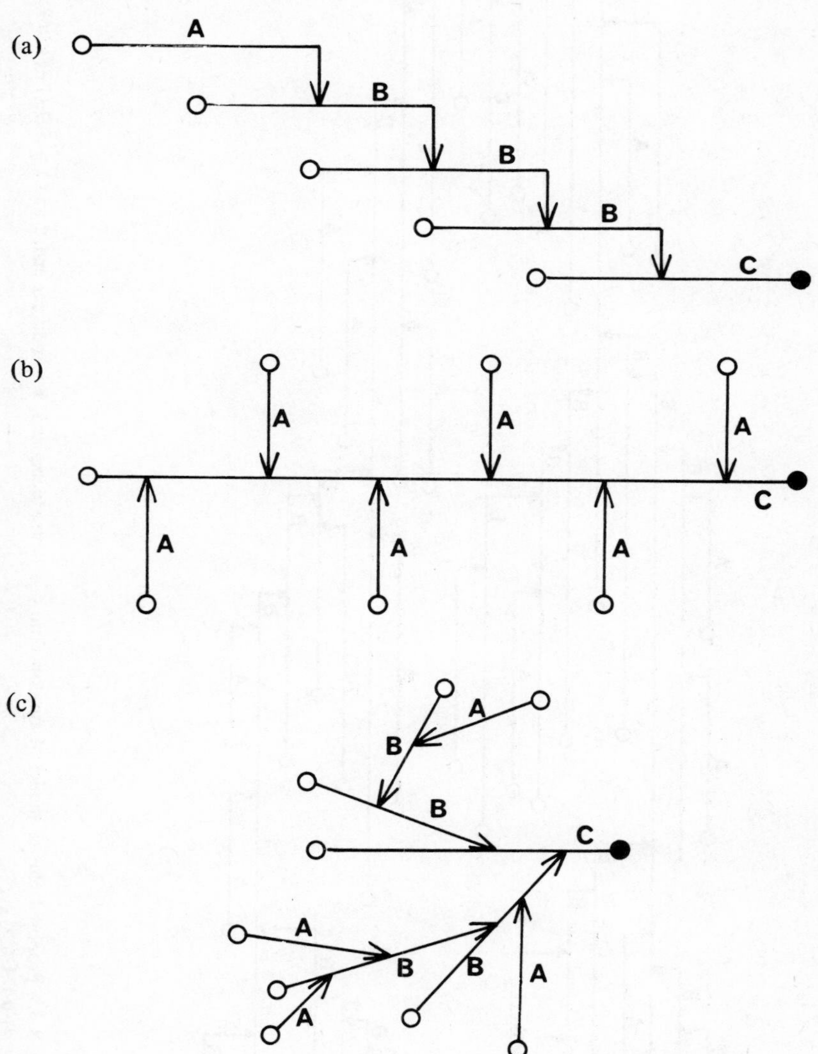

Fig. 8.2 – Standard models of amylopectin showing possible arrangements of the D-glucan chains of *ca.* 20 residues. ○ = nonreducing end, ● = reducing end, → = α-D-(1→6)-linkage, —— = $[(1\rightarrow4)\text{-}\alpha\text{-D-Glc}p\text{-}]_X$.

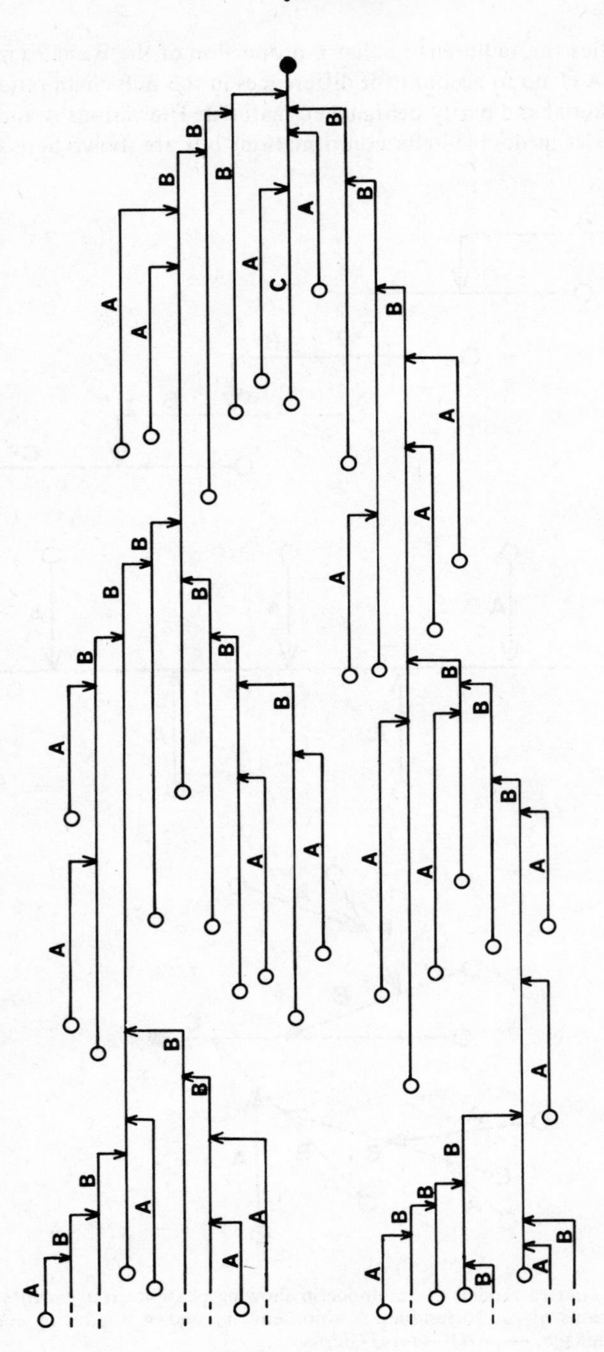

Fig. 8.3 – Proposed cluster model of amylopectin. ○ = nonreducing end, ● = reducing end, ↑ and ↓ = α-D-(1→6)-linkage, —— = $[(1\rightarrow4)\text{-}\alpha\text{-D-Glc}p]_x$.

The molecular size of amylopectin is almost too large to determine accurately; light scattering results indicate that there are of the order of 10^6 D-glucosyl residues per molecule making amylopectin one of the largest naturally occurring molecules. The size of the molecule has been shown to increase with increasing maturity of the parent starch in the plant.

The inability of amylopectin to bind much iodine, thereby producing the characteristic red colouration, is attributed to the large number of branch points in the molecule which disrupt any possible helix formation.

Cellulose

Cellulose is the most abundant organic substance found in nature. It forms the principal constituent of cell walls in higher plants, forming the main structural element (see review by Cabib and Shematek, 1981). It occurs in an almost pure form (98%) in cotton fibres and to a lesser extent in flax (80%), jute (60–70%) and wood (40–50%) and has also been found as a constituent of some algae and as a product of bacterial synthesis. Isolation of cellulose from most of its sources is difficult due to its insolubility in most solvents, and involves rather the solubilisation of the contaminating compounds such as hemicelluloses and lignin. The usual methods are based on alkaline pulping to remove the non-cellulosic polysaccharides, but it is difficult to remove all the contaminating monosaccharides and this has led to reports of the presence of monosaccharides other than in D-glucose in trace quantities.

The primary structure of cellulose was determined in the 1930s and little modification has subsequently had to be made to that original proposal. Methylation analysis yields over 90% of 2,3,6-tri-*O*-methyl-D-glucose showing that the molecule is essentially linear and, since partial hydrolysis yields cellobiose (71), it is a (1→4)-D-glucan. The β configuration of the inter D-glucosidic linkages was verified by enzymic studies. Determination of chain length by end-group analysis is inaccurate for the essentially linear molecule and has led to very low values of the order of 200 units being erroneously reported due to the occurrence of degradation during the analysis. Physical methods of chain length determination have shown that chains of up to 10,000 D-glucose units exist. Kinetic studies of the hydrolysis of cellulose have shown that over 99% of the linkages are of the same type and those which appear different do so as a result of physical effects. There has been no evidence for the presence of any other type of linkage.

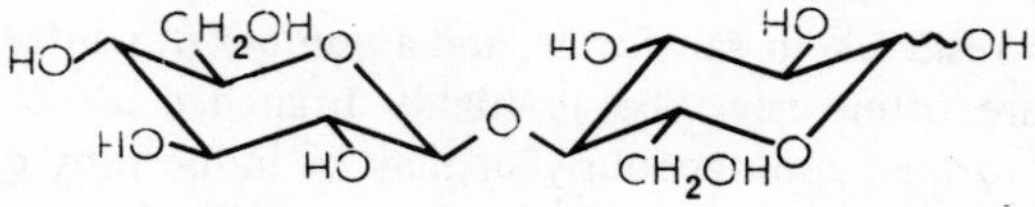

(71)

The characteristic properties of cellulose are due to the tendency of the individual chains to form microfibrils through inter- and intra-molecular hydrogen bonding to give a highly ordered structure. The microfibrils associate in a similar manner to give fibres but in this case the axis of the fibres is at an angle to the axis of the microfibril whereas the individual molecules lie parallel to the microfibril axis. This regular arrangement of molecules is sufficient to allow X-ray diffraction patterns to be obtained and such indicate that cellulose from almost every natural source has an essentially similar pattern but which is different from that regenerated from derivatives or from solution. Cellulose I is the crystalline arrangement found in nature and cellulose II is that of regenerated cellulose. Variants in this classification have been noted and four other forms of cellulose have been distinguished from each other by their X-ray patterns. Cellulose III_I and cellulose IV_I have similarities in the X-ray pattern with cellulose I whilst cellulose III_{II} and cellulose IV_{II} have similarities with cellulose II. Whilst all the polymers have different molecular conformations in the unit cell, the cellulose I family (I, III_I, and IV_I) have 'bent' chain conformations and the cellulose II family (II, III_{II} and IV_{II}) have 'bent-twisted' chain conformations. The term 'bent' chain refers to the cellobiose units which correspond to the repeat distance of the fibres. The original structures proposed were based on 'straight' chain conformations which had all the D-glucopyranosyl residues in the same plane. The adoption of 'bent' chain structures with alternate D-glucopyranosyl residues in the same plane is a more satisfactory conformation because it eliminates the disadvantages of overlap of the van der Waals radii of the O-2 and C-6 atoms and of the repulsions between the H-1 and H-4 atoms and retains the intermolecular hydrogen bonding which is consistent with polarised infrared studies. Cellulose II appears to be thermodynamically more stable than cellulose I and transformation of cellulose I into cellulose II can take place in sodium hydroxide, or preferably in a mixture of sodium hydroxide and sodium sulphide.

Heteropolysaccharides

Heteropolysaccharides derived from the bark, seeds, roots and leaves of plants fall into several structural types which can be divided into two distinct groups. These are the acidic polysaccharides described as gums, mucilages and pectins, and the neutral polysaccharides known as hemicelluloses.

Gums

The plant gums or exudate gums are essentially polysaccharides containing hexuronic acid residues, in salt forms, and a number of neutral monosaccharides units, which are often esterified in highly branched structures. These gums, which may be formed spontaneously or may be induced by deliberately cutting the bark or fruit, are exuded as viscous liquids which become hard nodules on dehydration to seal the site of the injury and so provide protection from microorganisms. Many of these gums find industrial applications as thickening agents or emulsion stabilisers (see Chapter 14 and Davidson, 1980). Determination of the structure of these gums utilises the differing rates of hydrolysis of the various glycosidic linkages to produce, by selected degradation, oligosaccharides whose structures can be determined with greater ease and certainty. The most common conditions for hydrolysis include autohydrolysis (heating the acidic gum in aqueous solution) and mild hydrolysis with 0.01 M acid to cleave L-arabinofuranosyl glycosidic linkages, acid hydrolysis with 0.1–0.5 M acid or extended autohydrolysis to rupture the more labile hexopyranosyl glycosidic linkages, and strong acid hydrolysis with 0.5 M sulphuric acid to hydrolyse all but hexopyranosyluronic acid glycosidic linkages resulting in the isolation of acidic oligosaccharides, particularly of disaccharides containing one acidic residue.

Gum arabic, from various species of *Acacia*, is probably the best known example of plant gums and is typical of a number of gums which contain interior chains of (1→3)-linked β-D-galactopyranosyl residues to which chains comprised of L-arabinofuranosyl, L-rhamnopyranosyl and D-glucopyranosyluronic acid residues are attached. Autohydrolysis of arabic acid, the salt-free polysaccharide, yields L-arabinose, L-rhamnose and a D-galactosyl-L-arabinose disaccharide showing that these entities are linked to the main chains as shown in the generalised structure (72) which summarises the structural features, but does not place, uniquely, the glycosyl side chains which can be monosaccharide groups or disaccharide units attached to the outer chain. A number of other gums from *Acacia* species have been shown to have the same general D-galactan core with variations in the degree of branching and in the nature and attachment of the peripheral L-arabinosyl and L-rhamnosyl residues. Some gums, such as the gum from *A. karroo* also contain D-glucopyranosyluronic acid residues (1→4)-linked to α-D-galactopyranosyl residues. More recently an *Acacia* gum exuded by species of a subseries *Juli florae* has been shown to contain more acidic groups, to have a higher molecular weight, viscosity and methoxyl content, but a lower proportion of L-rhamnopyranosyl and L-arabinofuranosyl residues than the majority of other *Acacia* gums.

Gum ghatti from *Anogeissus latifolia* has interior chains of alternate D-glucopyranosyluronic acid and D-mannopyranosyl residues with a high proportion of L-arabinofuranosyl groups in the terminal nonreducing positions. A small proportion of L-arabinopyranosyl residues are present in the more acid stable part of the polysaccharide. The structure (73) has been formulated to show the major structural features.

Gum tragacanth has interior chains of 4-*O*-substituted α-D-galactopyranosyluronic acid residues and provides an example of a gum which is structurally related to the pectic acids. The gums from *Sterculia* and *Khaya* species also contain a D-galacturonic acid interior chain but with various amounts 2-*O*-substituted L-rhamnopyranosyl residues interspersed and a variety of constituent monosaccharides including D-glucopyranosyluronic acid groups which occur only as nonreducing end groups, unlike the D-galactopyranosyluronic acid

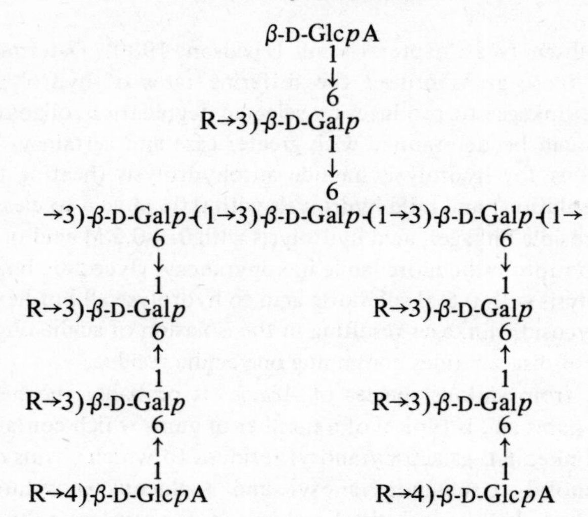

R = L-Ara*f*-(1→, L-Rha*p*-(1→, α-D-Gal*p*-(1→3)-L-Ara*f*-(1→, or,
less frequently, β-L-Ara*p*-(1→3)-L-Ara*f*-(1→

(72)

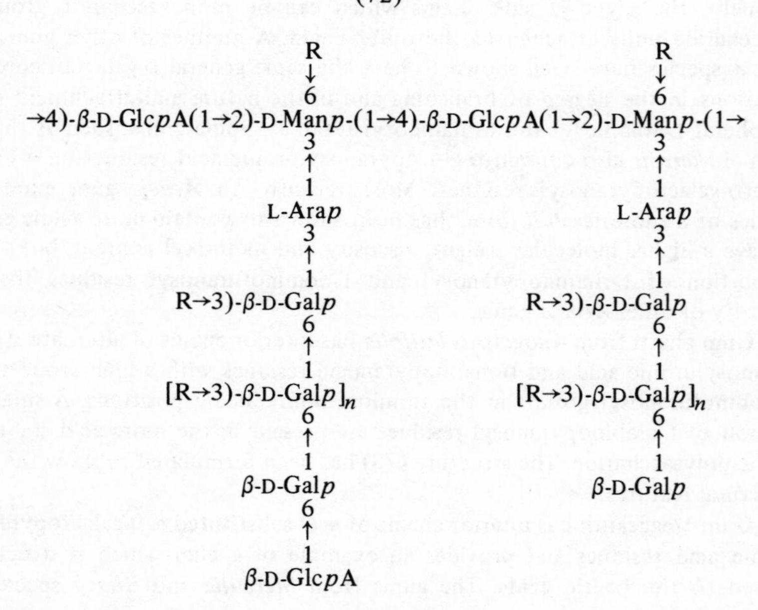

R = L-Ara*f*-(1→, or less frequently L-Ara*f*-(1→2)-L-Ara*f*-(1→,
L-Ara*f*-(1→3)-L-Ara*f*-(1→, or L-Ara*f*-(1→5)-L-Ara*f*-(1→

(73)

residues which are present mainly in the inner chains. Gum tragacanth is only partially soluble in water but its major component, tragacanthic acid, has been purified and found to contain D-galacturonic acid, D-xylose, L-fucose and D-galactose with small amounts of D-glucuronic acid, L-rhamnose and L-arabinose; the important structural features are summarised in (74). A minor component of the gum has been found to be a highly branched L-arabino-D-galactan.

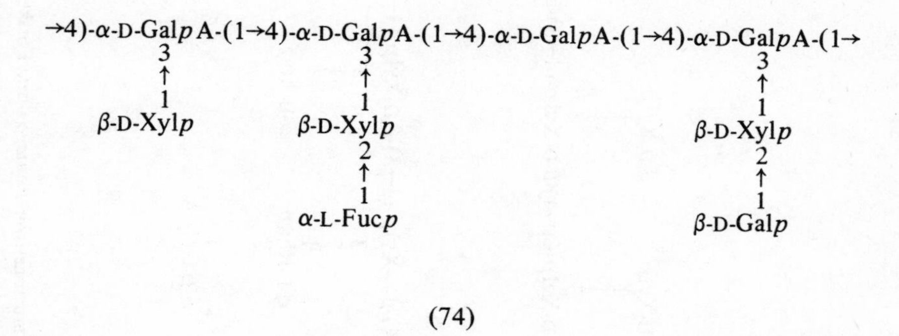

(74)

Mucilages

Mucilages have, in general, been less well characterised than many of the plant polysaccharides, but they are of a similar degree of structural complexity to the exudate gums. The presence of a particular acidic monosaccharide residue alone does not provide a basis for classification since many gums and mucilages contain both D-galactopyranosyluronic and D-glucopyranosyluronic acid residues but D-galacturonic acid is a characteristic component of many mucilages. Typical examples of mucilages include the acidic polysaccharides from slippery elm bark (75), cress seeds (76) and the husk from the seeds of *Plantago ovata* Forsk (Fig. 8.4). The role of mucilages is probably one of reservoirs for water retention to protect the seeds, etc., from desiccation. The mucilage from the seed husk of *Plantago ovata* Forsk is widely used as a prophylactic in treatment of large-bowel disorders, including diverticular disease. It is the ability of the mucilage to form a gel which retains many times its own weight of water over a wide range of concentrations that is responsible for the laxative action. Since D-xylan polymers are insoluble in water, it is proposed (Sandhu *et al.*, 1981) that the L-arabinofuranosyl and D-galactopyranosyluronic acid residues are responsible for binding water at the surface or within gels, whilst the parallel arrangement of the D-xylan chains is responsible for the gel formation.

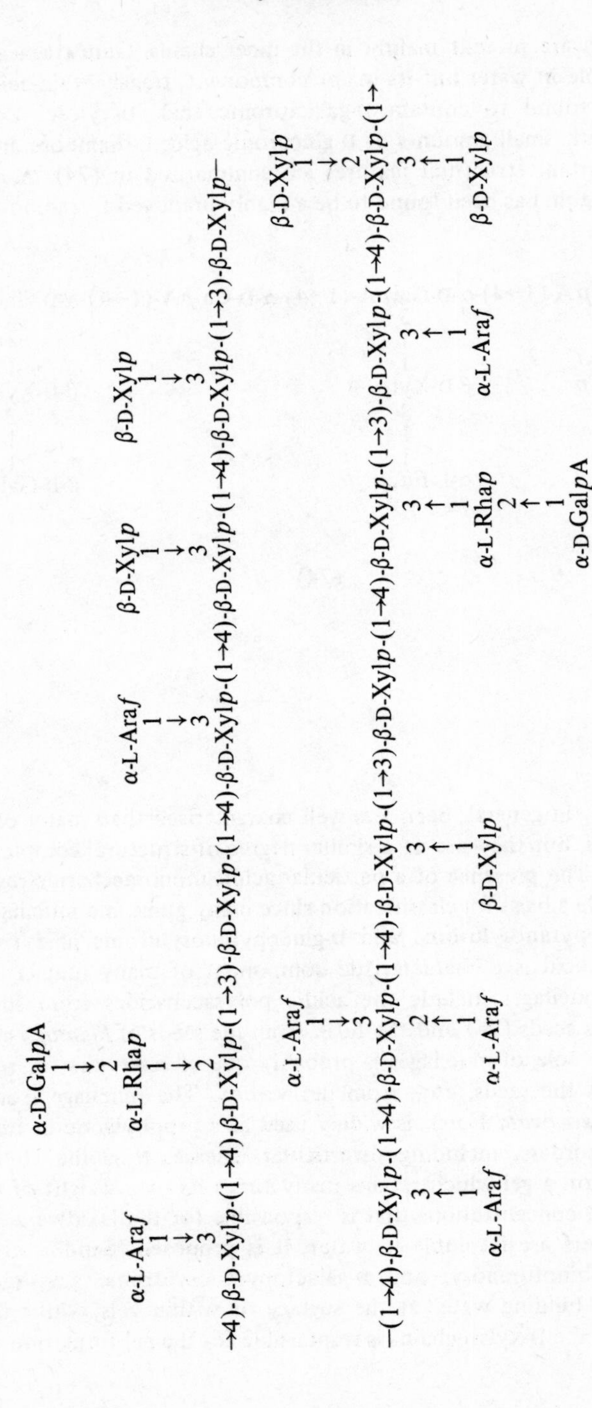

Fig. 8.4 – Representation of the main structural features of the mucilage from *Plantago ovata* Forsk.

```
→4)-α-D-GalpA-(1→2)-L-Rhap-(1→4)-α-D-GalpA-(1→2)-L-Rhap-(1→
                        3
                        ↑
                        1
                      D-Galp
                        4
                        ↑
                        1
                 3-O-Me-D-Galp
```

(75)

```
                        R
                        ↓
                        3
→4)-D-GalpA-(1→2)-L-Rhap-(1→4)-D-GalpA-(1→2)-L-Rhap-(1→
                        4                               4
                        ↑                               ↑
                        R                               R
```

R = L-Rhap-(1→, D-Galp-(1→, 4-O-Me-D-GlcpA-(1→4)-D-Galp-(1→ or D-Xylp-(1→4)-D-Galp-(1→

(76)

Pectins

The pectins are a group of substances found in the primary cell walls and intercellular layers in land plants in which the principal constituent is D-galactopyranose. They make up 15% of the dry weight of apples and 30% of the dry weight of the rinds of citrus fruits, but occur only in small proportions in woody tissues. A group of related polysaccharides in which D-galactopyranosyluronic acid residues and not D-galactopyranosyl residues are present are termed pectic acids whilst if the uronic acid residues are present as methyl esters they are termed pectinic acids. Pectinic acids are very easily extracted with water and possess considerable gelling powers which are used commercially for gelation of fruit juices to give jellies. In contrast, the pectic acids which frequently contain the calcium salts of the hexosyluronic acid residues, are less soluble and require the use of reagents such as bis[di(carboxymethyl)amino] ethane (H_4 edta, ethylenediaminetetraacetic acid) or sodium hexametaphosphate for their extraction. Among the pectic substances are those types of homopolysaccharides, D-galactans, L-arabinans, and D-galacturonans but the most common pectic substances are heteropolysaccharides containing both acidic and neutral sugars.

Structural studies carried out on these homopolysaccharidic pectins established the salient features of the more complex pectins. The increased recognition of the susceptibility of pectins to acidic and basic hydrolysis indicates that these homopolysaccharides may have arisen from degradation of more-complex natural polysaccharides. D-Galactans, containing (1→4)-linked β-D-galactopyranosyl residues in linear chains, have been isolated from white lupin seeds and from red spruce compression wood; a more complex, acidic D-galactan, which has the same major structural features, has been isolated from Norwegian spruce. Highly branched L-arabinofurans, containing no other monosaccharide residues, have been isolated from mustard seeds (77), and a D-galacturonan has been isolated from sunflower heads, although this type of acidic homopolysaccharide is of infrequent occurrence, D-galactopyranosyluronic acid being more frequently found in heteropolysaccharidic pectins. The structure of this D-galacturonan has linear chains of (1→4)-linked α-D-galactopyranosyluronic acid residues.

```
→5)-α-L-Araf-(1→5)-α-L-Araf-(1→5)-α-L-Araf-(1→5)-α-L-Araf-(1→5)-α-L-Araf-(1→
                    3                        3                        3
                    ↑                        ↑                        ↑
                    1                        1                        1
                α-L-Araf                 α-L-Araf                 α-L-Araf
```

(77)

A pectin from soybean (78) is characteristic of the L-arabino-D-galactans which are the only known neutral heteropolysaccharidic pectins which have structures containing chains of (1→4)-linked β-D-galactosyl residues to which various amounts of L-arabinosyl residues are linked (1→3).

```
→4)-β-D-Galp-(1→4)-β-D-Galp-(1→4)-β-D-Galp-(1→4)-β-D-Galp-(1→
                        3
                        ↑
                        1
                      L-Araf
                        5
                        ↑
                        1
                      L-Araf
```

(78)

The majority of the pectic and pectinic acids contain various proportions of neutral monosaccharides (usually between 10 and 25%) of which only L-rhamnose is found interrupting the D-galacturonic acid chains, the others being attached as side chains as shown in the generalised structure for pectic acids

(Fig. 8.5). A pectic acid isolated from soybean was the first example shown to contain D-xylopyranosyl residues which are present as mono- or di-saccharide block units (79) similar to those in tragacanthic acid.

$$\begin{array}{l}
\rightarrow 4)\text{-}\alpha\text{-D-Gal}p\text{A-}(1\rightarrow 2)\text{-}\beta\text{-L-Rha}p\text{-}(1\rightarrow 4)\text{-}[\alpha\text{-D-Gal}p\text{A}]_m\text{-}(1\rightarrow 4)\text{-}\alpha\text{-D-Gal}p\text{A-}(1\rightarrow \\
\quad\quad\quad 3 \uparrow R^1 \quad\quad\quad\quad\quad\quad\quad\quad\quad\quad\quad\quad\quad\quad\quad\quad\quad\quad\quad\quad 3 \uparrow R^2
\end{array}$$

R^1 = $[\alpha\text{-D-Gal}p]_n\text{-}(1\rightarrow$ or $[\alpha\text{-L-Ara}f]_n\text{-}(1\rightarrow$

R^2 = $\beta\text{-D-Xyl}p\text{-}(1\rightarrow$, $\alpha\text{-L-Fuc}p\text{-}(1\rightarrow 2)\text{-}\beta\text{-D-Xyl}p\text{-}(1\rightarrow$, or $\beta\text{-D-Gal}p\text{-}(1\rightarrow 2)\text{-}\beta\text{-D-Xyl}p\text{-}(1\rightarrow$

Fig. 8.5 – Generalised structure for pectic acids.

$$\rightarrow 4)\text{-}[\text{D-Gal}p\text{A-}(1\rightarrow 2)\text{-L-Rha}p\text{-}(1\rightarrow 4)]_n\text{-}\alpha\text{-D-Gal}p\text{A-}(1\rightarrow 4)\text{-}\alpha\text{-D-Gal}p\text{A-}(1\rightarrow 4)\text{-}\alpha\text{-D-Gal}p\text{A-}(1\rightarrow$$

$$\text{(first } \alpha\text{-D-Gal}p\text{A: O-3} \leftarrow 1\text{-}\beta\text{-D-Xyl}p\text{; last } \alpha\text{-D-Gal}p\text{A: O-3} \leftarrow 1\text{-}\beta\text{-D-Gal}p\text{, whose O-4} \leftarrow 1\text{ of:)}$$

$$\text{L-Ara}f\text{-}(1\rightarrow 2 \text{ or } 3)\text{-}\beta\text{-D-Gal}p\text{-}(1\rightarrow 4)\text{-}\beta\text{-D-Gal}p\text{-}(1\rightarrow 4)\text{-}\beta\text{-D-Gal}p$$

(79)

Hemicelluloses

The hemicelluloses are found in close association with cellulose in plant cell walls and were originally thought to be precursors of cellulose. They have now been shown to have no part in cellulose biosynthesis but represent a chemically distinct, separate group of polysaccharides which contain residues and linkages different from those found in cellulose. It is usual to limit the term hemicellulose to cell wall polysaccharides from land plants with the exclusion of cellulose and pectins, and to classify them according to the type of monosaccharide residues present. The majority of the hemicelluloses are relatively small molecules consisting of between 50 and 200 monosaccharide residues, whilst those from hardwoods are larger molecules (150–200 residues). Some of these compounds have crystalline structures and it has been found that the backbone of the molecule of the D-xylans consists of D-xylopyranosyl residues with a rotation of 120° between successive residues.

D-Xylans. D-Xylans are the most common of the hemicelluloses, occurring in all parts of all land plants in proportions of between 7 and 30%. The backbone of the molecule is an essentially linear chain of (1→4)-linked β-D-xylopyranosyl residues. Homoglycans are neither common nor abundant, but a typical example is the D-xylan from Esparto grass which has been found to contain (1→4)-linked β-D-xylopyranosyl residues. As indicated by the isolation of small amounts of 2,3,4-tri-*O*-methyl-D-xylose from methylation analysis, these chains contain a single branch point located at O-3 of a D-xylopyranosyl residue. The most common hemicellulose in soft woods is an (*O*-acetyl-L-arabino)-(4-*O*-methyl-D-glucurono)-D-xylan but the most common side chain is a (1→2)-linked, or less frequently, a (1→3)-linked 4-*O*-methyl α-D-glucopyranosyluronic acid, although annual plants have been shown to contain unmethylated residues. The number of D-glucuronic acid side chains varies considerably, with the hard wood D-xylans having, on average, one acidic residue for every ten D-xylopyranosyl residues. Another common side chain in D-xylans is (1→3)-linked α-L-arabinose but these residues do not always occur as nonreducing end groups. A typical example of non-terminal L-arabinopyranosyl residues occurs in barley husk D-xylan (80).

→4)-β-D-Xyl*p*-(1→
3
↑
1
→4)-β-D-Xyl*p*-(1→2)-α-L-Ara*f*

(80)

L-Arabino-D-xylans. The L-arabino-D-xylan group of neutral polysaccharides which occur in association with acidic polysaccharides come from cereal gums. These are highly branched polysaccharides in which the β-D-xylan chains are (1→4)-linked (81) and to which are attached, in an irregular manner, single α-L-arabinofuranosyl groups *via* (1→3)-links. The polysaccharide isolated from cress seeds has been found to contain an L-arabino-D-xylan comprising chains of (1→5)-linked α-L-arabinofuranosyl residues to which are attached L-arabinofuranosyl residues and D-xylosyl groups (82).

→4)-β-D-Xyl*p*-(1→4)-β-D-Xyl*p*-(1→4)-β-D-Xyl*p*-(1→4)-β-D-Xyl*p*-(1→
3 3
↑ ↑
1 1
α-L-Ara*f* α-L-Ara*f*

(81)

```
→5)-α-L-Araf-(1→5)-α-L-Araf-(1→5)-α-L-Araf-(1→
                 3                   3
                 ↑                   ↑
                 1                   1
            α-L-Araf            α-L-Araf
                 3                   3
                 ↑                   ↑
                 1                   1
            α-D-Xylp            α-D-Xylp
```

(82)

D-Mannans, D-galacto-D-mannans and D-gluco-D-mannans. D-Mannose is a common constituent of plant polysaccharides, occurring in homopolysaccharides, or in heteropolysaccharides in conjunction with D-galactose or D-glucose. D-Mannans occur in ivory nuts, green coffee beans and a number of other plant sources and have a common carbohydrate structure consisting of linear chains of (1→4)-linked β-D-mannopyranosyl residues, which differ in chain length depending on the source; such polysaccharides are insoluble in water. The presence of traces of D-galactose in the ivory nut D-mannans may indicate that they are in fact D-galacto-D-mannans at the extreme end of the whole spectrum of D-galacto-D-mannans which occur in the seeds of leguminous plants. The structure of the polysaccharides vary, according to the source of the material, in the relative amounts of the component monosaccharide ranging from a D-galactose:D-mannose ratio of 1:1 to 1:5, but all have the same structural feature, namely chains of (1→4)-linked β-D-mannopyranosyl residues to which side chains consisting of single α-D-galactopyranosyl groups attached by (1→6)-linkages (83) at various intervals, for example, guaran contains a D-galactopyranosyl residue on every other D-mannopyranosyl residue. These D-galacto-D-mannans are readily soluble in water and find commercial uses as sizes and beater additives in paper making and as gelling agents in the food industry due to the highly viscous solutions they form. The position of the side chains is such that regions with and without side chains assist to give the appearance of 'hairy' and 'smooth' regions. It is the smooth regions which interact with, for example, xanthan (see also pages 48 and 171).

```
→4)-β-D-Manp-(1→4)-β-D-Manp-(1→4)-β-D-Manp-(1→4)-β-D-Manp-(1→
        6                               6
        ↑                               ↑
        1                               1
    α-D-Galp                        α-D-Galp
```

(83)

D-Gluco-D-mannans have been isolated from hard woods in which the ratio of D-glucose to D-mannose is 1:2. Partial hydrolysis has shown the existence of disaccharides of D-glucose and D-mannose together with cellobiose, mannobiose, and mannotriose which suggests that these hemicelluloses have a random distribution of 1,4-substituted β-D-glucopyranosyl and 1,4-substituted β-D-mannopyranosyl residues in linear chains. Similar compounds have been isolated from seeds of certain iris, orchid and lily bulbs but they contain different proportions of D-glucose to D-mannose in the range 1:1 to 1:4. D-Galacto-D-gluco-D-mannans occur together with D-gluco-D-mannans in coniferous woods. Structural studies have shown that these have the same backbone as the D-gluco-D-mannans with single (1→6)-linked α-D-galactopyranosyl groups as side chains. There is also evidence that D-glucose and D-mannose occur as nonreducing end groups. These side chains are responsible for the greater solubility of these compounds compared to that of the D-gluco-D-mannans. This solubility is probably brought about by the side chains preventing the macromolecules from forming strong, intermolecular hydrogen bonds and act in the same way as the residues and groups, such as 4-*O*-methyl-D-glucuronic acid, L-arabinose, and *O*-acetylated residues, of the other hemicelluloses.

L-Arabino-D-galactans. Another group of hemicelluloses is the water-soluble L-arabino-D-galactans which are highly branched molecules with branched backbone chains of (1→3/6)-linked β-D-galactopyranosyl residues to which are attached side chains which may contain L-arabinofuranosyl, L-arabinopyranosyl, L-rhamnopyranosyl, etc., residues. Typical examples are the L-arabino-D-galactans from maritime pine (84) and from sycamore (85).

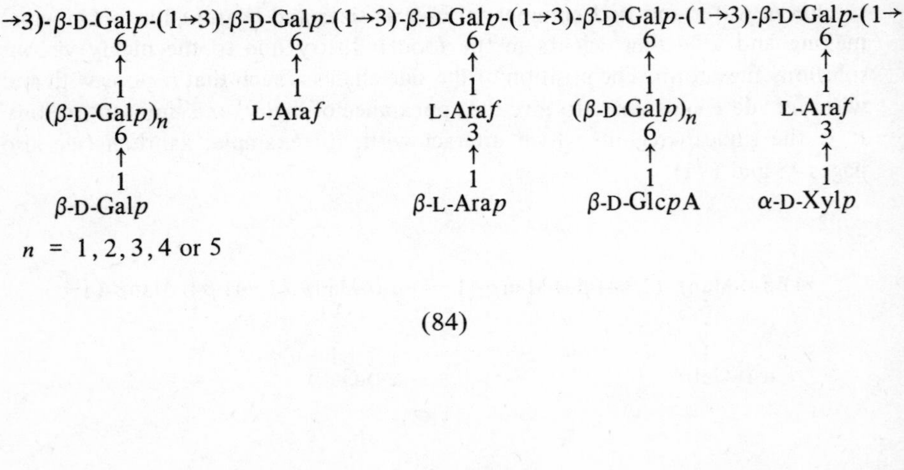

```
→3)-β-D-Galp-(1→3)-β-D-Galp-(1→3)-β-D-Galp-(1→3)-β-D-Galp-(1→3)-β-D-Galp-(1→
      6              6              6              6              6
      ↑              ↑              ↑              ↑              ↑
      1              1              1              1              1
  (β-D-Galp)n      L-Araf         L-Araf      (β-D-Galp)n       L-Araf
      6                             3              6              3
      ↑                             ↑              ↑              ↑
      1                             1              1              1
   β-D-Galp                      β-L-Arap      β-D-GlcpA       α-D-Xylp

 n = 1,2,3,4 or 5
```

(84)

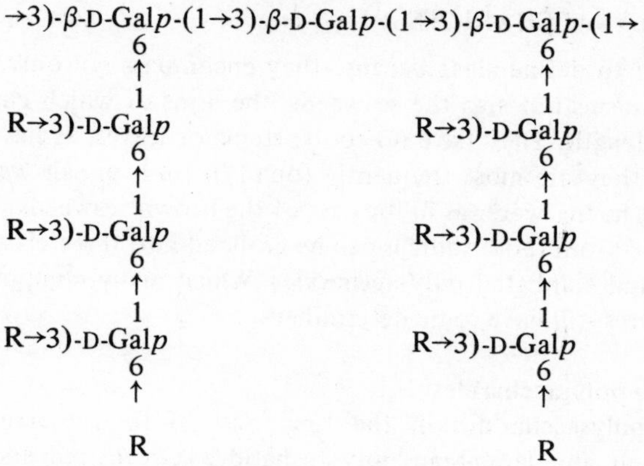

```
→3)-β-D-Galp-(1→3)-β-D-Galp-(1→3)-β-D-Galp-(1→
         6                               6
         ↑                               ↑
         1                               1
R→3)-D-Galp                     R→3)-D-Galp
         6                               6
         ↑                               ↑
         1                               1
R→3)-D-Galp                     R→3)-D-Galp
         6                               6
         ↑                               ↑
         1                               1
R→3)-D-Galp                     R→3)-D-Galp
         6                               6
         ↑                               ↑

         R                               R
```

R = L-Ara*f*-(1→, β-L-Ara*p*-(1→5)-L-Ara*f*-(1→, or L-Rha*p*-(1→

(85)

Miscellaneous plant polysaccharides

Lichenan is a D-glucan from Iceland moss and contains a random mixture of (1→4)- and (1→3)-linked β-D-glucopyranosyl residues in linear chains, with a higher proportion (about 70%) of (1→4) linkages. Iceland moss also contains a polysaccharide which is the α-D-analog of lichenan. This so-called isolichenan contains (1→4)- and (1→3)-linked α-D-glucopyranosyl residues with a higher proportion of the latter linkage. It also has a lower chain length of about 40–45 residues compared to 60–360 for lichenan. These polysaccharides have been important in the understanding of the mode of action of glycanases.

There are a number of glycans which occur in plants. Nigeran is an α-D-glucan with approximately equal proportions of alternating (1→3)- and (1→4)-linkages. The β-D-glucans found in plants include pustulan which contains (1→6)-linkages. Fructans also commonly occur in plants, acting as reserve carbohydrates either alone, or in conjunction with starch. They contain β-D-fructofuranosyl residues which are linked (2→1) in inulins and (2→6) in levans. Most D-fructans contain D-glucopyranosyl residues as nonreducing end groups and have relatively low molecular weights – of the order of 8,000. Grass levans have chain lengths of 20–30 units and are essentially linear.

Amyloids are a group of water soluble polysaccharides found in the seeds of a number of plants including tamarind and nasturtium and are so-called because of their staining reactions with iodine. They contain (1→4)-linked β-D-glucopyranosyl residues to which are linked single α-D-xylopyranosyl groups and 2-*O*-β-D-galactopyranosyl-α-D-xylopyranosyl units as side chains *via* (1→6)-linkages.

The unusual monosaccharide, L-glucose, has been reported to be among the hydrolysis products of the leaves of jute and *Grindelia* species.

ALGAL POLYSACCHARIDES

It is difficult to define algae because they encompass not only microscopic unicellular organisms but also the seaweeds, the arms of which can extend to over 150 feet in length. They have no roots, stems or leaves, as have the higher land plants, and they are most frequently found in fresh or salt water, occasionally free floating in the water as in the case of the brown seaweeds. The polysaccharides obtained from those sources can be grouped into three classes: food-reserve, structural, and sulphated polysaccharides. Whilst many compounds are known, their structures still have to be determined.

Food-reserve polysaccharides

The main polysaccharides in the first class of food-reserve polysaccharides include starch and laminaran polysaccharides. Green, red and blue-green seaweeds, and fresh water algae contain starch polysaccharides which can be fractionated into amylose and amylopectin. The absence of amylose from some extracts can be explained in terms of its destruction during isolation with acidic and alkaline solutions. The major differences between plant and algal starches are the lower intrinsic viscosities and lower iodine binding powers of the algal starches, which indicates smaller molecules. The presence of smaller molecules is also demonstrated by X-ray diffraction studies which show that the starch granules are less organised but still exhibit the characteristics of the plant starches. Algal starches are more susceptible to amylolytic enzymes. The average chain lengths vary between 10 and 19 units and small proportions of (1→3)-linked α-D-glucopyranosyl residues have been found.

Laminaran is a water soluble D-glucan which occurs as the principal reserve polysaccharide in a number of brown algae, particularly from *Laminaria* species. The main structural features of this polysaccharide are (1→3)- and (1→6)-linked β-D-glucopyranosyl residues forming chains with some (1→6)-linkages as branch points which are terminated by reducing D-glucopyranosyl or nonreducing D-mannitol groups. The ratio of these so-called G- and M- chains varies, but is about 1:1 for the majority of samples. The average chain lengths vary between 7 and 11, with molecular weights corresponding to about 30 units for soluble forms of laminaran, which indicates an average of 2–3 branch points per molecule, whilst insoluble forms contain 16–24 units with average chain lengths of 15–19 units and indicate essentially linear structures. Chains terminating in D-mannitol have not been found in the D-glucans isolated from members of the Chrysophycaae and Bacillariophycaae but they resemble laminaran in other respects.

Structural polysaccharides

Brown, red, and green algae have, as a structural polysaccharide, the second class of algal polysaccharides, a cellulose which is essentially similar to that of land plants and comprises about 10% of the dry weight. In very close association with this cellulose are a number of hemicelluloses and similar compounds which include D-mannans, D-xylans and a lichenan type of polysaccharide consisting of linear chains of (1→4)- and (1→3)-linked β-D-glucopyranosyl residues with a lower proportion of (1→3)-linkages than is found in plant lichenan. The D-mannans isolated from red seaweed resemble the ivory nut D-mannan in that they contain essentially linear chains of about 16 (1→4)-linked β-D-mannopyranosyl residues. The D-xylans found in seaweeds differ from those in land plants by being predominantly true D-xylans containing, in the case of the red seaweed *Rhodymenia palmata*, chains of (1→3)- and (1→4)-linked β-D-xylopyranosyl residues in an irregular distribution whereas the D-xylan from green seaweed, *Caulerpa filiformis*, contains only (1→3)-linked β-D-xylopyranosyl residues.

Alginic acid is a commercially important component of brown seaweeds and is the most important mucilaginous polysaccharide which prevents desiccation of the seaweed when exposed to the air, for example, at low tide. Industrially it is important as a thickening agent and emulsion stabiliser (see Chapter 14). It has been shown that the molecule readily produces fibres which according to X-ray diffraction are essentially linear. It was originally thought to be a (1→4)-linked β-D-mannuronan but alginic acid actually consists of D-mannopyranosyluronic and L-gulopyranosyluronic acid residues. Partial hydrolysis gives degraded polysaccharides containing entirely one or the other hexuronic acid residue and a number of oligosaccharides containing both residues, and it has been suggested that alginic acid contains crystalline regions of β-D-mannuronic acid and of α-L-guluronic acid and amorphous regions containing a mixture of both residues, all (1→4)-linked.

Sulphated polysaccharides

A group of polysaccharides isolated from algae are the sulphated polysaccharides which include carrageenan and agar. Carrageenan can be fractionated into a number of polysaccharides, proportions of which vary from species to species and with the season and environment. The major structural features of this group of polysaccharides are an alternating sequence of (1→3)-linked β-D-galactopyranosyl and (1→4)-linked α-D-galactopyranosyl residues containing various degrees and sites of sulphation. κ-Carrageenan, for instance, is an insoluble D-galactan containing alternating 3-*O*-substituted β-D-galactopyranosyl 4-sulphate and 4-*O*-substituted 3,6-anhydro-α-D-galactopyranosyl residues and small proportions of (1→4)-linked D-galactopyranosyl 6-sulphate residues whereas μ-carrageenan is soluble and contains only (1→3)-linked α-D-galactopyranosyl 4-sulphate and (1→4)-linked β-D-galactopyranosyl 6-sulphate. ι-Carrageenan has a repeating structure of (1→3)-linked α-D-galactopyranose 4-sulphate and (1→4)-linked 3,6-anhydro-β-D-galactopyranosyl 2-sulphate residues with approximately 1 in 10 of these residues being replaced by (1→4)-linked β-D-galactopyranosyl 2,6-disulphate.

ζ-Carrageenan is a polymer of D-galactopyranose 2-sulphate with β-(1→4)- and α-(1→3)-linkages. λ-And χ-carrageenans are similar, each containing (1→3)-linked α-D-galactopyranosyl 2-sulphate and (1→4)-linked β-D-galactopyranosyl 2,6-disulphate and 3,6-anhydro-β-D-galactopyranosyl 2-sulphate residues; χ-carrageenan is richer in 3,6-anhydro-D-galactopyranose units compared to λ-carrageenan.

Agar is considered to consist of chains having alternating (1→3)- and (1→4)-linkages with three extremes of structure: neutral agarose which contains (1→3)-linked β-D-galactopyranosyl and (1→4)-linked 3,6-anhydro-α-L-galactopyranosyl residues, a pyruvic acid acetal-containing agarose in which the D-galactopyranosyl residues are substituted with pyruvic acid acetals, and a sulphated galactan with few, or no, 3,6-anhydro-L-galactopyranosyl or pyruvic acid acetal-containing D-galactopyranosyl residues.

Miscellaneous algal polysaccharides

There are also a number of algal mucilages which have similar properties and whose structures have not yet been determined. They usually contain L-rhamnose, D-xylose, D-glucuronic acid, D- and L-galactose and D-mannose and are typified by the mucilage from fresh water red algae, which may contain a D-galactose and D-glucuronic acid repeating sequence, together with the above neutral residues and methylated L-rhamnosyl and D-galactosyl residues; and by a mucilage from brown seaweed which contains L-fucose, D-xylose and D-galactose as nonreducing terminal residues with D-mannose and D-galactose occupying branch positions.

MICROBIAL POLYSACCHARIDES

Microbes give rise to many polysaccharides and also to macromolecules, such as glycopeptides, glycoproteins and glycolipids (see Chapter 9 and 10), to the backbones of which carbohydrate chains are attached. Many of these compounds are unique to microbes, not being found in any other general areas such as plants and animals. The polysaccharidic material can occur as an integral part of the cell-wall, as capsules surrounding the cell, or can be elaborated in culture media (extracellular polysaccharides).

Teichoic acids

The name 'teichoic acids' has been given to a group of phosphate-containing polymers, isolated from cell-walls and membranes of Gram-positive bacteria. The first known members of the group were all polymers of either glycerol phosphate or ribitol phosphate repeating units joined together *via* phosphodiester linkages. With the discovery of monosaccharide- and oligosaccharide-1-phosphate polymers in the walls of some bacteria the definition has had to be extended to include these materials.

The structure of teichoic acids can be regarded as being a combination of two chemically different parts (which are biosynthesised by different mechanisms): (a) the main polymer chain and (b) the linkage region responsible for the attachment of the teichoic acid to a peptidoglycan. Several different types of main chain have been described which contain *O*-ribitol, sugar and phosphate residues. These are held together in repeating units of between 10 and 50 units by the phosphate residues. 1,5-Poly(ribitol phosphate) teichoic acids (86) occur in the walls of many bacilli, lactobacilli and staphylococci and most contain D-alanine residues attached *via* O-2 of the D-ribitol residues. In *Staphylococcus aureus* 2-acetamido-2-deoxy-α- or -β-D-glucopyranosyl residues are glycosidically attached by (1→2)-linkages, whilst in strains of *Bacillus subtilis* and *Bacillus licheniformis* β-D-glucopyranosyl residues are attached by (1→2)-linkages.

1,3-Poly(glycerol phosphate) teichoic acids (87) are more widespread than ribitol phosphate teichoic acids, occurring in the walls of several species of bacteria. The hydroxyl group at C-2 of each glycerol residue is substituted by D-alanine, D-glucose and 2-acetamido-2-deoxy-D-glucose and -D-galactose. A variation in the structural pattern is observed in some bacteria which contain a 1,2-poly(glycerol phosphate) in which the 3-position is substituted by glycosyl residues such as a disaccharide of D-galactopyranose and 2-acetamido-2-deoxy-D-glucopyranose as found in the teichoic acid from *Streptomyces antibioticus* (88).

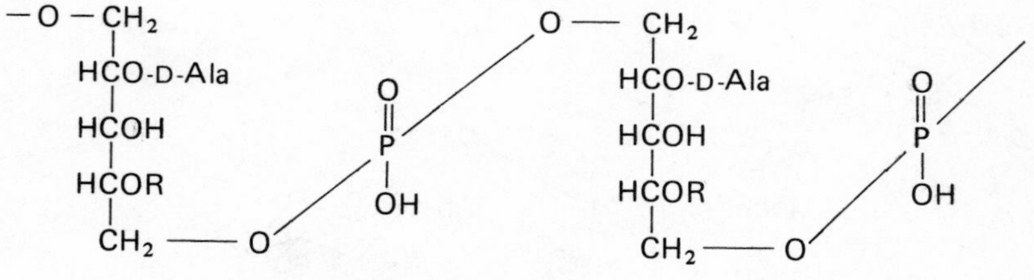

—O—CH₂ / HCO-D-Ala / HCOH / HCOR / CH₂—O—P(=O)(OH)—O—CH₂ / HCO-D-Ala / HCOH / HCOR / CH₂—O—P(=O)(OH)—

R = α-D-Glc*p*NAc-(1→, β-D-Glc*p*NAc-(1→, or β-D-Glc*p*-(1→

(86)

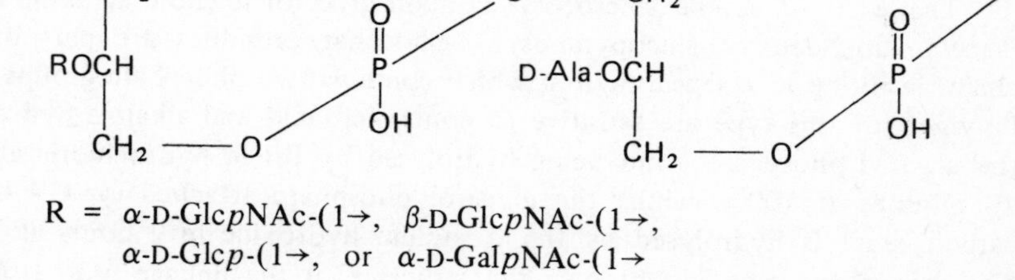

—O—CH₂ / ROCH / CH₂—O—P(=O)(OH)—O—CH₂ / D-Ala-OCH / CH₂—O—P(=O)(OH)—

R = α-D-Glc*p*NAc-(1→, β-D-Glc*p*NAc-(1→, α-D-Glc*p*-(1→, or α-D-Gal*p*NAc-(1→

(87)

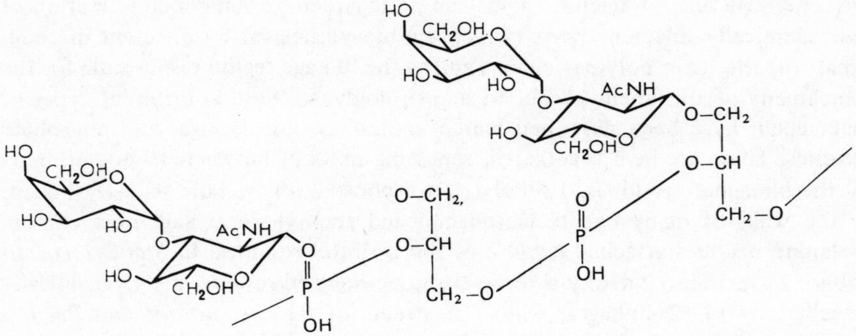

(88)

Some glycerol teichoic acids contain glycosidically-linked monosaccharides as part of the phosphodiester backbone. A poly(D-glucopyranosylglycerol phosphate) (89) has been isolated from the walls of *Bacillus stearothermophilus* whilst a strain of *Bacillus licheniformis* was found to contain more than one type of polymer chain teichoic acid, namely a mixture of poly(D-glucopyranosylglycerol phosphate), poly(D-galactopyranosylglycerol phosphate) and 1,3-poly(glycerol phosphate).

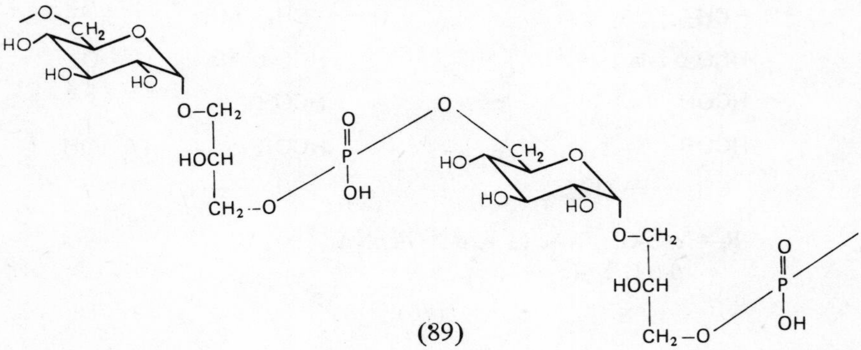

(89)

The walls of several micrococci contain glycerol teichoic acids in which 2-acetamido-2-deoxy-D-glucopyranosyl 1-phosphate residues are part of the chain, resulting in a repeating unit which contains two phosphate groups (90). Polymers of this type are sensitive to controlled acid and alkaline hydrolysis, the sugar 1-phosphate bond being hydrolysed by 0.1 M hydrochloric acid in 10 minutes at 100°C whilst the glycerol phosphate attached *via* C-4 of the carbohydrate is hydrolysed by 0.5 M sodium hydroxide in 2 hours at 20°C. This has given some insight into the structure of the linkage unit. D-Alanyl residues are frequently attached *via* O-6.

(90)

(91)

rest of teichoic acid chain

Mur = muramic acid residue of cell wall peptidoglycan

(92)

Sugar 1-phosphate polymers which are included within the teichoic acid definition have been isolated from the walls of various bacteria. The repeating units found include 2-acetamido-2-deoxy-α-D-glucopyranose 1-phosphate (91) and the β-isomer.

Teichoic acids are attached to muramic acid residues of cell-wall peptidoglycans through a phosphodiester linkage which is both acid and alkali labile (and therefore not a simple phosphodiester linkage). The discovery of a terminal 2-acetamido-2-deoxy-D-glucosyl residue at the reducing end of the polymer on hydrolysis led to the discovery of a linkage unit containing three glycerol phosphate residues as well as the aminosugar (92). Other linkage units have since been found, including substitution at O-6 rather than at O-4 of the amino sugar and a disaccharide linkage unit consisting of 2-acetamido-(4-*O*-2-acetamido-2-deoxy-D-mannopyranosyl)-2-deoxy-D-glucose. In the latter case the polymer chain is attached to the 2-acetamido-2-deoxy-D-mannopyranosyl residue *via* O-3 or O-4, whilst the 2-acetamido-2-deoxy-D-glucopyranosyl residue is attached to O-6 of the muramic acid residue in the peptidoglycan through an acid labile sugar 1-phosphate bond.

When Gram-positive organisms are grown under limiting conditions (that is, limited D-glucose or phosphate), their teichoic acids are replaced by teichuronic acids which are polymers containing D-glucuronic acid and 2-acetamido-2-deoxy-D-galactose. Originally these were thought to contain single disaccharide repeating units analogous to glycosaminoglycans (see Chapter 9) but recent studies have shown more-complex repeating tetrasaccharide repeating units (93) which can contain neutral sugar residues (94). This latter example, from *Bacillus megaterium* (Ivatt and Gilvarg, 1979), has a very large complex structure. Its molecular weight is 5×10^5, making it the largest ancillary polymer known to date (by a factor of 20). Unlike teichoic acids, which have random coil structures, this teichuronic acid has an ordered rod-like structure which is the most compressed reported for a complex polysaccharide, being 3 times as compressed as amylose. Reviews of teichoic acids (Baddiley, 1972, Sutherland, 1977, Munson and Glaser, 1981, and McArthur, 1981) give full descriptions of the function and biosynthesis of all these materials.

→4)-β-D-Glc*p*A-(1→4)-β-D-Glc*p*A-(1→3)-β-D-Gal*p*NAc-(1→6)-α-D-Gal*p*NAc-(1→

(93)

→4)-β-D-Glc*p*-(1→3)-α-L-Rha*p*-(1→3)-α-L-Rha*p*-(1→
3
↑
1
β-D-Glc*p*A

(94)

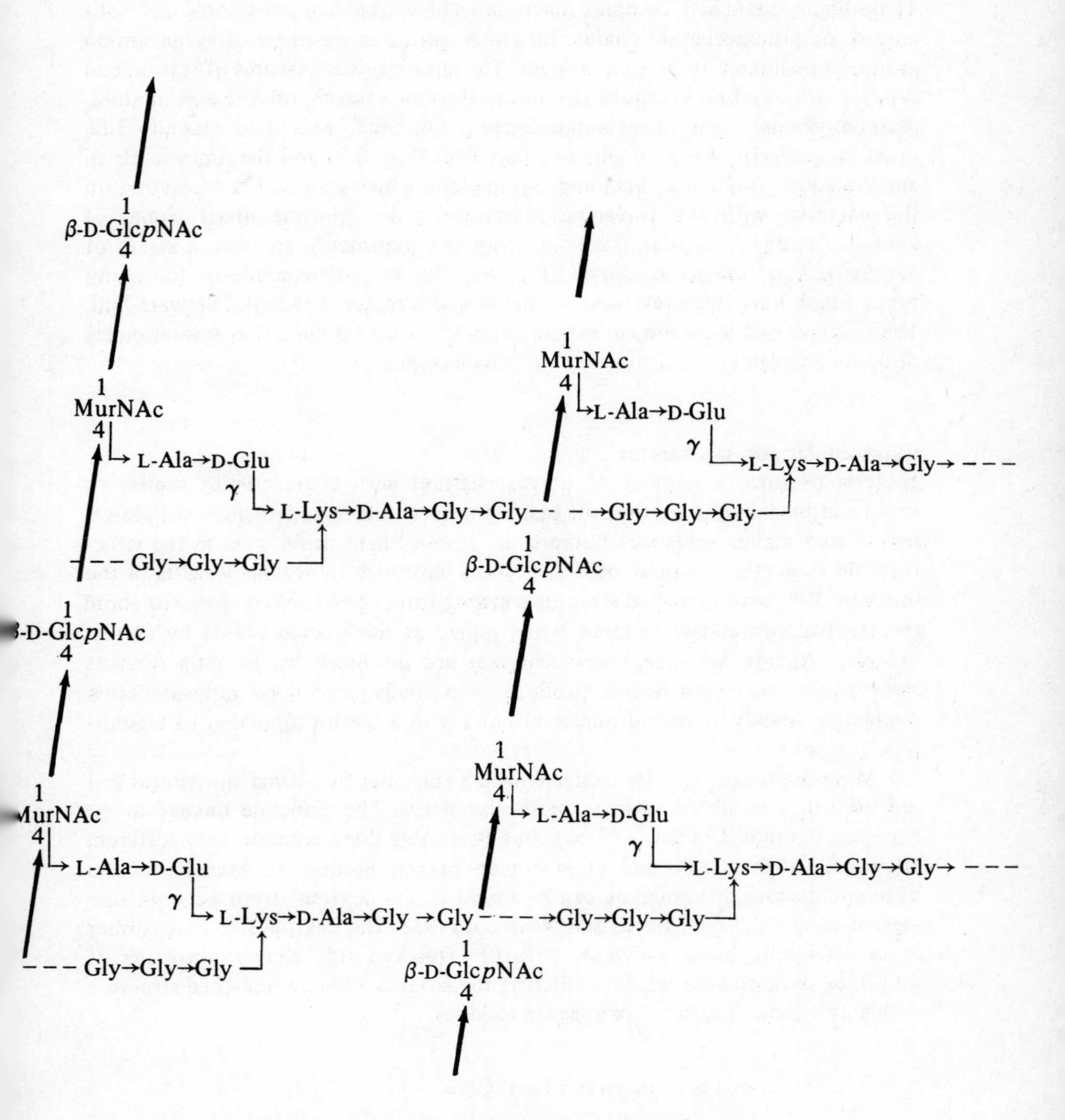

Fig. 8.6 – Schematic representation of the structure of the peptidoglycan from a strain of *Staphylococcus aureus*. (➔ inter carbohydrate linkage, → inter aminoacid linkages.)

Cell-wall peptidoglycans (mureins)

These highly branched, complex macromolecules occurring in bacterial cell walls consist of polysaccharide chains, in which individual residues carry an amino group, cross-linked by peptide bridges. The characteristic features of this unique type of structural polymer are the amino-type monosaccharides, 2-acetamido-2-deoxy-D-glucose and 2-acetamido-2-deoxy-muramic acid (2-acetamido-3-*O*-(1-carboxyethyl)-2-deoxy-D-glucose) (see later, Fig. 8.9) and the aminoacids D- and L-alanine, D-glutamic acid and L-lysine and other aminoacids depending on the bacteria, with the foregoing monosaccharides forming linear chains of alternate units. A typical fragment from the peptidoglycan from a strain of *Staphylococcus aureus* is shown (Fig. 8.6) by way of example of the many types which have been reviewed (Schleifer and Kandler, 1982, and Stewart-Tull, 1980). Since cell walls are *ca*. 20 nm thick it can be assumed that several sheets of peptidoglycan are held together by cross-linkages.

Extracellular polysaccharides

Bacteria produce a number of polysaccharides with characeristics similar to those outlined in the section on plant polysaccharides and include celluloses, levans, and alginic acids (see Sutherland, 1979). Slight differences in the structures do exist; for example, bacterial levans have high molecular weights of the order of 10^6, with branched structures arising from (2→6)-linked chains of about ten β-D-fructofuranosyl residues being joined at the branch points by (2→1)-linkages. Alginic acid-like polysaccharides are produced by bacteria such as *Azotobacter vinelandii* which produces a partially acetylated polysaccharide containing mainly D-mannuronic acid units with a small proportion of L-guluronic acid units.

More important are the dextrans which find uses as plasma substitutes and are used in a modified form as molecular sieves. The principle linkage in the α-D-glucopyranosyl chain is (1→6), but branching does occur to very different degrees through (1→3)- and (1→4)-linked branch points. An example of the different degrees of branching can be found in the dextrans from *Leuconostoc mesenteroides* and *Betacoccus arabinosaceous* (95). The dextran from the former is an essentially linear molecule with (1→3)-linked side chains consisting of mono- or di-saccharide residues whereas the latter is a highly branched structure with a unit chain length of six or seven residues.

```
→6)-α-D-Glcp-(1→6)-α-D-Glcp
                          1
                          ↓
                          3
→6)-α-D-Glcp-(1→6)-α-D-Glcp-(1→6)-α-D-Glcp-(1→
```

(95)

A family of related extracellular polysaccharides, many of which have not yet been fully characterised, are produced by the various strains of *Xanthomonas*, the most important for industrial purposes (see Chapter 14) being xanthan gum, which is obtained from *X. campestris*. Xanthan gum has a structure consisting of linear chains of (1→4)-linked β-D-glucosyl residues to which are attached trisaccharide side chains consisting of D-mannose and D-glucuronic acid in the molar ratio of 2:1 (see Fig. 8.7). The α-D-mannosyl residues which are linked *via* (1→3)-linkages to the D-glucan backbone are usually acetylated at C-6 whilst approximately half of the terminal α-D-mannosyl groups carry pyruvic acid acetal groups attached *via* O-4 and O-6. Variation in the amount and distribution of acetate, pyruvic acid acetal and trisaccharide side chains occurs with different strains of bacteria, Fig. 8.7 representing only an average situation rather than a specific case.

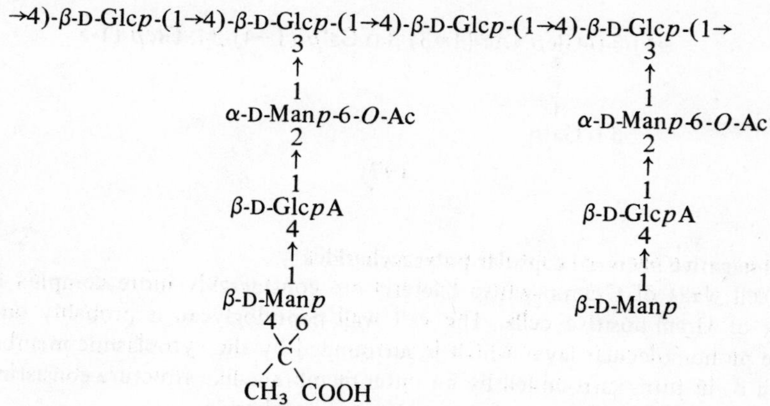

Fig. 8.7 – Repeating unit of xanthan gum shown in the acidic form. Distribution of side chains and pyruvic acid acetal reflect average values.

Gram-positive bacterial capsular polysaccharides

The Gram-positive bacteria produce polysaccharides which have complex structures and frequently contain aminosugar residues as well as neutral and acidic monosaccharides. Many of the aminosugars have been found to have rare structures (for example, diamino-trideoxysugars, etc.) which were previously regarded as 2-amino-2-deoxy-D-glucose. This was due to the inadequate specificity of the degradation method used to identify the aminosugar and the incorrect assumption that no diaminosugars etc. could exist. Mild, physicochemical methods of analysis have shown that many other aminosugars do, in fact, exist (see Fig. 8.9 and Horton and Wander, 1980a). They frequently have immunological properties which are characteristic of a particular chemical structure. The structures of a number of these polysaccharides have been found to consist of a variety of polysaccharide types.

Polysaccharides from strains of *Streptococcus pneumoniae*, which were the first non-protein materials shown to be antigenic, are produced externally to the cell wall and form capsules that cover the cells. Over 70 different types of polysaccharide have been identified and the structures of many have been established. Some, such as types 13 and 34, contain ribitol phosphate plus a number of monosaccharides and these are considered to be teichoic acids. Other polysaccharides contain simple repeating units such as type 3 which contains a disaccharide repeating unit (96) or type 14 which contains a tetrasaccharide repeating unit (97).

→3)-β-D-Glc*p*A-(1→4)-β-D-Glc*p*-(1→

(96)

→6)-β-D-Glc*p*NAc-(1→3)-β-D-Gal*p*-(1→4)-β-D-Glc*p*-(1→
4
↑
1
β-D-Gal*p*

(97)

Gram-negative bacterial capsular polysaccharides

The cell walls of Gram-negative bacteria are considerably more complex than those of Gram-positive cells. The cell wall peptidoglycan is probably only a single monomolecular layer which is surrounded by the cytoplasmic membrane which is, in turn, surrounded by an outer membrane-like structure consisting of protein, phospholipid, lipoprotein and lipopolysaccharides.

LIPOPOLYSACCHARIDES

The lipopolysaccharides (LPS) do not appear to be covalently linked to the peptidoglycan (unlike teichoic acids in Gram-positive bacteria), but have been shown to be located on the outer surface of the cell from which it can be removed, intact, by reagents such as phenol and propan-2-ol. These lipopolysaccharides are responsible for type specific immunological reactions. Bacteria can have more than one antigenic characteristic, many of which are sponsored by a structural feature of a/the capsular polysaccharide of the bacterium. Most of the antigenic specificity of the polysaccharides arises from the nonreducing end groups and ranges of antisera have been used in the structural analysis of such polysaccharides. The structural basis for these serological reactions can be shown by the following example. The acidic polysaccharides from *Klebsiella aerogens* and *Enterobacter 349* cross react to the extent of about 50% with the heterologous antibodies (that is, from the antiserum to the opposite polysaccharide). Structural analysis has shown that these polysaccharides both contain (1→3)-linked D-galactopyranosyl and (1→3)- and (1→4)-linked D-mannopyranosyl residues, but that the *Klebsiella aerogenes* polysaccharide contains D-mannopyranosyluronic acid residues forming disaccharide repeating units with the D-mannopyranosyl residues, whereas the *Enterobacter* polysaccharide contains disaccharide repeating units of D-mannopyranosyl residues attached to D-galactopyranosyluronic acid and to D-glucopyranosyluronic acid residues. A number of less common monosaccharides are found in the polysaccharides from Gram-negative bacteria, including 3,6-dideoxy derivatives of D-glucose, D-galactose and D-mannose and a number of aldoheptoses, the structures of which are given in Fig. 8.9.

The part of the lipopolysaccharides not responsible for antigenic activity is known as the core polysaccharide and this is attached to the lipid material (see Fig. 8.8). Compared to the complexity of the antigenic outer chains, the core polysaccharide of nearly all lipopolysaccharides consists of D-glucose, D-galactose, 2-acetamido-2-deoxy-D-glucose and a heptose which are linked to an octose. The heptose is usually L-*glycero*-D-*manno*-heptose and the octose always 3-deoxy-D-*manno*-octopyranulosonic acid (KDO). Lipid A is a common constituent of all lipopolysaccharides and has the repeating structure (98) consisting of a disaccharide of 2-acetamido-2-deoxy-D-glucopyranosyl residues to which KDO residues are attached as shown.

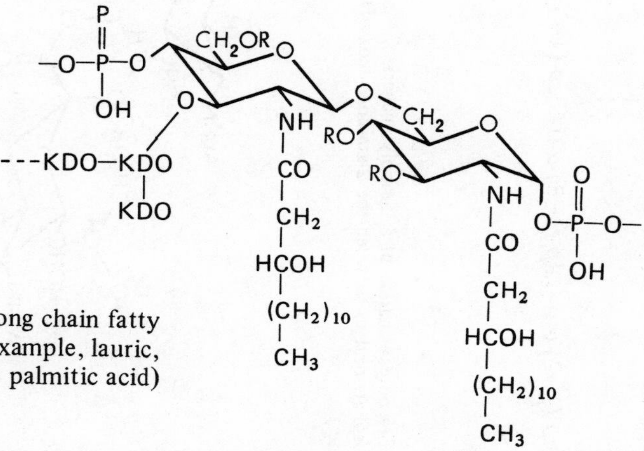

(98) R = long chain fatty acids (for example, lauric, myristic or palmitic acid)

A unique homopolysaccharide from Gram-negative bacteria is a polymer of 5-acetamido-3,5-dideoxy-D-*glycero*-D-*galacto*-2-nonulopyranonic acid called colominic acid which contains (2→8)-linkages and also in some samples (1→9)-internal ester linkages between adjacent residues (99).

The structures of the less common monosaccharides found in bacterial polysaccharides are given in Fig. 8.9 but their inclusion in this listing does not preclude their existence in other carbohydrate-containing macromolecules.

```
                       α-D-GlcpNAc                      α-D-Galp
                            1                               1
                            ↓                               ↓
                            2                               6
Antigenic chain →4)-α-D-Glcp-(1→2)-α-D-Galp-(1→3)-α-D-Glcp-(1→3)-α-Hep-(1→3)-α-Hep-(1→5)-KDO-(2→7 or 8)-KDO→ Lipid A
                                                                                 |                          5
                                                                                 P                          ↑
                                                                                                            2
                                                                                                           KDO
```

Fig. 8.8 – Structure of core polysaccharide of a lipopolysaccharide. A number of chains are crosslinked by phosphate bridges between heptose residues and by pyrophosphate bridges between the 2-amino-2-deoxy-D-glucopyranosyl residues of lipid A. (Hep = L-*glycero*-D-*manno*-heptopyranose, Ⓟ = phosphate.)

(99)

Deoxysugars

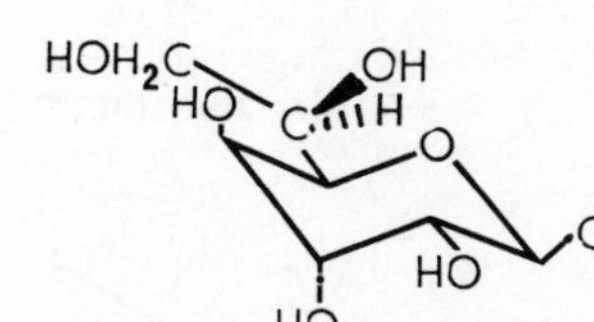
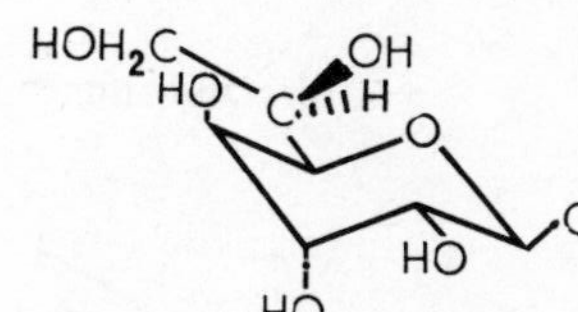
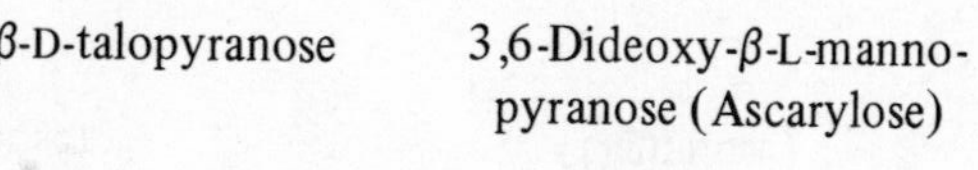
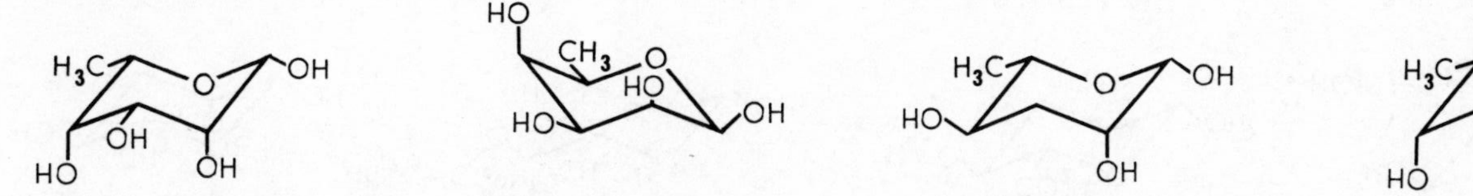
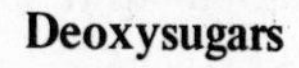

6-Deoxy-β-L-talopyranose

6-Deoxy-β-D-talopyranose

3,6-Dideoxy-β-L-manno-pyranose (Ascarylose)

3,6-Dideoxy-β-L-galacto-pyranose (Colitose)

3,6-Dideoxy-β-D-gluco-pyranose (Paratose)

3,6-Dideoxy-β-D-manno-pyranose (Tyvelose)

3,6-Dideoxy-β-D-galacto-pyranose (Abequose)

Heptoses

L-*glycero*-β-D-*manno*-Heptopyranose

D-*glycero*-β-D-*galacto*-Heptopyranose

D-*glycero*-β-D-*manno*-Heptopyranose

D-*glycero*-β-D-*gulo*-Heptopyranose

Fig. 8.9 – Structures of less common monosaccharides found in bacterial polysaccharides (*continued next page*)

Octose

3-Deoxy-β-D-*manno*-octopyranulosonic acid (KDO)

Uronic acids

β-L-Gulopyranuronic acid

β-D-Ribofuranuronic acid

Ether derivatives

3-*O*-Methyl-6-deoxy-β-L-talopyranose (Acovenose)

3-*O*-Methyl-β-L-rhamnopyranose (Acofriose)

3-*O*-Methyl-β-D-fucopyranose (Digitalose)

2,3-Di-*O*-methyl-β-L-rhamnopyranose

2,3-Di-*O*-methyl-β-D-rhamnopyranose

4-*O*-(1-Carboxyethyl)-β-D-glucopyranose

4-*O*-Methyl-β-D-glucopyranuronic acid

2-Amino-3-*O*-(1-carboxyethyl)-2-deoxy-β-D-glucopyranose (Muramic acid)

Aminosugars

2-Amino-2-deoxy-β-D-mannopyranose (D-Mannosamine)

2-Amino-2-deoxy-β-L-fucopyranose (L-Fucosamine)

2-Amino-2-deoxy-β-D-fucopyranose (D-Fucosamine)

2-Amino-2,6-dideoxy-β-L-talopyranose (Pneumosamine)

2-Amino-2,6-dideoxy-β-D-glucopyranose (Quinovosamine)

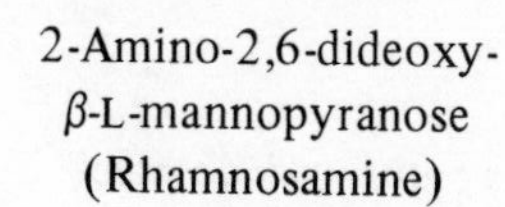

2-Amino-2,6-dideoxy-β-L-mannopyranose (Rhamnosamine)

3-Amino-3,6-dideoxy-β-D-galactopyranose

3-Amino-3,6-dideoxy-β-D-glucopyranose

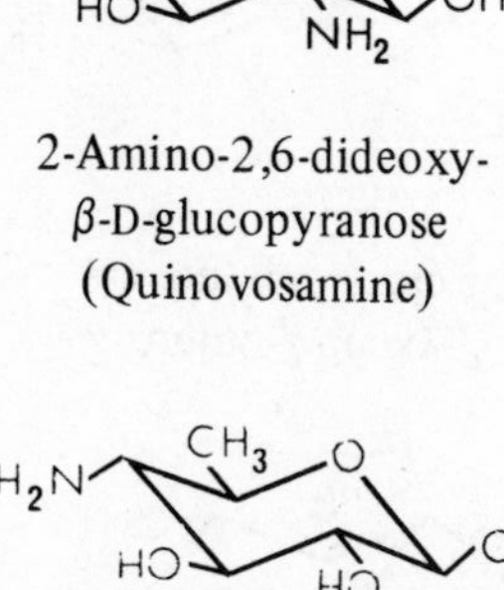

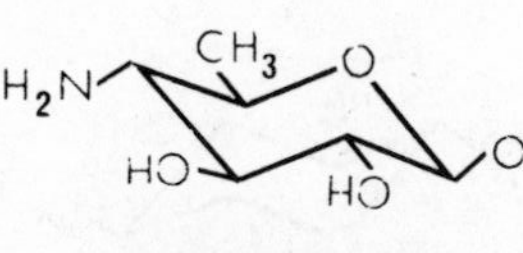

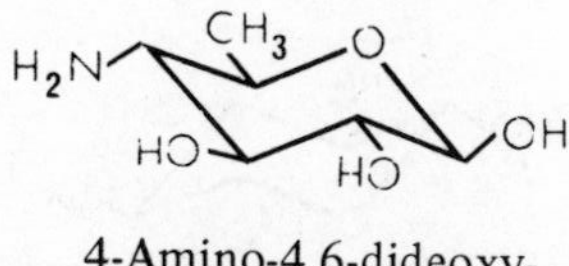

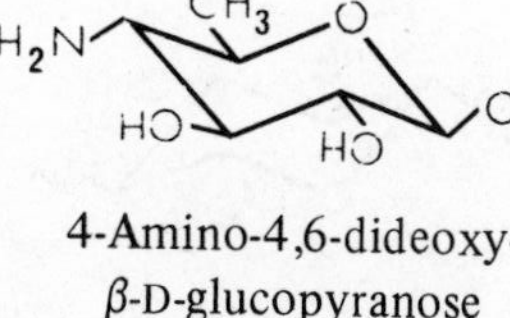

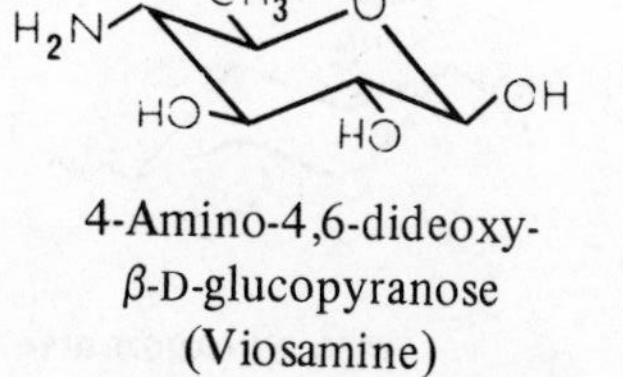

4-Amino-4,6-dideoxy-β-D-glucopyranose (Viosamine)

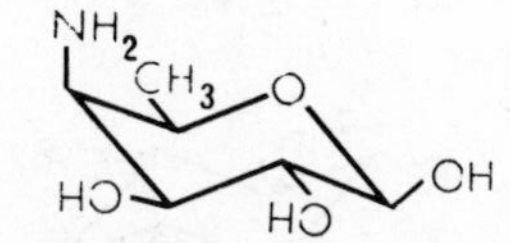

4-Amino-4,6-dideoxy-β-D-galactopyranose (Thomasamine)

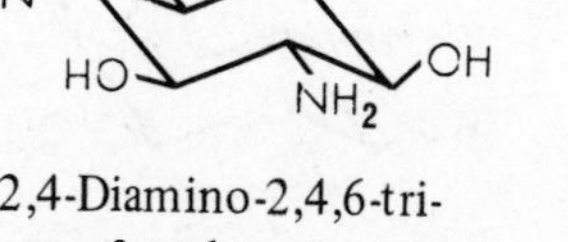

2,4-Diamino-2,4,6-trideoxy-β-D-glucopyranose (Bacillosamine)

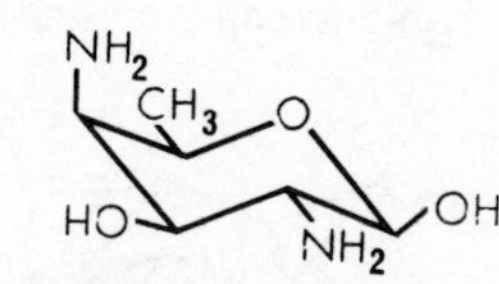

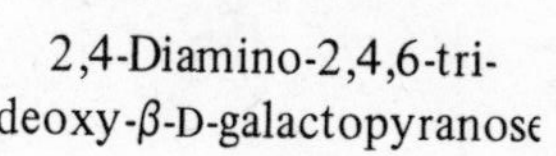

2,4-Diamino-2,4,6-trideoxy-β-D-galactopyranose

Fig. 8.9 – Structures of less common monosaccharides found in bacterial polysaccharides (*continued next page*)

Aminouronic acids

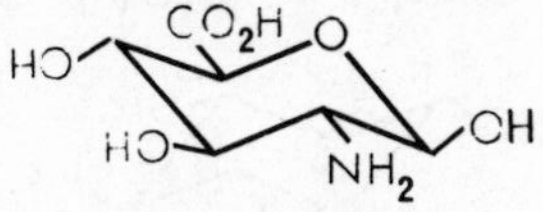

2-Amino-2-deoxy-β-D-glucopyranuronic acid

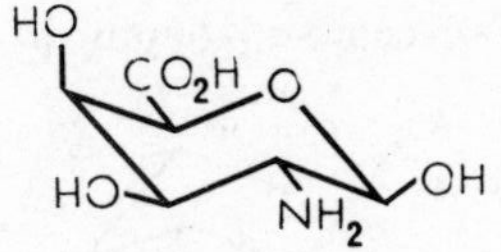

2-Amino-2-deoxy-β-D-galactopyranuronic acid

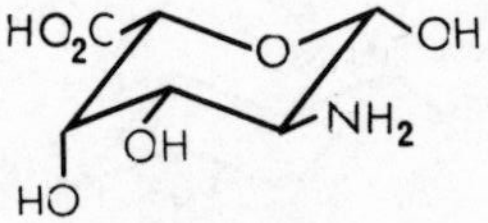

2-Amino-2-deoxy-β-L-galactopyranuronic acid

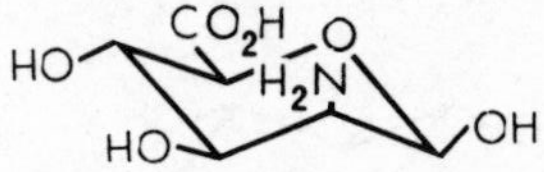

2-Amino-2-deoxy-β-D-mannopyranuronic acid

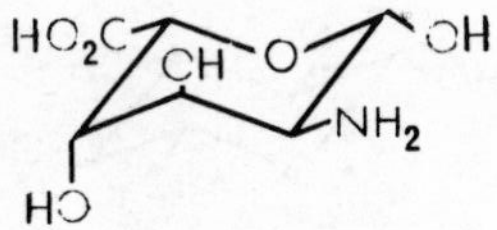

2-Amino-2-deoxy-β-L-gulopyranuronic acid

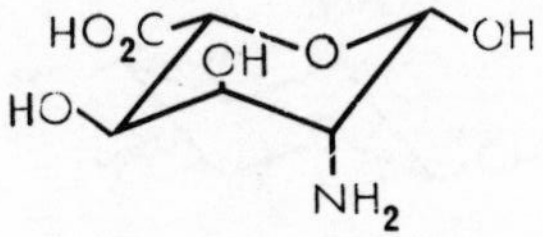

2-Amino-2-deoxy-β-L-altropyranuronic acid

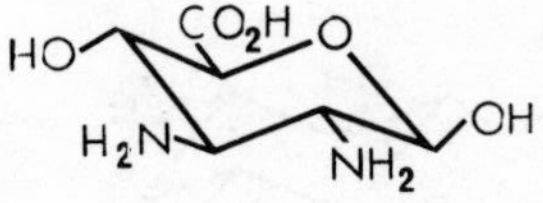

2,3-Diamino-2,3-dideoxy-β-D-glucopyranuronic acid

Fig. 8.9 – Structures of less common monosaccharides found in bacterial polysaccharides (shown in one anomeric form only). See Figs. 2.8 and 12.1 for other monosaccharide structures.

FUNGAL POLYSACCHARIDES

The major polysaccharides elaborated by fungi include: α-D-glucans, such as one containing only (1→3)-linked residues obtained from cell walls of *Polyporus tumulosus*; β-D-glucans, such as pachyman from *Porin cocos*, which consists of (1→3)-linked residues, and luteose which is a (1→6)-linked polymer from *Penicillium luteum*; and a number of storage products of the laminaran type which contain (1→3)-, (1→4)- and (1→6)-linked residues in linear and branched structures. The other major type of fungal polysaccharides are the D-mannans typified by the (1→6)-lined D-mannan from *Saccharomyces rouxii* and the (1→2)-linked D-mannan from *Saccharomyces cerevisiae*. The latter D-mannan has also been shown to contain phosphorylated side chains and has the structure (100), arising from the addition of (1→3)-linked D-mannopyranosyl groups to the D-mannopyranosyl phosphorylmannotriose chains. A D-galactan, galactocarolose, from *Penicillium charlesii*, is a low molecular weight polysaccharide consisting of unbranched chains of approximately ten (1→5)-linked β-D-galactofuranosyl residues.

α-D-Man*p*-(1→3)-α-D-Man*p*-(1→2)-α-D-Man*p*-(1→2)-α-D-Man*p*-(1→
|
α-D-Man*p*-(1→3)-α-D-Man*p*→Ⓟ

→Ⓟ = orthophosphate diester linkage

(100)

A number of heteropolysaccharides have also been isolated and characterised, including the D-galacto-D-mannan from *Cladosporium herbarum* which contains D-galactosyl residues in both pyranose and furanose forms, and a D-xylo-D-mannan from *Cryptococcus leuerentii*.

ANIMAL POLYSACCHARIDES

Glycogen

The principal reserve polysaccharide in the animal kingdom is glycogen, being found in most tissues of which the most convenient source for the purpose of extraction is liver or muscle. Human liver contains glycogen in up to 10% of its dry weight. Unlike starch, isolation and purification of glycogen is not simple. The classical method was to use strongly alkaline solutions at 100°C for about 3 hours to dissolve the tissue and then to precipitate the glycogen with ethanol, but with the development of understanding of alkaline degradation the use of milder techniques had to be sought. The use of cold dilute trichloroacetic acid for the extraction procedure results in a product the molecular size of which is some 10 times larger than that obtained with the traditional method. Methods are now available which avoid more-complete destruction during isolation so that it is possible to investigate realistically molecular weights of the isolated polysaccharide. It has been found that, for example, glycogen from the liver in cases of general glycogen storage disease has a lower molecular weight than normal. Classical analytical methods such as methylation have shown the structure of glycogen to be chains of (1→4)-linked α-D-glucopyranosyl residues with 1,4,6-trisubstituted residues as branch points.

The use of amylolytic enzymes for determination of the fine structure has indicated a tree-like structure, similar to that of amylopectin, with a unit chain length of 12 D-glucopyranosyl residues. This short chain length in a molecule which can possess a high molecular weight in the range 10^7–10^8 necessarily results in a highly branched structure, a consequence of which is the extremely limited uptake of iodine by the molecule compared even with that of amylopectin. Regions of dense branching that are resistant to the action of α-amylase are randomly distributed throughout the molecule. With the availability of paracrystalline glycogen it should be possible to use physical methods to examine the structure in greater detail.

Chitin

The most abundant of polysaccharides containing amino-type monosaccharides is chitin which occurs as a structural polysaccharide in the shells of crustacea. It also occurs in fungi and some green algae. The chemical evidence for its structure has been based on the isolation and characterisation of oligosaccharides obtained as a result of partial hydrolysis. This shows that chitin is a homopolymer of 2-acetamido-2-deoxy-β-D-glucopyranose, each residue being (1→4)-linked to form linear chains (101). The polysaccharide may be considered as an analogue of cellulose, the hydroxyl groups at C-2 positions of cellulose being replaced by acetamido groups.

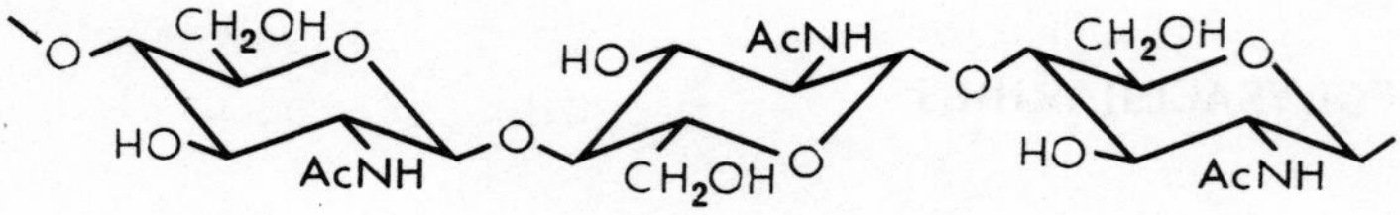

(101)

Chitin rarely occurs alone in nature, it being complexed or covalently bound to protein in the shells of crabs and lobsters. This property may be attributed to the more recently discovered fact that not all the amino groups of the majority of chitins are *N*-acetylated. Accordingly they can operate as basic groups and thereby complex with other molecules of a suitable ionic disposition. Chitin is insoluble in water and many organic solvents, which has made its structural determination difficult, its insolubility being reflected in its low reactivity towards methylation. The majority of samples obtained as a result of treatment with mineral acid have a degree of de-*N*-acetylation and are also of lower molecular weight than chitin in its native state. X-ray diffraction studies of crystalline chitin has shown that the unit cell contains two chains with bent conformations with inter- and intra-molecular hydrogen bonding in a similar manner to cellulose.

A group of animal polysaccharides, the glycosaminoglycans, which normally exist covalently bound to protein, are discussed in the chapter on glycoproteins and proteoglycans (Chapter 9).

The occurrence, isolation, structure and chemistry of all the polysaccharides mentioned in this chapter, and of many other polysaccharides, are regularly reviewed in an ongoing series of articles (Various Authors, 1968 onwards). More detailed reviews of cell-surface carbohydrates can be found in Cook and Stoddart, 1973, Hughes, 1975, and Sutherland, 1977.

CHAPTER 9

Glycoproteins and proteoglycans

CLASSIFICATION

Many macromolecules were originally believed to consist entirely of protein and it was formerly believed that any carbohydrate found in the presence of these biological macromolecules from such sources as human red blood cells and mucous secretions was an impurity. The chemical evidence, reported in 1865, that elemental analysis of a purified mucin yielded values for carbon and nitrogen significantly lower than the values required for pure protein was inconsistent with this, but not understood at the time.

Originally the carbohydrate which was released on acidic hydrolysis was thought to be glucose. Gradually the picture emerged that a number of natural macromolecules (that is, glycoproteins) exist in which carbohydrate forms only part of the total structure. The difficulty in isolating undegraded carbohydrate moieties from the protein (except in the case of the glycosaminoglycans), and the evidence that the heterosaccharides present in a single species of glycoprotein are often not identical but show minor variations in their composition, have made the progress in structural elucidation of glycoproteins much slower, but reviews published on this subject (Gottschalk, 1972, Cook and Stoddart, 1973, Sharon, 1975, Sharon and Lis, 1979, and Various Authors, 1968 onwards) show that a vast amount of work has already been pursued. Undoubtedly improved methods of compositional analysis for carbohydrate together with the above-mentioned developments in structural analysis have led to the discovery that carbohydrate is an integral part of many molecules hitherto described as proteins. This necessary readjustment of description is one which is currently being repeated over and over again, but many authors continue to refer to some glycoproteins as proteins. Some proteins have been shown to contain no carbohydrate using the most recent methods of analysis and are therefore designated as proteins *per se*. An example of such a compound is concanavalin A, a plant lectin (see p. 209).

'Glycoproteins' is a term used (often too generally) to apply to any macromolecule which contains carbohydrate and protein, and in such loose areas of use the term really applies to molecules which if properly classified come under the headings of glycoproteins, proteoglycans and carbohydrate–protein complexes. Glycoproteins contain a protein chain which consists of about 200 or so aminoacid units which are any of the 20 naturally occurring L-α-aminoacids (Fig. 9.1). Covalently attached to this protein backbone and pendent to it is/are the carbohydrate part(s) of the molecule, consisting of hetero-oligosaccharide chains. These are usually branched (Fig. 9.2(a)) and can contain neutral monosaccharides (D-glucose, D-galactose, D-mannose or L-fucose), basic monosaccharides (2-amino-2-deoxy-D-glucose or 2-amino-2-deoxy-D-galactose) and the unique nine-carbon sugar, 5-amino-3,5-dideoxy-D-*glycero*-D-*galacto*-2-nonulopyranonic acid. The basic residues are usually *N*-acetylated and the 5-amino-3,5-dideoxy-D-*glycero*-D-*galacto*-2-nonulopyranonic acid residues may be *N*-glycolyated or *N*-acetylated, and in some cases, certain hydroxyl groups are also *O*-acetylated. Thus oligosaccharide chains are rendered mildly acidic on account of the free carboxyl group in the nine-carbon sugar.

Neutral aminoacids (one amino and one carboxyl group)

H
NH_2—CH—COOH
Glycine (g)

CH_3
NH_2—CH—COOH
Alanine (g)

CH—$(CH_3)_2$
NH_2—CH—COOH
Valine (g, e)

CH_2—$CH(CH_3)_2$
NH_2—CH—COOH
Leucine (g, e)

$CH(CH_3)$—C_2H_5
NH_2—CH—COOH
Isoleucine (g, e)

$(CH_2)_3$—CH_3
NH_2—CH—COOH
Norleucine (l)

CH_2—C_6H_5
NH_2—CH—COOH
Phenylalanine (g, e)

CH_2—C_6H_4—OH
NH_2—CH—COOH
Tyrosine (g)

NH_2—CH—COOH
CH_2—S
CH_2—S
NH_2—CH—COOH
Cystine (g)

CH_2—SH
NH_2—CH—COOH
Cysteine (g)

CH_2—OH
NH_2—CH—COOH
Serine (g)

$CH(CH_3)$—OH
NH_2—CH—COOH
Threonine (g, e)

Fig. 9.1 – Structures of the L-α-aminoacids (*continued next page*)

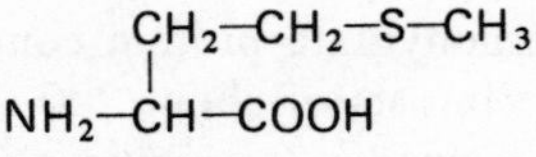

$CH_2-CH_2-S-CH_3$
$NH_2-CH-COOH$
Methionine (g, e)

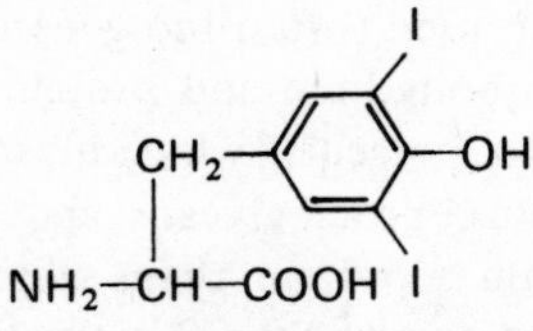

I
CH_2—[ring]—OH
I
$NH_2-CH-COOH$
Di-iodotyrosine (l)
(Iodogoric acid)

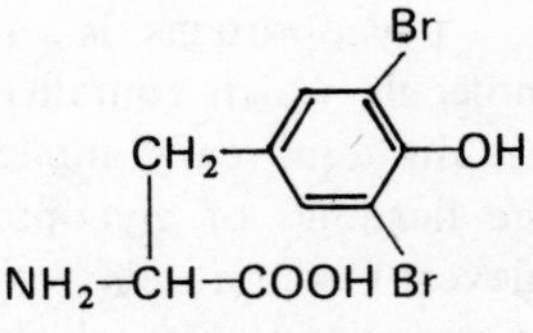

Br
CH_2—[ring]—OH
Br
$NH_2-CH-COOH$
Dibromotyrosine (l)

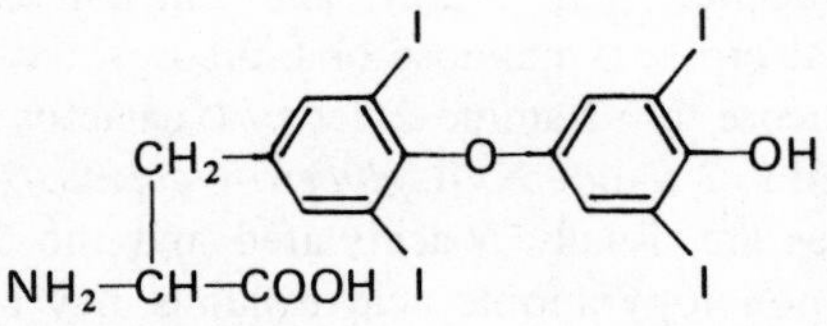

I I
CH_2—[ring]—O—[ring]—OH
I I
$NH_2-CH-COOH$
Thyroxine (l)
(Tetraiodothyronine, T_4)

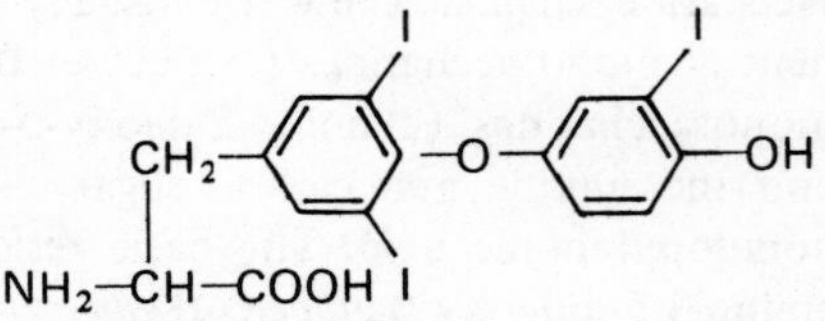

I I
CH_2—[ring]—O—[ring]—OH
I
$NH_2-CH-COOH$
Tri-iodothyronine (l)
(T_3)

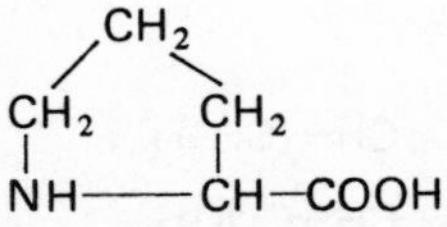

CH_2
CH_2 CH_2
NH——$CH-COOH$
Proline (g)

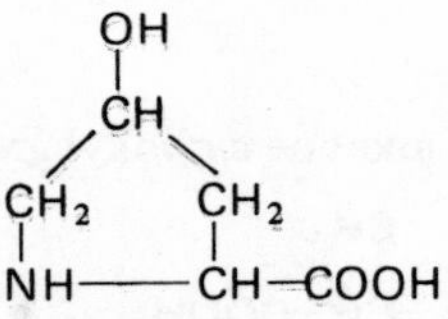

OH
CH
CH_2 CH_2
NH——$CH-COOH$
Hydroxyproline (l)

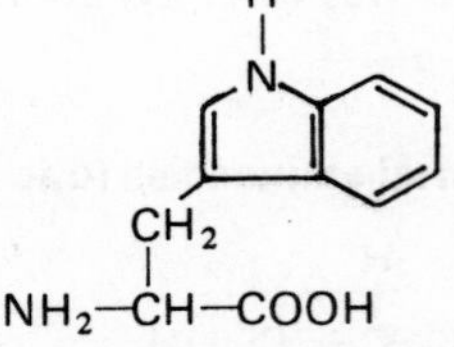

H
N [indole ring]
CH_2
$NH_2-CH-COOH$
Tryptophan (g, e)

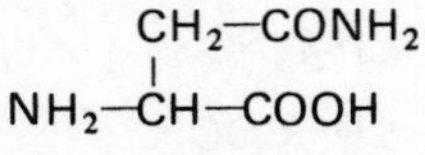

CH_2-CONH_2
$NH_2-CH-COOH$
Asparagine (g)

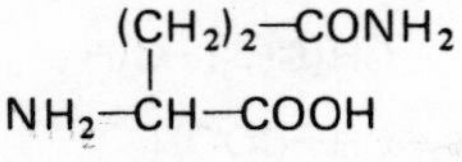

$(CH_2)_2-CONH_2$
$NH_2-CH-COOH$
Glutamine (g)

Acidic aminoacids (one amino and two carboxyl groups)

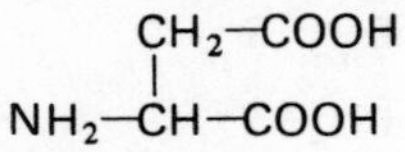

CH_2-COOH
$NH_2-CH-COOH$
Aspartic acid (g)

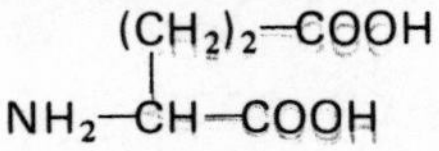

$(CH_2)_2-COOH$
$NH_2-CH-COOH$
Glutamic acid (g)

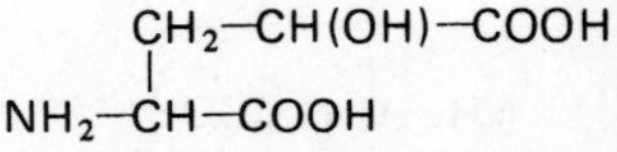

$CH_2-CH(OH)-COOH$
$NH_2-CH-COOH$
γ–Hydroxyglutamic acid[a] (l)

Basic aminoacids (two amino and one carboxyl groups)

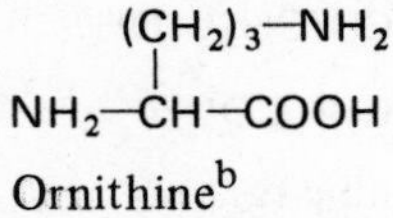

$(CH_2)_3-NH_2$
$NH_2-CH-COOH$
Ornithine[b]

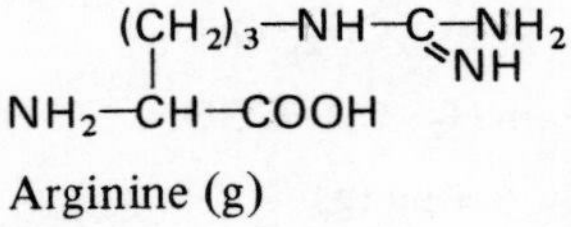

$(CH_2)_3-NH-C(=NH)-NH_2$
$NH_2-CH-COOH$
Arginine (g)

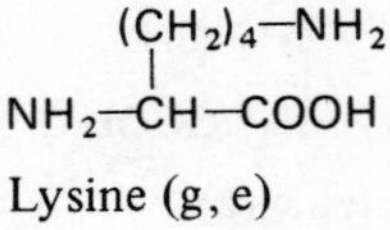

$(CH_2)_4-NH_2$
$NH_2-CH-COOH$
Lysine (g, e)

Fig. 9.1 – Structures of the L-α-aminoacids (*continued next page*)

Histidine (g): imidazol-4-yl–CH_2–$CH(NH_2)$–COOH

δ-Hydroxylysine (l): NH_2–$CH(COOH)$–$(CH_2)_2$–$CH(OH)$–CH_2–NH_2

General stereochemistry for L-aminoacids: H_2N–C(COOH)(R)–H

[a] Occurrence in proteins uncertain

[b] Ornithine is probably not present in proteins, but is formed by the hydrolysis of arginine.

(e) Essential in man (g) General occurrence (l) Less common occurrence

Fig. 9.1 – Structures of the L-α-aminoacids.

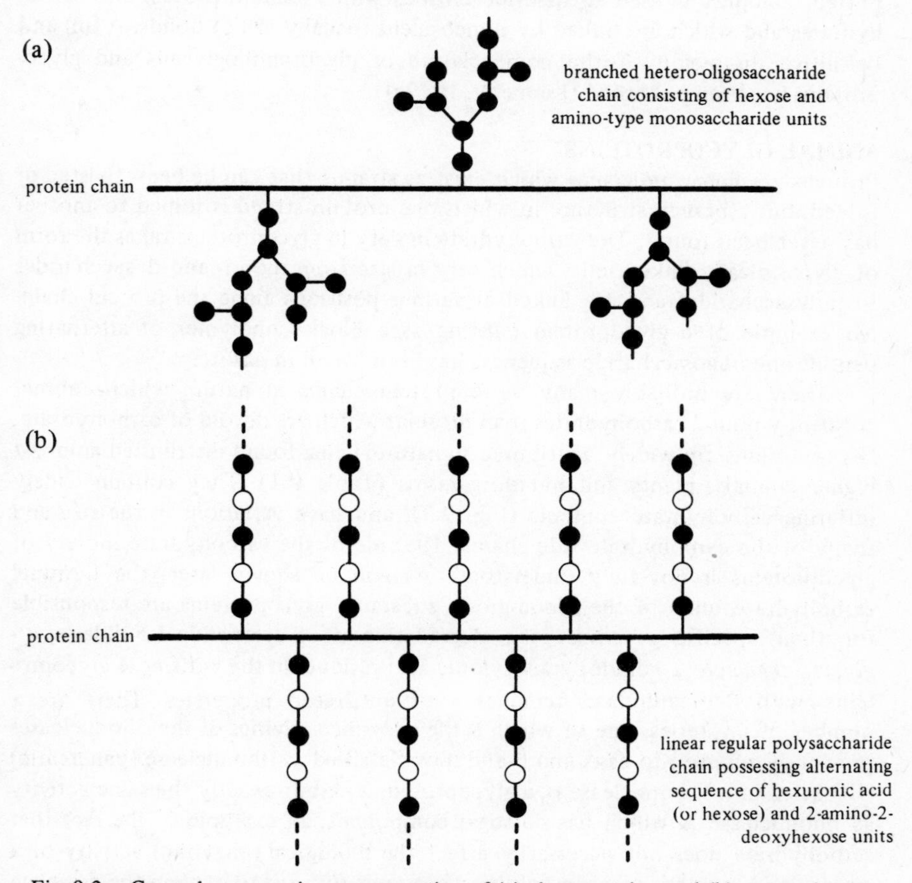

Fig. 9.2 – General comparative representation of (a) glycoproteins, and (b) proteoglycans

Proteoglycans also contain a backbone of protein, but the carbohydrate residue takes the form of linear chains possessing regular alternating monosaccharides (Fig. 9.2(b)) involving acidic monosaccharide (D-glucuronic acid or L-iduronic acid) and basic monosaccharide (2-amino-2-deoxy-D-galactose and 2-amino-2-deoxy-D-glucose). The basic units are usually *N*-acetylated and sometimes *N*-sulphated and the acidic units are sometimes *O*-sulphated. This results in the chains being strongly acidic, a factor recognized in their former names of 'acidic mucopolysaccharides'. The systematic name for these chains is 'glycosaminoglycans'.

A factor distinguishing glycoproteins from proteoglycans, as the names suggest, is the number of carbohydrate units per unit length (or molecular weight) of protein backbone, with protein being predominant in glycoprotein and carbohydrates predominating in proteoglycans. The term carbohydrate-protein complex is used to describe entities which contain protein and carbohydrates and which are linked by noncovalent (usually ionic) bonds. A full and definitive discussion of the nomenclature of glycosaminoglycans and glycoproteins has been published (Kennedy, 1979a).

ANIMAL GLYCOPROTEINS

Proteins are linear molecules which exist as strands that can be bent, twisted or folded, but a branch structure in which one protein strand is joined to another has never been found. The carbohydrate moiety in glycoproteins takes the form of glycosidically linked units which vary in size from mono- and di-saccharides to polysaccharides and are linked at various positions along the protein chain. No example of a glycoprotein existing as a block copolymer of alternating peptide and oligosaccharide sequences has been found in nature.

There are probably many more protein chains in nature which contain covalently bound carbohydrates than proteins which are devoid of carbohydrate. Glycoproteins are widely distributed in nature, being found distributed amongst higher animals, plants and microorganisms (Table 9.1). They contain widely differing carbohydrate contents (Fig. 9.3), and have variations in the size and shape of the carbohydrate side chains. The role of the carbohydrate moiety of glycoproteins is not fully understood. As will be shown later, the terminal carbohydrate units of the blood-group substance glycoproteins are responsible for their specificity, whilst the degradation of 5-acetamido-3,5-dideoxy-D-*glycero*-D-*galacto*-2-nonulopyranosylonic acid residues in the antifreeze glycoproteins, with neuraminidase, removes their antifreeze properties. There are a number of mysteries, one of which is the enzymic activities of the ribonucleases previously referred to as A and B and now classified as ribonuclease (pancreatic) (EC 3.1.27.5). Ribonuclease B, a glycoprotein, exhibits exactly the same activity as ribonuclease A which has no sugar component, an example of the fact that carbohydrate does not necessarily affect the biological (enzymic) activity of a molecule. A number of proposals have been put forward to explain the presence of carbohydrate, ranging from providing components for intercellular communication and handles for transportation of proteins from one part to another of a cell, to rendering proteins resistant to enzymatic degradation but allowing them to be recognized by certain receptor sites (Ryle *et al.*, 1970).

Table 9.1 Distribution and function of some glycoproteins

Presumed function	Name
Enzyme	Acetyl cholinesterase
	Bromelain
	Ficin
	Porcine α-amylase
	Porcine deoxyribonuclease
	Ribonuclease (pancreatic)
	Taka-amylase (a fungal α-amylase)
	Yeast invertase (a β-D-fructofuranosidase)
Food reserve	Casein
	Endosperm glycoproteins
	Ovalbumin
	Pollen allergens
Hormone	Erythropoietin
	Follicle-stimulating hormone
	Human chorionic gonadotrophin
	Luteinizing hormone
	Thyroglobulin
Plasma and body fluids	α-, β-, and γ-Glycoproteins
Protective	Fibrinogen
	Immunoglobulin
	Interferon
	Mucins
Structural	Bacterial cell wall
	Collagen
	Extensin (plant cell wall)
Toxin	Fungal phytotoxins
	Ricin
Transport	Ceruloplasmin
	Haptoglobin
	Transferrin
Unknown	Avidin (egg white)
	Blood-group substances

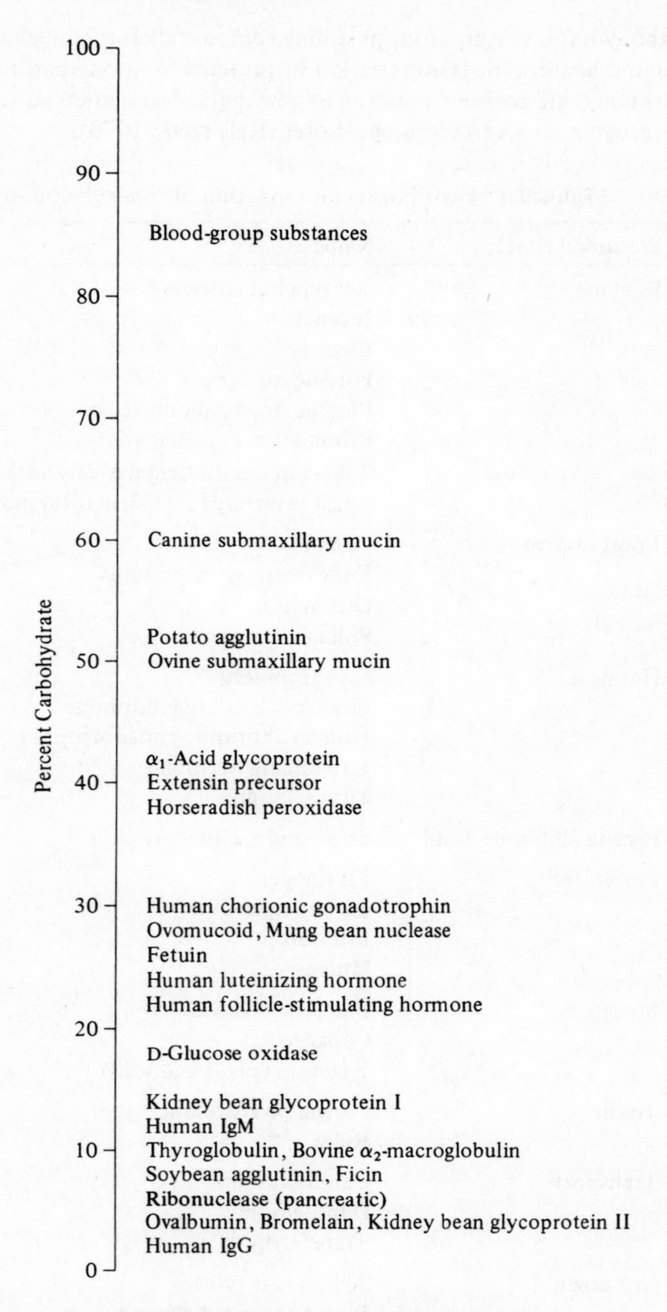

Fig. 9.3 – Carbohydrate content of some glycoproteins.

As many of the glycoproteins contain only a small proportion of carbohydrate, structural investigations are facilitated by the use of proteolytic enzymes (for example, pronase EC 3.4.24.4) to give glycopeptides which contain a small number of aminoacid residues to which are attached the intact carbohydrate chains. These glycopeptides may be analysed, not without problems, by the classical methods of periodate oxidation and methylation, or by sequential enzyme hydrolysis methods (see Chapter 4) to characterise the component monosaccharide residues and ideally give a single aminoacid residue linked to a monosaccharide unit. Only two types of linkage have been found to occur, namely *O*-glycosidic linkages to L-serine, L-threonine, hydroxy-L-proline and hydroxy-L-lysine and *N*-glycosylamide linkages to L-asparagine. Only five monosaccharide units have been shown to be involved in these linkages, L-arabinose, D-xylose, D-galactose, 2-acetamido-2-deoxy-D-glucose and 2-acetamido-2-deoxy-D-galactose. By way of example, a number of glycoproteins will be discussed to show the structure of their carbohydrate moieties since no classification can readily be made in terms of structural features.

Hormonal glycoproteins

This group of glycoproteins includes follicle-stimulating hormone (FSH), luteinizing hormone (LH), human chorionic gonadotrophin (hCG), human menopausal gonadotrophin (hMP), pregnant-mare serum gonadotrophin, and thyroid-stimulating hormone (TSH). The determination of primary structures of the carbohydrate moieties of these glycoproteins (particularly the human form) is still in its infancy due to the problems of isolating quantities of a pure hormone from closely similar (in chemical and physical nature) hormones and other macromolecules, including glycoproteins present in the media or origin of the sample. The purification of a particular hormone involves a complex series of purification stages in which advantage is taken of the properties peculiar to the hormone molecule, namely acidic and basic groups from certain aminoacid residues and acidic groups of 5-acetamido-3,5-dideoxy-D-*glycero*-D-*galacto*-nonulopyranonic acid. It is essential to monitor purification stages by determination of biological activity because no chemical (strictly abiological) method is hormone specific, and loss of a few residues can remove all biological activity; thus conditions which do not deviate markedly from the physiological have to be used. As the purification proceeds, it becomes more difficult as the remaining impurities become predominantly more and more those that are most similar to the hormone. Microheterogeneity complicates the separations relying on molecular change due to variation in, for example, the 5-amino-3,5-dideoxy-D-*glycero*-D-*galacto*-2-nonulopyranonic acid content, particularly as it now appears that some carbohydrate and aminoacid residues, particularly in certain terminal positions, are not essential for hormonal activity.

An example of a typical glycoprotein purification process is that applied to human follicle-stimulating hormone which involves a number of chromatographic separations on calcium phosphate, ion-exchange resins and gel filtration media but which has achieved a purification of 5000 fold (Butt *et al*., 1972). Immunoadsorption methods of separation are showing great promise as the purification can theoretically take place in one stage particularly if the hormone can be isolated in very high purity initially in order to allow the specific antibody to be raised, although initial high purity is not an absolute requirement.

Immunological reactions of the hormonal glycoproteins have shown that considerable similarities exist between these hormones; for example, the antiserum to follicle-stimulating hormone cross-reacts with luteinizing hormone, human chorionic gonadotrophin and thyroid-stimulating hormone, indicating the presence of a number of common antigenic groups in these hormones as well as their specific antigenic groups. These hormones have been shown to contain subunits, which can be formed by the action of trypsin, 1 M propionic acid, 8M urea, or sodium dodecyl sulphate. According to tests carried out on human glycoproteins a pattern has emerged and it seems quite certain that the β-subunits are hormone specific whereas the α-subunits are interchangeable. Such a finding is in keeping with the similarities of the aminoacid sequences of the α-subunits but the unique characters of the β-subunits. Recombination of subunits can be effected by incubation together under physiological conditions, the resulting biological activity being greater than the sum of the biological activities of the separate subunits. It has also been found that the subunits of a particular hormone from various species are interchangeable as far as activity is concerned. More important, however, is the fact that hybrid molecules can be produced by combining subunits from different hormonal glycoproteins, the type of hormonal activity of the hybrid glycoprotein being designated by the activity in which the β-subunit used was originally involved.

The application of glycoside hydrolases to the intact human follicle-stimulating hormone for the purpose of carbohydrate sequence determination has proved disappointing due to the inhibition of release of the monosaccharide units by adjacent parts of the molecule, but it has identified the nonreducing terminal groups as 5-acetamido-3,5-dideoxy-D-*glycero*-D-*galacto*-2-nonulopyranosylonic acid which are adjacent to D-galactosyl residues. Methylation and identification of the hydrolysis products by gas-liquid chromatography amd mass spectrometry demonstrated that L-fucose groups occupy nonreducing terminal positions, D-galactosyl residues are linked in the 1- and 2-positions, the D-mannosyl residues are present as nonreducing terminal groups, 1,6-linked residues and 1,3,4-linked branch points, with the 2-acetamido-2-deoxy-D-glucosyl residues 1,6-linked, all the sugars being in the pyranose form.

The use of glycoside hydrolases and chemical methods in the analysis of human chorionic gonadotrophin has been more successful with some studies on the linkage and sequence of the carbohydrate units on the intact molecule.

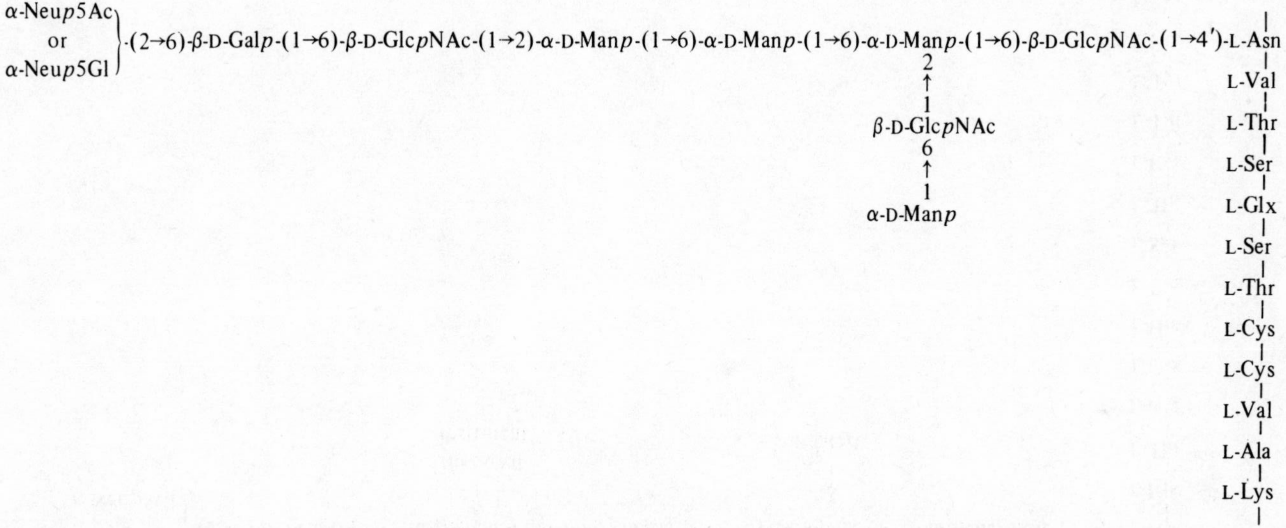

```
α-Neup5Ac⎫
    or   ⎬-(2→6)-β-D-Galp-(1→6)-β-D-GlcpNAc-(1→2)-α-D-Manp-(1→6)-α-D-Manp-(1→6)-α-D-Manp-(1→6)-β-D-GlcpNAc-(1→4')-L-Asn
α-Neup5Gl⎭                                                                  2                           |
                                                                            ↑                         L-Val
                                                                            1                           |
                                                                       β-D-GlcpNAc                    L-Thr
                                                                            6                           |
                                                                            ↑                         L-Ser
                                                                            1                           |
                                                                        α-D-Manp                      L-Glx
                                                                                                        |
                                                                                                      L-Ser
                                                                                                        |
                                                                                                      L-Thr
                                                                                                        |
                                                                                                      L-Cys
                                                                                                        |
                                                                                                      L-Cys
                                                                                                        |
                                                                                                      L-Val
                                                                                                        |
                                                                                                      L-Ala
                                                                                                        |
                                                                                                      L-Lys
                                                                                                        |
```

(102)

```
                                                                                                                              |
                                                                                                                            L-Val
                                                                                                                              |
                                                                                                                            L-Glx
α-Neup5Ac)                                                                                                                    |
   or     }-(2→6)-β-D-Galp-(1→2)-α-D-Manp-(1→6)-β-D-GlcpNAc-(1→6)-α-D-Manp-(1→6)-β-D-GlcpNAc-(1→4')-L-Asn
α-Neup5Gl)                          6                                  2                                     |
                                    ↑                                  ↑                                   L-His
                                 Unknown                               1                                     |
                                substituent                         β-D-Glcp                               L-Thr
                                                                                                             |
                                                                                                           L-Ala
                                                                                                             |
                                                                                                           L-Cys
                                                                                                             |
                                                                                                           L-His
                                                                                                             |
                                                                                                           L-Cys
                                                                                                             |
                                                                                                           L-Ser
                                                                                                             |
                                                                                                           L-Thr
                                                                                                             |
                                                                                                           L-Cys
                                                                                                             |
                                                                                                           L-Tyr
                                                                                                             |
                                                                                                           L-Tyr
                                                                                                             |
                                                                                                           L-His
                                                                                                             |
                                                                                                           L-Lys
                                                                                                             |
                                                                                                           L-Ser
                                                                                                             |
                                                                                                             OH
```

(103)

Studies on the pure α-subunit revealed the presence of two glycopeptide structures (102) and (103).

The structural requirements essential for the biological activity of the hormonal glycoproteins have been investigated, using a variety of techniques. Photo-oxidation has been used to modify selectively the side-chain functional groups of certain aminoacid residues, namely L-histidine, L-tryptophan and L-tyrosine, by destruction of the aromatic ring systems. Other chemical methods used to modify selectively the protein or carbohydrate residues include periodate oxidation, mild acid hydrolysis, methylation, and reaction with maleic anhydride, *N*-acetylimidazole, t-butyl azidoformate and formaldehyde. The use of highly purified enzymes, including trypsin, chymotrypsin, β-D-galactosidase, α-D-mannosidase, neuraminidase, etc., has also been made. It has been shown, by the application of such techniques, that terminal 5-acetamido-3,5-dideoxy-D-*glycero*-D-*galacto*-2-nonulopyranosylonic acid residues are essential for the transport, but not the biological activity of the hormones. The carbohydrate moiety, L-tryptophan residues and the guanido group of L-arginine residues have been shown to be essential for the hormonal activity of both follicle-stimulating hormone and luteinizing hormone.

Serum and plasma glycoproteins

Many of the 'proteins' in plasma and serum have now been identified as glycoproteins, only serum albumin and prealbumin having no carbohydrate moieties. The structure and functions of many of these have been reviewed, including the role of the carbohydrate in regulating the survival time of plasma glycoprotein in sera (Ashwell and Morell, 1974). Examples of these glycoproteins include α_1-acid glycoprotein (orosomucoid), transferrin, and fetuin; aspects of their carbohydrate structures are discussed by way of examples.

α_1-Acid glycoprotein is the serum glycoprotein having the highest carbohydrate content (*ca*. 40%) and much work has been carried out on its structure since variations in its content in serum occurs in some diseases. Carbohydrate analyses have shown the components to be 2-acetamido-2-deoxy-D-glucose (12.2 to 15.3%), D-mannose (4.7 to 6.5%), D-galactose (6.5 to 12.2%), L-fucose (0.7 to 1.5%) and 5-acetamido-3,5-dideoxy-D-*glycero*-D-*galacto*-2-nonulopyranonic acid (10.8 to 14.7%) in which L-fucosyl and 5-acetamido-3,5-dideoxy-D-*glycero*-D-*galacto*-2-nonulopyranosylonic acid groups are the nonreducing terminal groups linked principally to D-galactosyl residues. A number of possible structures have been postulated for the glycoprotein, after removal of the 5-acetamido-3,5-dideoxy-D-*glycero*-D-*galacto*-2-nonulopyranosylonic acid groups of which (104) is representative.

Transferrin is a glycoprotein which forms complexes with iron and is responsible for transporting iron from the storage form in tissues, especially in liver, to the metabolically functioning iron in haemoglobin. The carbohydrate chain has been characterised by periodate oxidation, methylation and glycoside hydrolase digestion to give a structure (105) for the oligosaccharide chains.

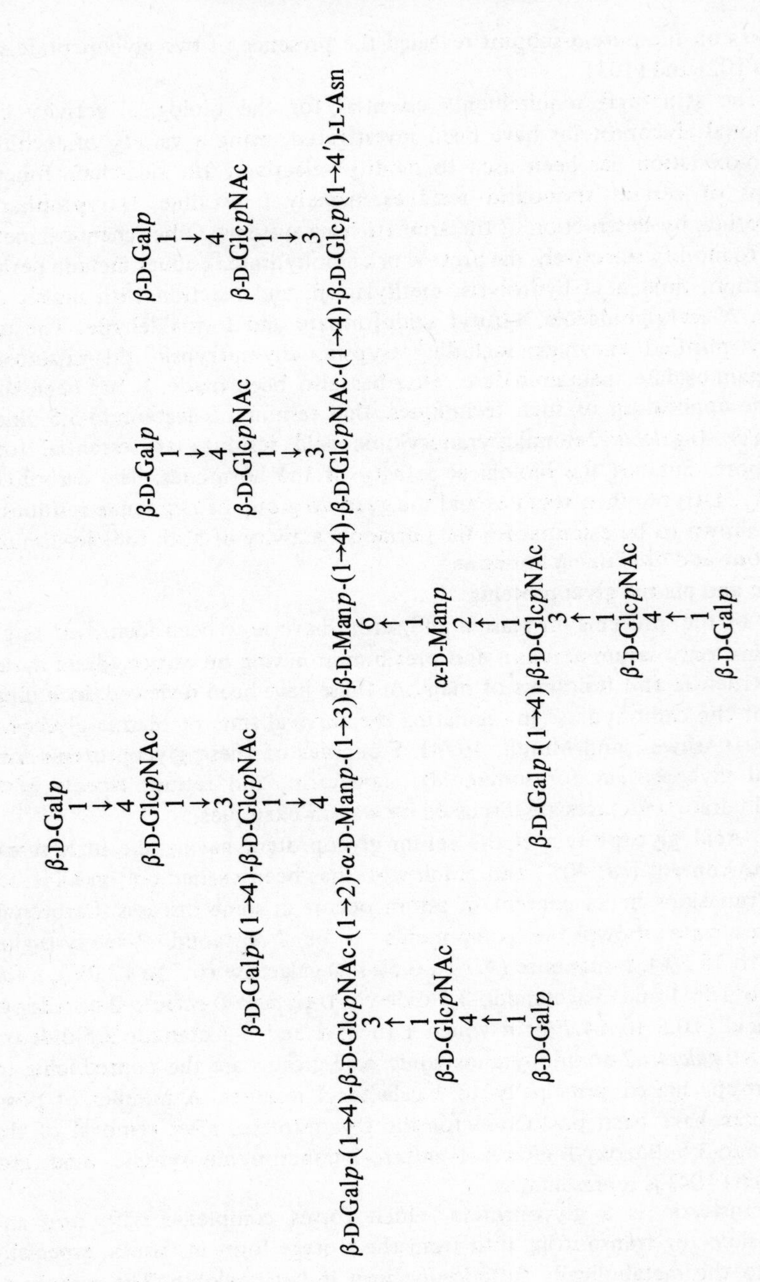

(104)

```
Neup5Ac-(2→6)-D-Gal-(1→3 or 4)-D-GlcNAc-(1→3)-D-Man-(1→2 or 4)-
        D-Man-(1→2 or 4)-D-Man-(1→2, 4 or 6)-D-Man-(1→3 or 4)-D-GlcNAc-(1→4′)-L-Asn
                                                  3
                                                  ↑
                                                  1
   Neup5Ac-(2→6)-D-Gal-(1→3 or 4)-D-GlcNAc-(1→3 or 4)-D-GlcNAc
```

(105)

```
 α-Neup5Ac         α-Neup5Ac        α-Neup5Ac
     2                 2                2
     ↓                 ↓                ↓
     3                 3                3
 β-D-Galp          β-D-Galp         β-D-Galp
     1                 1                1
     ↓                 ↓                ↓
     4                 4                4
 β-D-GlcpNAc       β-D-GlcpNAc      β-D-GlcpNAc
     1                 1                1
     ↓                 ↓                ↓
     2                 3 or 4           2, 4 or 6
α-D-Manp-(1→2 or 6)-α-D-Manp-(1→3)-β-D-Manp-(1→4)-β-D-GlcpNAc-(1→4)-β-D-GlcpNAc-(1→4′)-L-Asn
```

(106)

α-Neup5Ac-(2→3)-β-D-Galp-(1→3)-α-Neup5Ac-(2→6)-α-D-GalpNAc-(1→3′)-L-Ser (or L-Thr)

(107)

Fetuin is the principle glycoprotein in foetal calf serum, but becomes gradually replaced by the 'normal' glycoproteins of adult serum. Fetuin resembles α_1-acid glycoprotein in many of its properties and the structure of an oligosaccharide chain (106) has been shown to have similarities in its core structure (the carbohydrate units nearest the aminoacid unit). However, oligosaccharides have been found which are glycosidically linked to L-serine and L-threonine (107), some of which lack the internal 5-acetamido-3,5-dideoxy-D-*glycero*-D-*galacto*-2-nonulopyranosylonic acid residues (Spiro and Bhoyroo, 1974) showing the presence of a different type of oligosaccharide chain attached to the glycoprotein.

Immunoglobulins

Immunoglobulins are a group of serum glycoproteins that have antibody activity and are produced in response to stimuli by antigens. There are, at present, five classes of immunoglobulins known, which are designated IgG, IgA, IgM, IgD and IgE and those that have been examined all contain, in their monomeric form, the same fundamental structure of four polypeptides chains linked by interchain disulphide bonds. These chains are of two kinds, so-called light and heavy chains, with each monomer containing two of each (see Fig. 9.4). The light chains are of two kinds, kappa (κ) and lambda (λ), which differ in the allotypic specificity they carry. These are common to all immunoglobulin classes and an individual immunoglobulin monomer will contain two light chains, either two κ or two λ chains. The heavy chains are specific to, and determine the class of, the immunoglobulin. Each class of immunoglobulin has its own characteristic carbohydrate content which varies from about 22 monosaccharide units in IgG to about 82 units in monomeric IgM. Polymeric forms of immunoglobulin have been reported and include dimeric IgA and pentameric IgM. Macromolecular IgM is thought to contain five monomer units joined in the form of a ring from which radiate the five arms.

Normal IgG contains three glycopeptides representing two types of oligosaccharide units. Periodate oxidation analysis has shown that some of the 2-acetamido-2-deoxy-D-glucopyranosyl and D-mannopyranosyl residues are (1→3)-linked and other results based on mild acid hydrolysis suggest that L-fucosyl and 5-acetamido-3,5-dideoxy-D-*glycero*-D-*galacto*-2-nonulopyranosylonic acid residues are present at nonreducing terminal positions, linked to D-galactose, with D-mannopyranosyl residues substituted at O-3 or occurring at branch points, and some 2-acetamido-2-deoxy-D-glycosyl residues, those oxidized by periodate, linked *via* O-6 or in a terminal position. These results, plus information obtained from glycoside hydrolase degradation studies, led to the proposal of structure (108) for human IgG. Similar structures have been found for human IgE and IgA, all having various amounts of nonreducing terminal 5-acetamido-3,5-dideoxy-D-*glycero*-D-*galacto*-2-nonulopyranosylonic acid and D-galactosyl residues indicating various stages of completion of the outer chains or microheterogeneity. Bovine IgG contains a glycopeptide (109) which is identical to human IgG minus the 5-acetamido-3,5-dideoxy-D-*glycero*-D-*galacto*-2-nonulopyranosylonic acid residues.

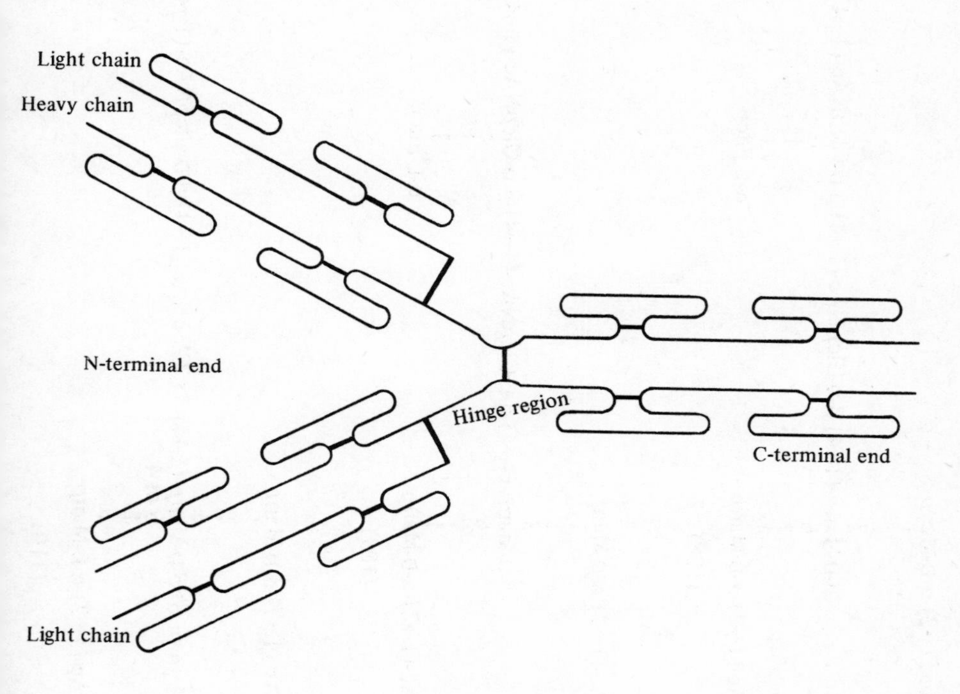

Fig. 9.4 – Schematic representation of an immunoglobin molecule.
(**—** represents a disulphide bond).

Characteristically different variations in glycopeptide structures to those described above have been found in human IgA myeloma glycopeptide 11A (110), which lacks L-fucose and has an extra 2-acetamido-2-deoxy-D-glucosyl residue on the branch point D-mannosyl residue. Glycopeptide B-3 from human myeloma IgG also has a difference in the core structure (111). The high D-mannose containing glycopeptide C-1 from human IgE has a very different core structure (112) containing alternating D-mannosyl and 2-acetamido-2-deoxy-D-glucosyl residues. One of the 2-acetamido-2-deoxy-D-glucosyl residues uniquely has the α-configuration. These changes in the core structure are to be distinguished from the microheterogeneity resulting from incomplete outer chains, in that changes in the core structure can reflect genetic differences (for example, myeloma).

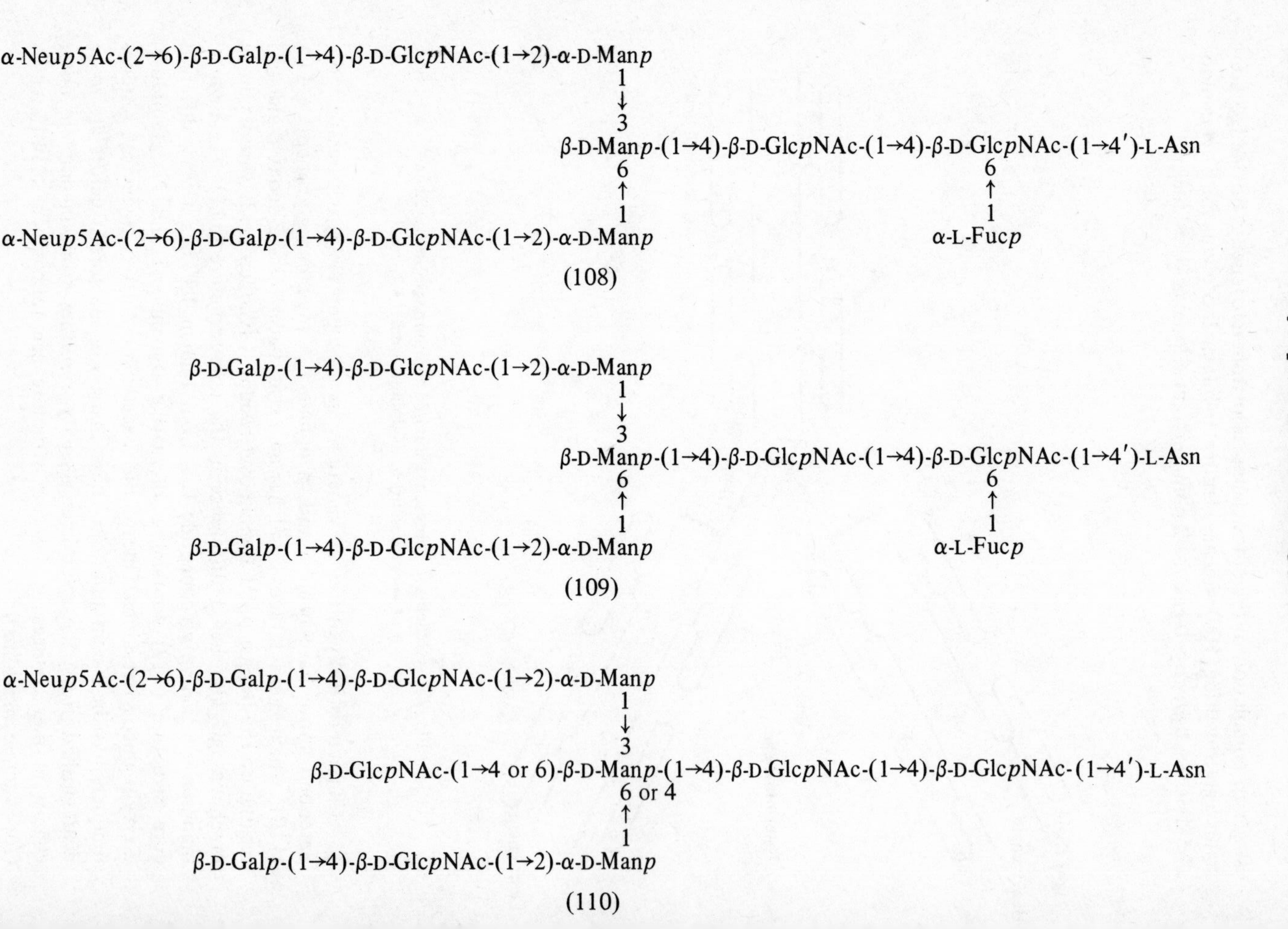
α-Neup5Ac-(2→6)-β-D-Galp-(1→4)-β-D-GlcpNAc-(1→2)-α-D-Manp
1
↓
3
β-D-Manp-(1→4)-β-D-GlcpNAc-(1→4)-β-D-GlcpNAc-(1→4′)-L-Asn
6
↑
1
α-Neup5Ac-(2→6)-β-D-Galp-(1→4)-β-D-GlcpNAc-(1→2)-α-D-Manp
6
↑
1
α-L-Fucp
(108)
β-D-Galp-(1→4)-β-D-GlcpNAc-(1→2)-α-D-Manp
1
↓
3
β-D-Manp-(1→4)-β-D-GlcpNAc-(1→4)-β-D-GlcpNAc-(1→4′)-L-Asn
6
↑
1
β-D-Galp-(1→4)-β-D-GlcpNAc-(1→2)-α-D-Manp
6
↑
1
α-L-Fucp
(109)
α-Neup5Ac-(2→6)-β-D-Galp-(1→4)-β-D-GlcpNAc-(1→2)-α-D-Manp
1
↓
3
β-D-GlcpNAc-(1→4 or 6)-β-D-Manp-(1→4)-β-D-GlcpNAc-(1→4)-β-D-GlcpNAc-(1→4′)-L-Asn
6 or 4
↑
1
β-D-Galp-(1→4)-β-D-GlcpNAc-(1→2)-α-D-Manp
(110)

```
α-Neup5Ac-(2→6)-β-D-Galp-(1→4)-β-D-GlcpNAc-(1→2)-α-D-Manp
                                                     1
                                                     ↓
                                                     3
                         β-D-GlcpNAc-(1→4)-β-D-Manp-(1→4)-β-D-GlcpNAc-(1→4)-β-D-GlcpNAc-(1→4′)-L-Asn
                                                     6                                6
                                                     ↑                                ↑
                                                     1                                1
α-Neup5Ac-(2→6)-β-D-Galp-(1→4)-β-D-GlcpNAc-(1→2)-α-D-Manp                          α-L-Fucp
```

(111)

```
β-D-Manp-(1→3)-β-D-Manp-(1→4)-α-D-GlcpNAc-(1→3)-β-D-Manp-(1→4)-β-D-GlcpNAc-(1→4′)-L-Asn
                    6               6
                    ↑               ↑
                    1               1
                α-D-Manp        α-D-Manp
                                    2
                                    ↑
                                    1
                                α-D-Manp
```

(112)

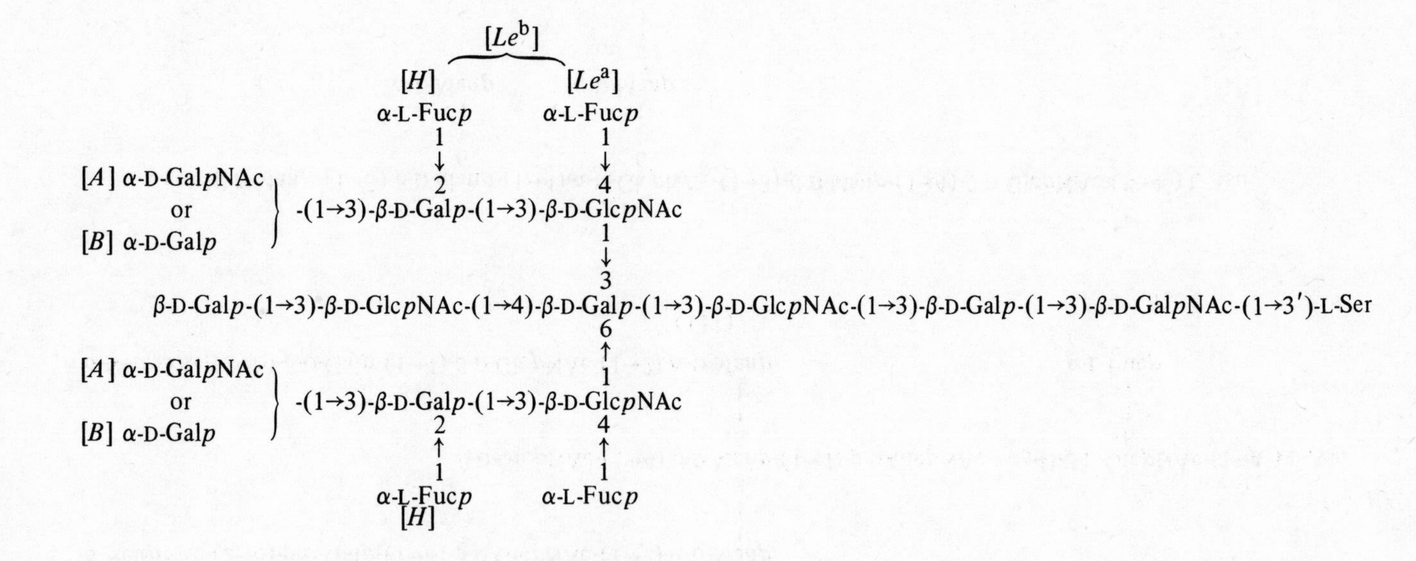

Fig. 9.5 – Composite structure of the carbohydrate chain of blood-group glycoproteins, indicating the residues which confer blood-group A, B, H(O), Le[a] and Le[b] specificity.

Blood-group substances

Blood-group substances are a group of structurally similar glycoproteins which occur on the surface of red blood cells and are responsible for the determination of particular blood-group types. The importance of the oligosaccharide moiety of the glycoprotein in this specificity has been demonstrated by enzymic removal of terminal 2-acetamido-2-deoxy-D-galactopyranosyl groups from type A erythrocytes or D-galactopyranosyl groups from type B erythrocytes which resulted in both being converted to type O erythrocytes. Confirmation of these results is found in the determination of specific enzymes (the relevant glycosyl-transferases involved in the blood-group substance biosynthesis) in the blood. Type A blood has an enzyme which will transfer 2-acetamido-2-deoxy-D-galactose to a core oligosaccharide whilst type B blood contains an enzyme catalysing transfer of D-galactose to the same core. Type O blood has neither enzyme.

The structure of the carbohydrate chains has been analysed using immunological methods to determine changes in serological activity on hydrolysis by acid or specific enzymes. This gives information on the specific terminal carbohydrate units responsible for the immunological specificity. Alkaline degradation has shown that linkage of the oligosaccharide chain to the protein molecule is *via* glycosidic linkages to L-serine or L-threonine. The structures of these glycoproteins with respect to their immunological properties have been reviewed (Hakomori and Kobata, 1974), and a composite structure has been proposed to indicate the residues which confer blood-group specificity for A, B, H(O), Le^a and Le^b (Fig. 9.5).

Involvement in diseases

Since glycoproteins are probably involved in coding and decoding of the intercellular messages required for the correct movement and assembly of cells that will eventually form the mature central nervous system, a genetic defect in their biosynthesis or regulation could account for many of the mental retardation syndromes that result from developmental impairment. However, this field remains relatively unexplored and no conclusions have yet been made.

The mechanisms by which glycoproteins are catabolised is perhaps better understood, and, with the possible exception of the removal of 5-acetamido-3,5-dideoxy-D-*glycero*-D-*galacto*-2-nonulopyranosylonic acid residues, the carbohydrate units are degraded by the sequential action of a group of lysosomal glycoside hydrolases (see Table 9.2). As will be seen later (Chapter 10), some of these enzymes are also involved in glycolipid catabolism. In some of the lipidoses impaired glycoprotein catabolism is a secondary effect but in other cases, namely fucosidosis, α-D-mannosidosis and aspartylglucosaminuria, the glycoprotein-derived storage material due to a deficiency in the required catabolic enzyme can be considered to be a major cause of the observed severe mental retardation. The structures of the storage material in many cases has been determined (see reviews

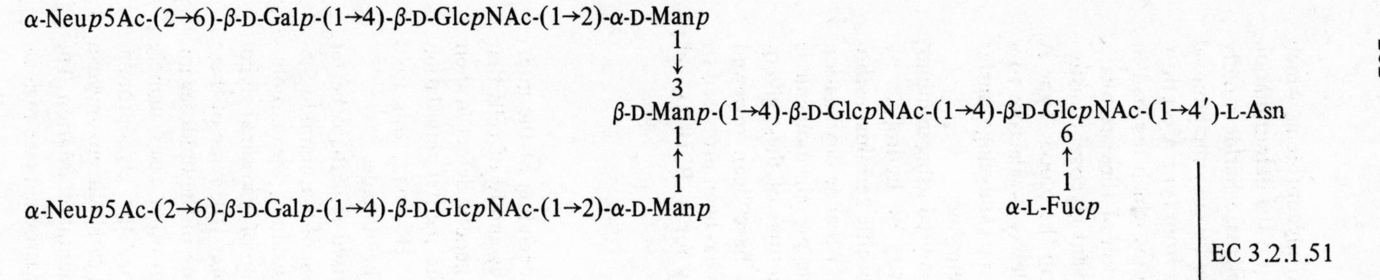
α-Neup5Ac-(2→6)-β-D-Galp-(1→4)-β-D-GlcpNAc-(1→2)-α-D-Manp
1
↓
3
β-D-Manp-(1→4)-β-D-GlcpNAc-(1→4)-β-D-GlcpNAc-(1→4')-L-Asn
1
↑
1
α-Neup5Ac-(2→6)-β-D-Galp-(1→4)-β-D-GlcpNAc-(1→2)-α-D-Manp
6
↑
1
α-L-Fucp
EC 3.2.1.51
?
EC 3.2.1.18

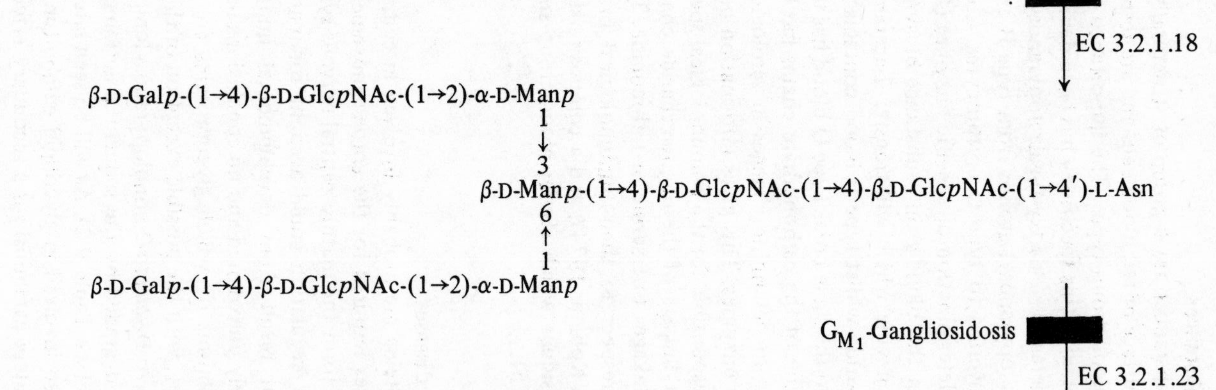
β-D-Galp-(1→4)-β-D-GlcpNAc-(1→2)-α-D-Manp
1
↓
3
β-D-Manp-(1→4)-β-D-GlcpNAc-(1→4)-β-D-GlcpNAc-(1→4')-L-Asn
6
↑
1
β-D-Galp-(1→4)-β-D-GlcpNAc-(1→2)-α-D-Manp
G_{M1}-Gangliosidosis
EC 3.2.1.23

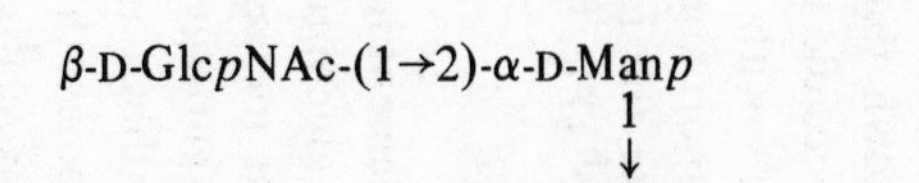
β-D-GlcpNAc-(1→2)-α-D-Manp
1
↓
3
β-D-Manp-(1→4)-β-D-GlcpNAc-(1→4)-β-D-GlcpNAc-(1→4')-L-Asn
6
↑
1

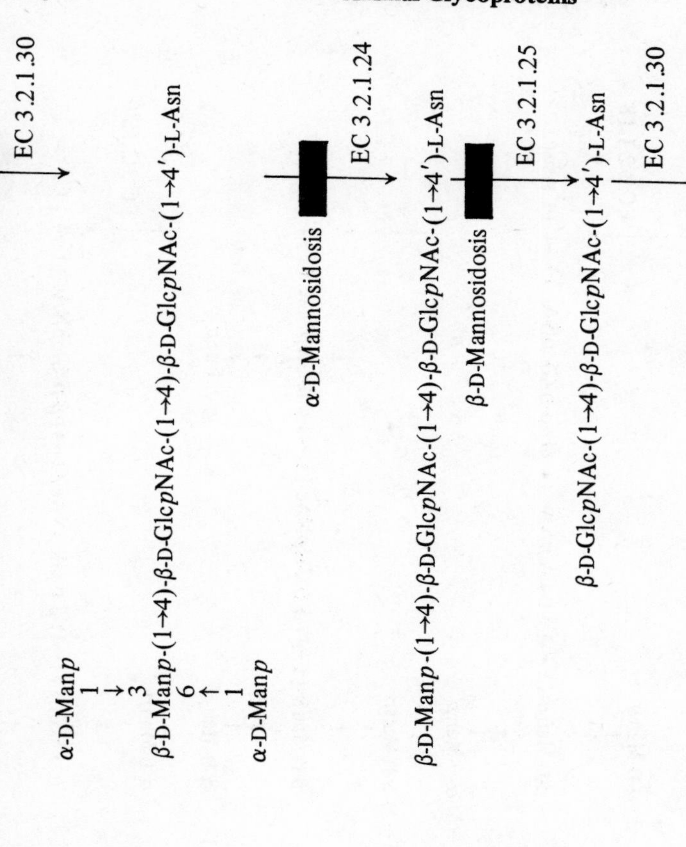

Fig. 9.6(a) – Proposed schematic catabolic pathway for immunoglobulin glycoproteins showing possible sites for blockages in known inborn errors of glycoprotein catabolism.

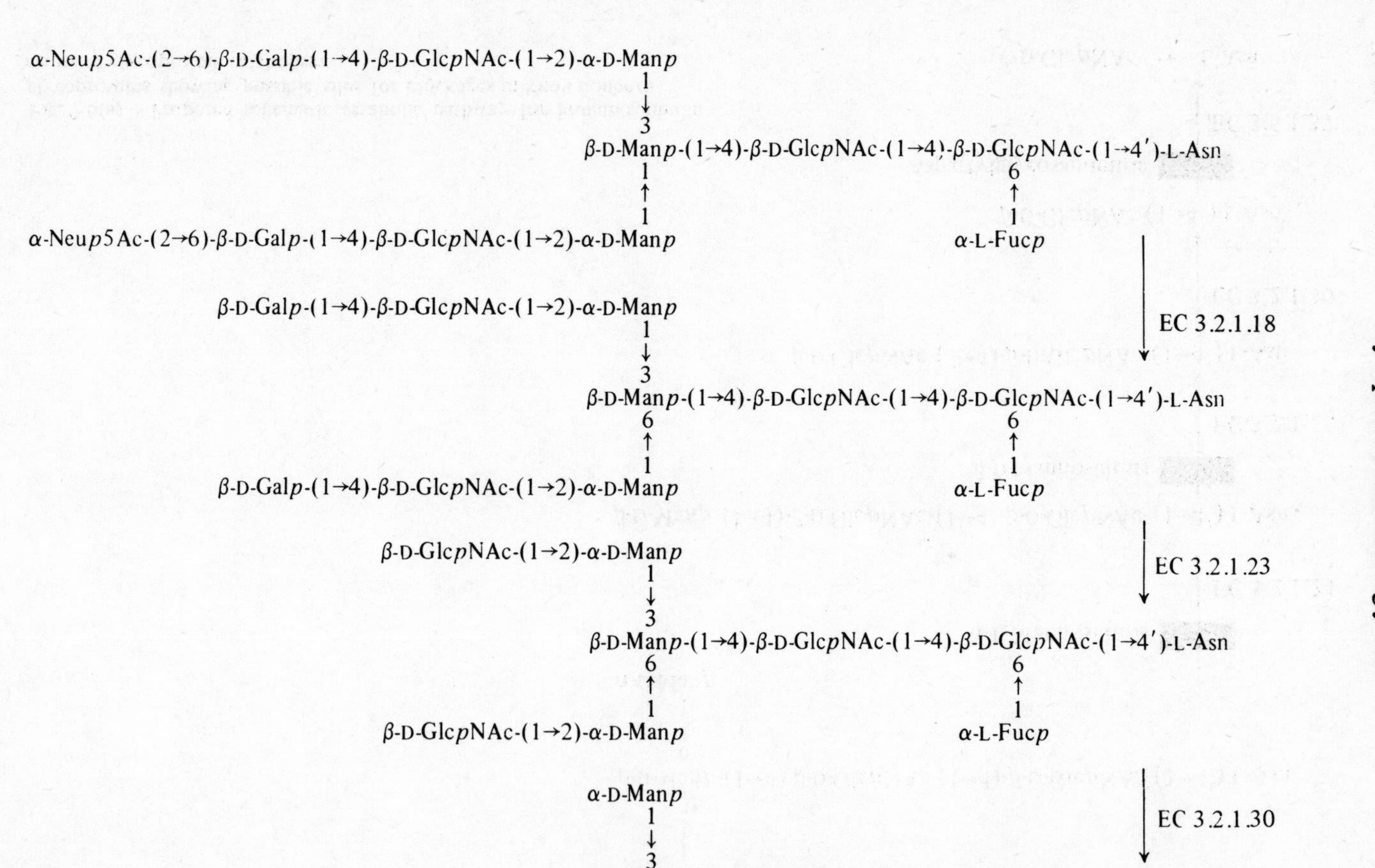
α-Neup5Ac-(2→6)-β-D-Galp-(1→4)-β-D-GlcpNAc-(1→2)-α-D-Manp
β-D-Manp-(1→4)-β-D-GlcpNAc-(1→4)-β-D-GlcpNAc-(1→4′)-L-Asn
α-Neup5Ac-(2→6)-β-D-Galp-(1→4)-β-D-GlcpNAc-(1→2)-α-D-Manp
α-L-Fucp
EC 3.2.1.18
β-D-Galp-(1→4)-β-D-GlcpNAc-(1→2)-α-D-Manp
β-D-Manp-(1→4)-β-D-GlcpNAc-(1→4)-β-D-GlcpNAc-(1→4′)-L-Asn
β-D-Galp-(1→4)-β-D-GlcpNAc-(1→2)-α-D-Manp
α-L-Fucp
EC 3.2.1.23
β-D-GlcpNAc-(1→2)-α-D-Manp
β-D-Manp-(1→4)-β-D-GlcpNAc-(1→4)-β-D-GlcpNAc-(1→4′)-L-Asn
β-D-GlcpNAc-(1→2)-α-D-Manp
α-L-Fucp
EC 3.2.1.30
α-D-Manp
β-D-Manp-(1→4)-β-D-GlcpNAc-(1→4)-β-D-GlcpNAc-(1→4′)-L-Asn

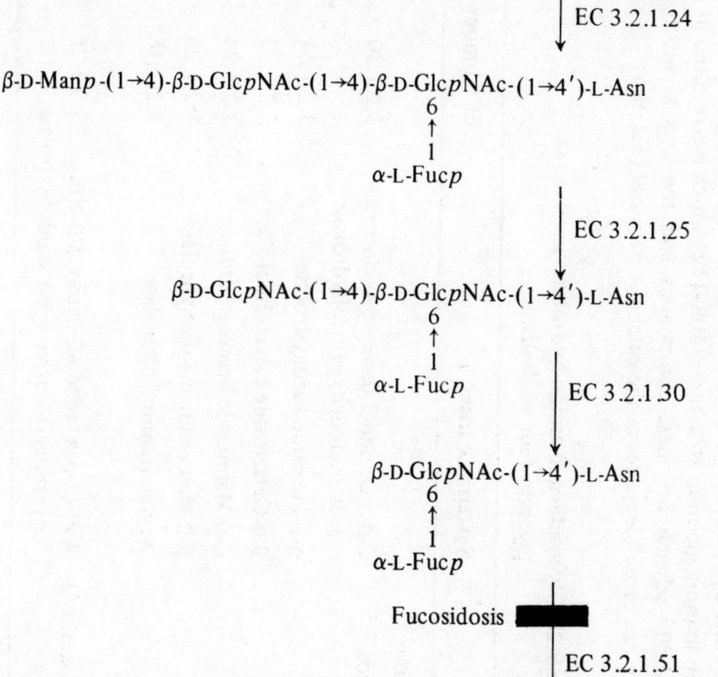

Fig. 9.6(b) – Proposed schematic catabolic pathway for immunoglobulin glycoproteins showing the site for blockage in fucosidosis, an inborn error of glycoprotein catabolism.

by Kornfeld and Kornfeld, 1976, and Dawson and Tsay, 1977), and was found to be related to the immunoglobulin structures (108–112) which has resulted in the proposed catabolic scheme for these compounds and the sites at which enzyme deficiencies can cause blockages of the catabolic pathway (Fig. 9.6).

Table 9.2 Glycoside- and amido-hydrolases involved in glycoprotein catabolism

Trivial name	Systematic name	EC number
Glycoside hydrolases		
β-*N*-Acetyl-D-glucosaminidase	2-Acetamido-2-deoxy-β-D-glucoside acetamidodeoxyglucohydrolase	3.2.1.30
α-L-Fucosidase	α-L-Fucoside fucohydrolase	3.2.1.51
β-D-Galactosidase	β-D-Galactoside galactohydrolase	3.2.1.23
α-D-Mannosidase	α-D-Mannoside mannohydrolase	3.2.1.24
β-D-Mannosidase	β-D-Mannoside mannohydrolase	3.2.1.25
Neuraminidase	Acylneuraminyl hydrolase	3.2.1.18
Amidohydrolase		
4-*N*-(2-β-D-Glucosaminyl)-L-asparaginase	4-*N*-(2-Acetamido-2-deoxy-β-D-glucopyranosyl)-L-asparagine amidohydrolase	3.5.1.26

PLANT AND ALGAL GLYCOPROTEINS

It was originally believed that glycoproteins were restricted to animal origins, but the isolation of an L-asparaginyl-oligosaccharide from soya bean agglutinin (Lis *et al*., 1964) was the first proof of their existence in plants. It is now well established that glycoproteins are widely distributed in plants and have similar functions to the animal glycoproteins and include enzymes, structural materials, etc. The carbohydrate residues in plant glycoproteins are similar to those found in animal glycoproteins with the exceptions that 5-acetamido-3,5-dideoxy-D-*glycero*-D-*galacto*-2-nonulopyranonic acid does not occur and 2-amino-2-deoxy-D-galactose has only rarely been found (in the leaf glycoproteins of *Cannabis* species) whilst D-xylose and L-arabinose, which are uncommon in animal glycoproteins, are found in many plant glycoproteins.

The number of well characterised plant glycoproteins is still very small (see Sharon and Lis, 1979) with only the glycoprotein lectins as distinct from protein lectins (see later p. 209) being investigated in sufficient detail to allow the structure of their carbohydrate moieties to be determined. However, a number of characteristic features have been discovered and these include a number of glycopeptide linkages which differ from those found in animal glycoproteins (see Fig. 9.7). Only N^4-(2-acetamido-2-deoxy-β-D-glucopyranosyl) hydrogen L-asparaginate is common to both plant and animal glycoproteins.

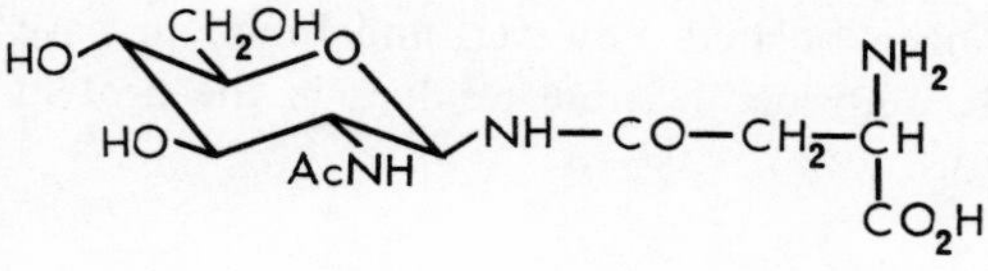

N^4-(2-Acetamido-2-deoxy-β-D-gluco-pyranosyl) hydrogen L-asparaginate

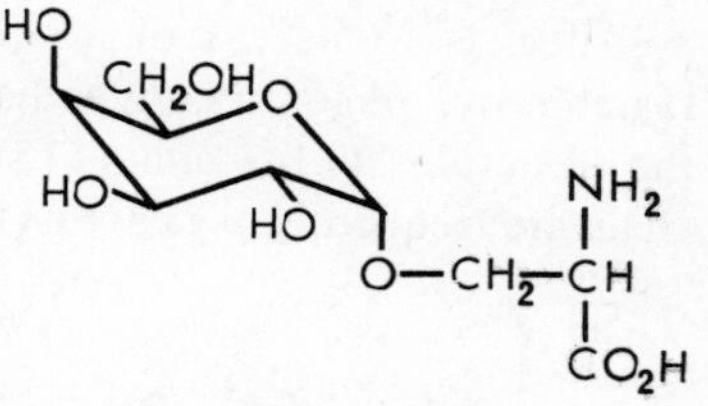

O^3-α-D-Galactopyranosyl-L-serine

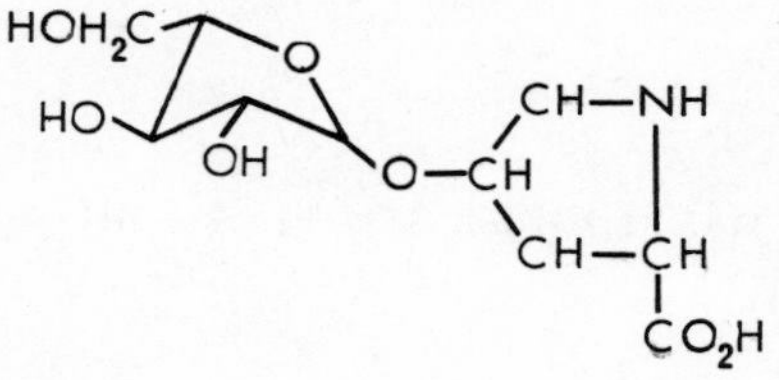

O^4-β-L-Arabinofuranosyl-hydroxy-L-proline

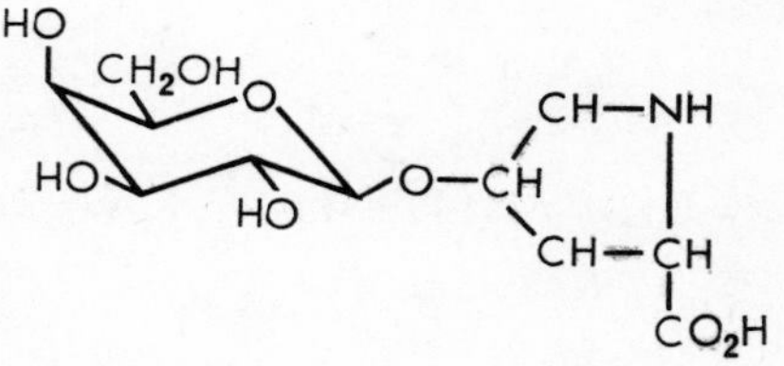

O^4-β-D-Galactopyranosyl-hydroxy-L-proline

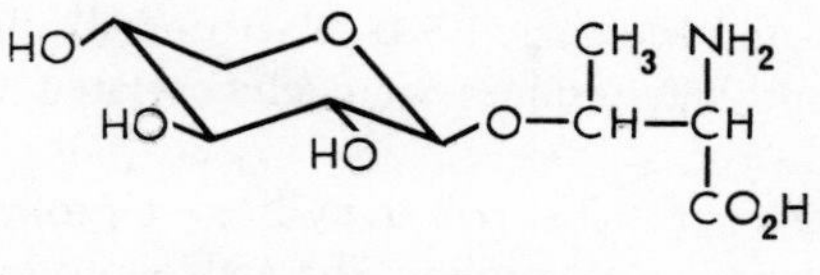

O^3-β-D-Xylopyranosyl-L-threonine

Fig. 9.7 – Glycopeptide linkages in plant glycoproteins

Bromelain, which was identified as a glycoprotein in 1967, has, since that time, become one of the better characterised plant glycoproteins and shown to contain the core structure (113) which is common to many of the animal glycoproteins. It has been reported (Fukuda *et al.*, 1976) that the D-mannopyranosyl residue in the core region is uniquely substituted at C-2 and C-6, unlike the animal glycoproteins in which the D-mannopyranosyl residue is substituted at C-3 and C-6.

Extensin is an insoluble glycoprotein found attached to the cell walls of many plants. It is believed to have a structural function analogous to that of collagen in mammals. The insolubility of extensin, thought to be a result of covalent crosslinking to cell-wall polysaccharides, has meant that degradative techniques are used in its isolation which results in limited information being available as to its carbohydrate moiety, molecular weight, etc. A number of L-arabinosyl oligosaccharides have been isolated, however, and found to have the structures (114) and (115). In addition, L-serine residues in the protein chain are frequently D-galactosylated.

β-D-Man*p*-(1→4)-β-D-Glc*p*NAc-(1→4)-β-D-Glc*p*NAc-(1→4′)-L-Asn
(113)

β-L-Ara*f*-(1→2)-β-L-Ara*f*-(1→2)-β-L-Ara*f*-(1→4′)-L-Hyp
(114)

β-L-Ara*f*-(1→3)-β-L-Ara*f*-(1→2)-β-L-Ara*f*-(1→2)-β-L-Ara*f*-(1→4′)-L-Hyp
(115)

A number of other hydroxy-L-proline-containing plant glycoproteins have been identified and isolated from other sources, including the leaves of broad beans and cannabis plants and as a component of rice bran and corn pericarp. Many of these are poorly characterised but that obtained from sandal (*Santalum album*) leaves contains 16% L-arabinose and less than 1% D-galactose. Alkaline hydrolysis showed that all hydroxy-L-proline residues were glycosylated by either L-arabinosyl di- or tri-saccharides.

A number of soluble extracellular glycoproteins, rich in hydroxy-L-proline, are found in the medium in which plant tissues are grown. These glycoproteins contain 80–95% carbohydrate which consists of D-galactose and L-arabinose in approximately equal proportions and are frequently (but incorrectly) referred to as the 'L-arabino-D-galacto-proteins'. The extracellular glycoprotein released by sycamore callus tissue contains L-arabino-D-galactan polysaccharides attached to about 50% of the hydroxy-L-proline residues. In addition, about 30% of the hydroxy-L-proline residues carry short L-arabinosyl oligosaccharides, with the trisaccharides predominating.

L-Arabino-D-galacto-glycoproteins have also been isolated from aqueous extracts of the seeds, leaves, stem and fruits of a variety of plants, a number of which exhibit lectin-like behaviour towards the β-D-glycosyl dyes (such as 1,3,5-tri-[4-*O*-β-D-glucopyranosyloxy-phenylazo]-2,4,6-trihydroxybenzene (116)) and have been designated the β-lectins (Jermyn and Yeow, 1975).

(116)

Evidence exists to indicate that many major plant polysaccharides (see Chapter 8), including starch and cellulose, contain small amounts of covalently bound protein and should be considered as glycoproteins.

Hydroxy-L-proline-rich glycoproteins have been found in green algae. These algal glycoproteins contain O^4-β-L-arabinofuranosyl-hydroxy-L-proline and O^3-α-D-galactopyranosyl-L-serine residues similar to those in extensin and, in addition, O^4-β-D-galactosyl-hydroxy-L-proline residues which are found in the L-arabino-D-galacto-glycoproteins but not in extensin.

Lectins

The large group of plant lectins (previously known as phytohaemagglutinins), which have the ability to bind carbohydrates and carbohydrate-containing macromolecules in an analogous manner to the antigen–antibody interaction, contains many glycoproteins, the structures of some of which have been elucidated (Lis and Sharon, 1977). Some plant lectins, including concanavalin A, contain no carbohydrate and are classified as protein lectins.

Soya bean agglutinin was first isolated in purified form in 1952 but was not known to be a glycoprotein until 1964. Since then much work has been expended on determination of its structure. Like most other lectins, it is now prepared by affinity chromatography (see Chapter 13, p. 302) using an immobilised form of the monosaccharide for which it possesses a specific affinity (in this case 2-amino-2-deoxy-D-galactopyranose). The carbohydrate content of soya bean agglutinin of 6% consists of D-mannose and 2-acetamido-2-deoxy-D-glucose in the ratio of 3:1. The structure of the carbohydrate unit varies due to micro-

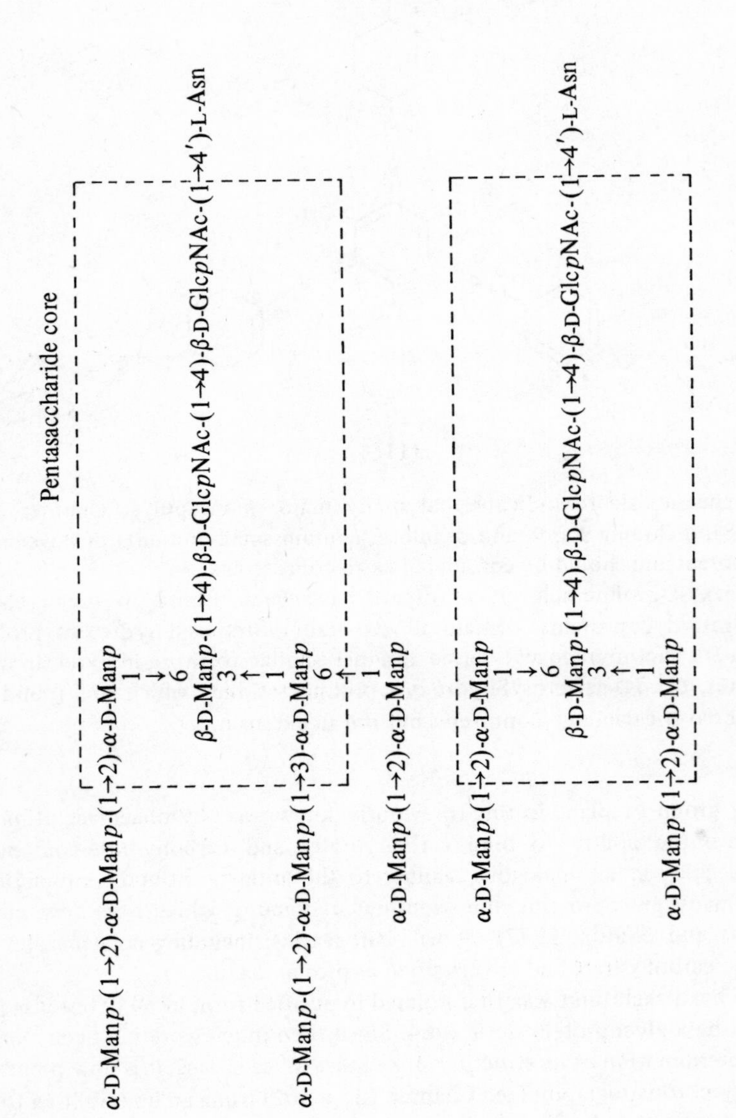

Fig. 9.8 – Structures of the carbohydrate moieties of soya bean agglutinin showing the common pentasaccharide core.

heterogeneity, but consists of a common pentasaccharide core to which additional D-mannosyl residues are attached, with a hepta- and undeca-saccharide having been identified (see Fig. 9.8). The structure of the pentasaccharide core region is common to that of many animal glycoproteins (for example, immunoglobulins, structures 108–110) indicating a common biosynthetic pathway for all L-asparagine-linked carbohydrate units in animals, plants and microorganisms (Sharon and Lis, 1979). Lima bean agglutinin has a similar carbohydrate composition and structure with the exception that L-fucose is attached to one of the 2-acetamido-2-deoxy-D-glucopyranosyl residues.

Potato lectin contains 47% L-arabinose and 3% D-galactose and has a structure in which some L-serine residues are substituted with a single α-D-galactopyranosyl group whilst all the hydroxy-L-proline residues carry L-arabinofuranosyl oligosaccharides, mainly the tri- and tetra-saccharides. The aminoacid sequence is such that sequences of three or more hydroxy-L-proline residues are present and form a rigid, open, left-handed helix with a region (of about 163 aminoacid residues) which contains no glycosyl residues. This region which contains the carbohydrate-binding site (specific for (1→4)-linked oligosaccharides of 2-acetamido-2-deoxy-β-D-glucopyranose) of the lectin contains large amounts of L-glycine and L-cysteine, the latter being present at almost every sixth aminoacid residue and all of which are cross-linked *via* disulphide bonds. This high L-glycine and disulphide linkage region is also found in wheat germ agglutinin which has the same carbohydrate specificity but which is a nonglycosylated protein and not a glycoprotein. A similar structure has been found for the glycoprotein lectin from jimson weed which contains hydroxy-L-proline (6.3%), L-cysteine (16.3%), 2-amino-2-deoxy-D-glucose (4.5%) and neutral sugars (28%, predominantly L-arabinose) and which is also specific for (1→4)-linked 2-acetamido-2-deoxy-β-D-glucopyranosyl oligosaccharides.

The use of immobilised lectins for affinity chromatography is discussed in Chapter 13.

PROTEOGLYCANS AND GLYCOSAMINOGLYCANS

Having discussed the many varied types of glycoproteins, a more compact range of carbohydrate–protein macromolecules are now discussed. In correct nomenclature, these are proteoglycans, and their carbohydrate components are glycosaminoglycans. These compounds are the only source of hexuronic acids in animals and occur in nearly all parts of mammalian bodies and to lesser extents in fish and bacteria. They are amongst the essential building blocks of the macromolecular framework of connective and other tissues. Hyaluronic acid appears to act, on account of its viscosity in solution, as a lubricant and shock absorbing gel in limb joints. The size and charged nature of the proteoglycans may allow them to operate as filtration media for the exclusion of foreign matter from the body. Heparin is unique among the glycosaminoglycans in that it possesses biological activity as a blood anticoagulant. It is therefore used therapeutically.

Nomenclature

In view of the confusion and lack of consistency in the nomenclature in this field, it is necessary to discuss a number of other relevant terms. The purpose of the introduction in 1938 of the term 'mucopolysaccharide' was to describe collectively 2-amino-2-deoxyhexose-containing polysaccharide materials of animal origin occurring either as free polysaccharides or as their protein derivatives. However, with the various subsequent discoveries of other types of carbohydrate-containing macromolecules, the term has come to be used in so many ways that it is now, in a sense, quite vague. Since the 'glycosaminoglycans' have always come within the 'mucopolysaccharide' category irrespective of the way in which that term has been used, they were described widely as 'acidic mucopolysaccharides' on account of their highly cationic nature. However, this nomenclature arose at a time when it was not realized that the glycosaminoglycans, as we call them today, are attached covalently to protein, and at a time when the polysaccharide was isolated with some aminoacids units attached. Thus, 'acidic mucopolysaccharide' means the 'glycosaminoglycan of a proteoglycan plus (sometimes) a few aminoacid units', where 'glycosaminoglycan' means purely the polysaccharide part of proteoglycan. The term 'acid mucopolysaccharide', used to a lesser extent, is synonymous with 'acidic mucopolysaccharide', whereas 'acidic polysaccharide' applies to any polysaccharide containing acidic groups. The term 'aminopolysaccharides' was also coined some time ago to describe collectively blood-group-specific substances, acidic mucopolysaccharides and chitin.

The general terms 'polysaccharide–protein' and 'mucopolysaccharide–protein complex' and the specific terms, for example, 'chondroitin sulphate–protein complex', were introduced when it was realized that the glycosaminoglycan is associated with protein, but when it (soon) became apparent that the association is stronger than that of a noncovalent complex, the general name 'protein–polysaccharide' and specific name 'chondromucoprotein' were employed to replace the three foregoing terms. These terms were all forerunners of 'proteoglycan' (general) and, for example, chondroitin 4-sulphate proteoglycan (specific).

It is unnecessary to discuss the historical aspects of the development of all these terms. It is sufficient to conclude that the terms 'acid mucopolysaccharide', 'acidic mucopolysaccharide', 'aminopolysaccharide', 'mucopolysaccharide', 'mucopolysaccharide-protein complex', 'mucoprotein', and 'protein–polysaccharide' are misleading, historical, and redundant, and that they can well be done without.

On account of the apparent regularity of the polysaccharide chains in proteoglycans and the early belief that the protein present in preparations of the polysaccharide parts represented impurity, greatest attention has been given to the glycosaminoglycan chains themselves rather than to proteoglycans as a whole. Thus the glycosaminoglycans have been named individually, but not so much according to their component monosaccharides and their simplified disaccharide repeating structures (these were often unknown at the times of original isolation), but according to trivial reasoning, for example, by naming after the source. 'Hyaluronic acid' is derived from 'hyaloid' (vitreous) plus 'uronic acid', 'chondroitin sulphate' from the Greek χονδροζ (chondros – cartilage) plus 'sulphate', 'dermatan sulphate' from the root 'derm' (skin) plus 'sulphate', 'heparin' from the Greek ηπαρ (hepar – liver), 'heparan sulphate' from heparin with an altered amount of sulphate, etc., and keratan sulphate from the Greek κεραζ (keras – horn) after its discovery in cornea.

In all, eight glycosaminoglycans of essentially different chemical structures have been identified. Through the times, these glycosaminoglycans have been individually named in a number of ways, as shown in Table 9.3. Most of these names are used currently, and so the reader will find it useful to have this table available when consulting the primary literature. Where the term chondroitin sulphate appears in the more recent literature, this can mean chondroitin 4-sulphate or chondroitin 6-sulphate or a mixture of the two. The terms keratan sulphate I and keratan sulphate II are sometimes used to denote keratan sulphate of corneal and skeletal origin, respectively, there being some differences between the two.

However, it has been accepted by many that, pending a complete systematisation of polysaccharide nomenclature by the various nomenclature committees, only one name for each should be regarded as the one which is up-to-date. As will be seen, some of the names in being altered have been systematised to include the terminal 'an'. (The term 'chondroitin sulphate' will be used when no distinction between the 4-sulphate and 6-sulphate isomers is given.) It is recommended that only these names should be used and that use of the others, which are misleading to the uninformed, should be discontinued completely forthwith. As will be seen later, the glycosaminoglycans contain acidic (anionic) groups which are capable of forming salts. The salt type is not usually indicated when naming a particular glycosaminoglycan since it is not necessarily known which cation, of those available throughout the extraction and purification processes, was finally assimilated by the glycosaminoglycan. However, where the salt form is known this may be specified, for example, sodium hyaluronate, potassium chondroitin 4-sulphate. Also included in Table 9.3 is the extension of the semi-systematic naming system which denotes the monosaccharide units involved in the polysaccharide, but it must be recognized that in certain instances these may represent simplifications.

For the reasons of the lesser emphasis invested in, and the general irregularity of, the sequences already discussed above, the protein parts of proteoglycans have not been named, nor for that matter have they been adequately chemically identified for names to be applied to them. This is reflected in the absence of names for the proteoglycans themselves other than those which indicate the type of glycosaminoglycan(s) involved, for example, dermatan sulphate proteoglycan and chondroitin 4-sulphate–keratan sulphate proteoglycan, but such of course

Table 9.3 Nomenclature of the glycosaminoglycans

Modern name	Original name	Other names	Semi-systematic name denoting actual monosaccharide units involved in structure[a]	Semi-systematic name denoting classes of monosaccharide units involved in structures	General semi-systematic name
Chondroitin	Chondroitin	–	Galactosaminoglucuronan		
Chondroitin 4-sulphate	Chondroitin sulphate A[b]	Chondroitin sulphate	Galactosaminoglucuronan		
Chondroitin 6-sulphate	Chondroitin sulphate C[b]	Chondroitin sulphate	Galactosaminoglucuronan		
Chondroitin 6-sulphate	Chondroitin sulphate D[c]	–	Galactosaminoglucuronan		
Dermatan sulphate[d]	Chondroitin sulphate B[b]	Chondroitin sulphate Dermatan[e] β-Heparin	Galactosaminoiduronan		
Heparin	Heparin	–	Glucosaminoglucurono-iduronan	Glycosaminoglycuronan	Glycosaminoglycan
Heparan sulphate	Heparitin sulphate	Acetylheparan sulphate Heparan[e] Heparin monosulphate Heparin monosulphuric acid Heparin sulphate Heparitin[e]	Glucosaminoglucurono-iduronan		
Hyaluronic acid	Hyaluronic acid	–	Glucosaminoglucuronan		
Keratan sulphate	Keratosulphate	Keratan[e]	Glucosaminogalactan	Glycosaminoglycan	

a As known so far.
b Known originally collectively as chondroitin sulphuric acid.
c Regarded by some as a separate glycosaminoglycan, but is really a super-sulphated chondroitin 6-sulphate and is therefore classified as such.
d Dermatan, the non-sulphated polysaccharide equivalent to chondroitin, is only known to be produced by laboratory desulphation of dermatan sulphate.
e Names used incorrectly irrespective of system used.

are nondescriptive of the protein. The alternative nomenclature, in the form proteochondroitin 4-sulphate, etc., has been used infrequently, but is not to be recommended. The attachment of only peptide to a glycosaminoglycan is designated by, for example, chondroitin 4-sulphate peptide.

Structures

The glycosaminoglycans can be distinguished by their compositions and primary structure, since the repeating disaccharide structures which have been discovered as general features of all of them are individually characteristic. Much of the earlier work on structure elucidation has been reviewed (Muir and Hardingham, 1975). Hyaluronic acid has the simplest of the glycosaminoglycan structures with a repeating unit composed of the monosaccharides D-glucuronic acid and 2-acetamido-2-deoxy-D-glucose linked such that there is a (1→3) linkage between the β-D-glucopyranosyluronic acid and the 2-acetamido-2-deoxy-D-glucopyranosyl residues, that is, the uronic acid has the β-D-anomeric configuration at C-1 from which it is linked glycosidically to position 3 of the 2-acetamido-2-deoxy-D-glucopyranosyl residue. Between the 2-acetamido-2-deoxy-D-glucopyranosyl and the β-D-glucopyranosyluronic acid residues the linkage is (1→4) which results in the repeating structure [→4)-*O*-(β-D-glucopyranosyluronic acid)-(1→3)-*O*-(2-acetamido-2-deoxy-β-D-glucopyranosyl)-(1→] (117). Chondroitin is the only other nonsulphated glycosaminoglycan which occurs naturally and is isomeric with hyaluronic acid. The repeating unit has the structure [→4)-*O*-(β-D-glucopyranosyluronic acid)-(1→3)-*O*-(2-acetamido-2-deoxy-β-D-galactopyranosyl)-(1→] (118). This makes the only difference between hyaluronic acid and chondroitin the orientation of one hydroxyl group on every other monosaccharide unit along the chain.

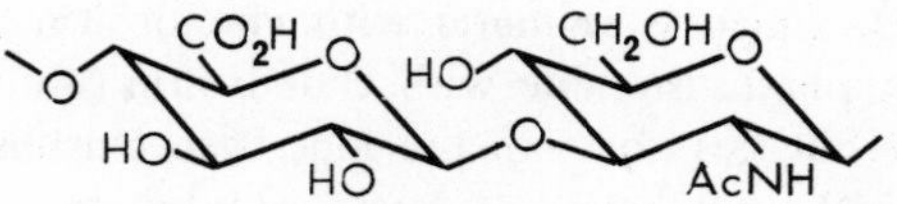

(117)

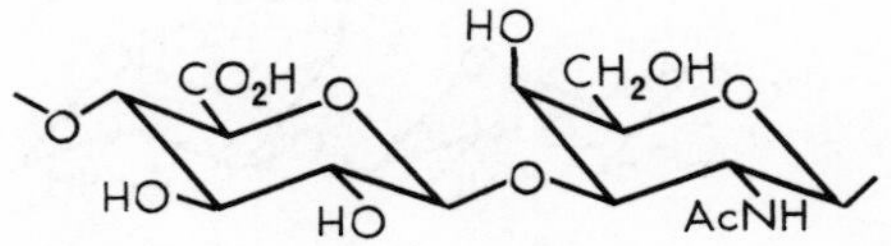

(118)

Chondroitin 4-sulphate and chondroitin 6-sulphate are varieties of chondroitin, the sulphate group being located on the 4- and 6-positions of the 2-acetamido-2-deoxy-D-galactosyl residues resulting in the repeating structures (119) and (120) respectively. Variants of chondroitin sulphate with sulphate contents greater than 1 mole per mole of disaccharide unit are known to occur but generally have no specific name. The so-called chondroitin sulphate D, obtained from shark cartilage, is, in fact, a chondroitin 6-sulphate containing, on average, more than one sulphate group per disaccharide repeating unit (sometimes referred to as 'oversulphation').

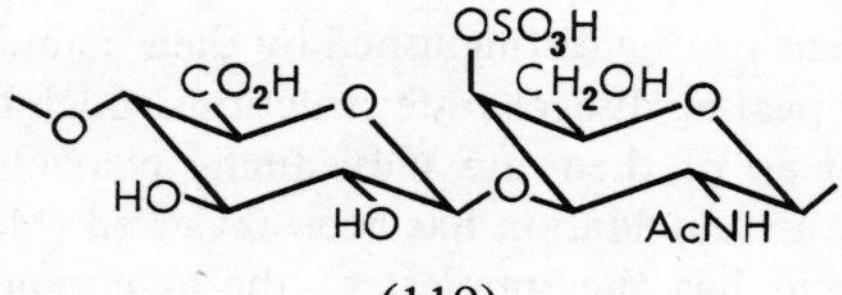

(119)

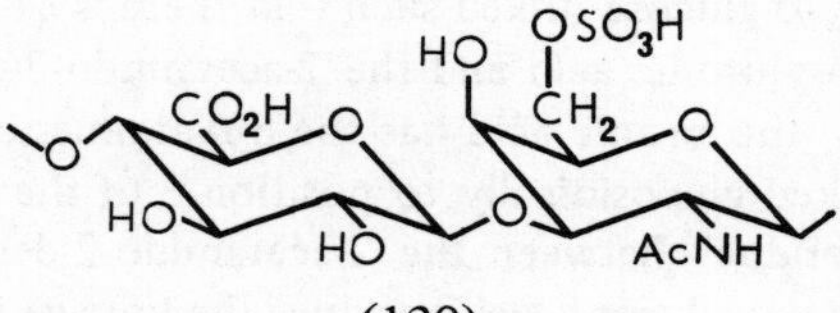

(120)

Dermatan sulphate is an isomer of chondroitin 4-sulphate with L-iduronic acid residues replacing D-glucuronic acid residues with the linkage positions and absolute orientations of the linkages remaining the same, resulting in a repeating structure (121) which differs from chondroitin 4-sulphate only in the orientation of the carboxyl group at C-5 on every other residue. Irregularities in the structure are known to exist, with one or two randomly spaced D-glucuronic acid residues replacing L-iduronic acid residues and the repeating disaccharide sometimes being di- or non-sulphated. Dermatan, the nonsulphated form of dermatan sulphate, which is isomeric with chondroitin and hyaluronic acid, and dermatan 6-sulphate, isomeric with chondroitin 6-sulphate, have not been detected in nature but this does not preclude their existence. Dermatan can be prepared by chemical desulphation of dermatan sulphate.

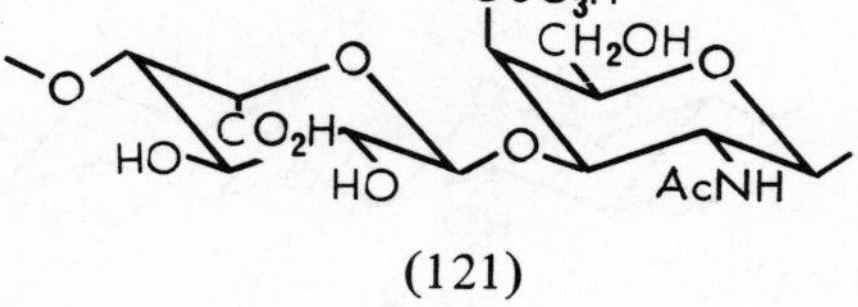

(121)

The repeating unit in heparin has been much more difficult to define and, in spite of extended efforts, it was not until the late 1960s that the location of the sulphate groups could be assigned with confidence. It was now recognized that the composition of heparin is a mixture of two disaccharide repeating units comprising L-iduronic acid 2-sulphate and 2-deoxy-2-sulphamido-D-glucose 6-sulphate (122) and D-glucuronic acid and 2-deoxy-2-sulphamido-D-glucose 6-sulphate (123). Although the monosaccharide residues are analogous to those in hyaluronic acid and dermatan sulphate, heparin is particularly different due to the anomeric configuration of the D-glucopyranosyl and 2-amido-2-deoxy-D-glucopyranosyl residues. The overall sulphate content is in the range 2–3 moles sulphate per mole disaccharide repeating unit. Heparan sulphate is not, as the name might suggest, a further sulphated version of heparin. The backbone carbohydrate structure of heparan sulphate is similar to that of heparin but the 2-amido-2-deoxy-D-glucopyranosyl residues are less frequently *O*-sulphated and may be *N*-acetylated.

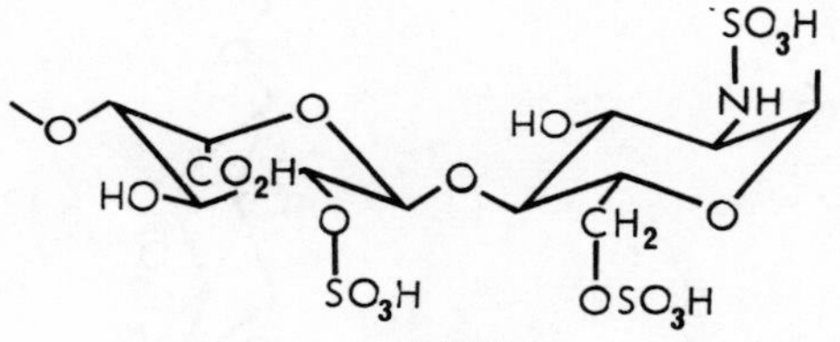

(122)

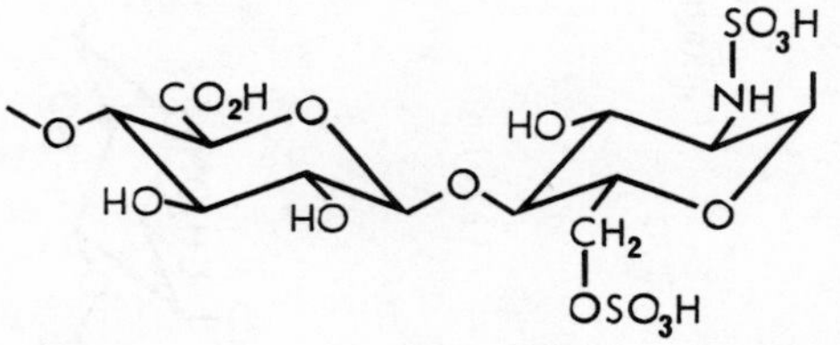

(123)

Keratan sulphate is unlike the other glycosaminoglycans in that it contains no (hex)uronic acid residues. It has a repeating disaccharide unit (124) consisting of D-galactopyranosyl and 2-acetamido-2-deoxy-D-glucopyranosyl 6-sulphate residues with the structure [→3)-*O*-β-D-galactopyranosyl-(1→4)-*O*-(2-acetamido-2-deoxy-β-D-glucopyranosyl 6-sulphate-(1→]. Variants of keratan sulphate are known in which other neutral sugars such as D-mannose, L-fucose and 5-acetamido-3,5-dideoxy-D-*glycero*-D-*galacto*-2-nonulopyranonic acid occur and in which there is more than 1 mole of sulphate per mole disaccharide repeating unit, additional sulphation occurring at C-6 of the neutral monosaccharide residues. Occasionally the neutral monosaccharides are non-sulphated.

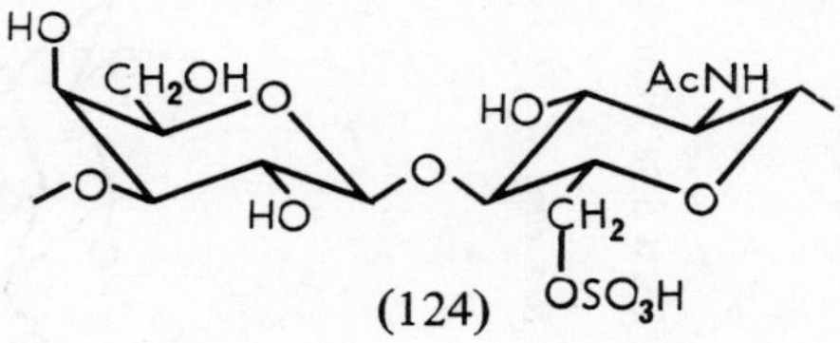

(124)

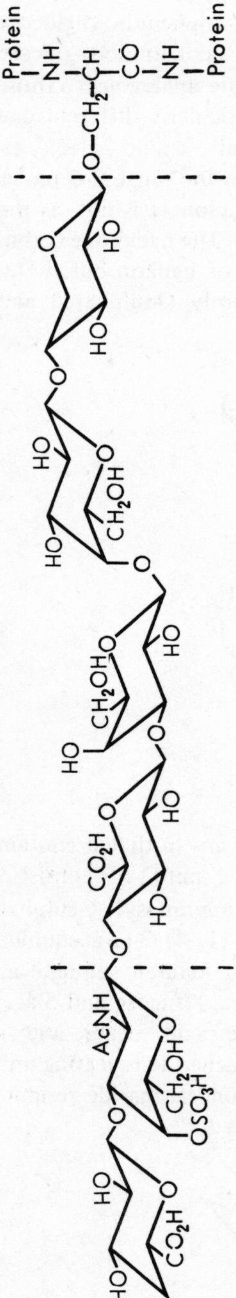

First disaccharide repeating unit

Linkage region

R = rest of glycosaminoglycan chain

(125)

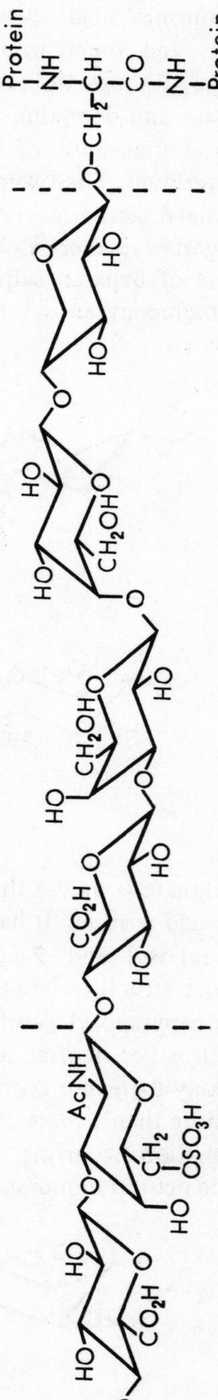

First disaccharide repeating unit

Linkage region

R = rest of glycosaminoglycan chain

(126)

Whereas the structures of the glycosaminoglycans are presented here largely as being based on repeating oligosaccharide units, it must be borne in mind that infrequent irregularities in the chains do exist, such irregularities being most apparent in the structures of heparin, heparan sulphate and keratan sulphate (for a full description see Kennedy, 1979a).

In the overall proteoglycan, the glycosaminoglycans are linked to the protein *via* a glycopeptide linkage which, in most cases, involves monosaccharide units different from those of the main polysaccharide chain. These linkage monosaccharides are glycosidically linked to one another with the reducing terminal residue being linked to a side chain of an aminoacid residue.

Chondroitin 4-sulphate-containing proteoglycan was afforded first attention in this area, and after exhaustive proteolysis, the only aminoacid remaining, in quantities large enough to account for all the glycopeptide linkages as one type, was L-serine. Partial acid hydrolysis yielded a number of oligosaccharides and glycopeptides which were fitted into an overall structure (125) in the glycopeptide linkage region and commencement of the repeating disaccharide unit. The actual linkage of chondroitin 6-sulphate to protein is identical to that for chondroitin 4-sulphate, and therefore the overall structure in the glycopeptide linkage region and commencement of the repeating disaccharide unit is structure (126).

Dermatan sulphate has also been found to conform to this general pattern in its glycopeptide linkage, including the D-glucopyranosyluronic acid residue (127). It is unclear whether heparin occurs as a proteoglycan, but all heparin samples contain aminoacids and again the glycopeptide linkage acid unit has the D-*gluco*-, not the L-*ido*-, configuration (128). The presence of D-galactose and D-xylose in preparations of heparan sulphate suggest that the glycopeptide linkage of this glycosaminoglycan may be analogous to the foregoing.

Skeletal keratan sulphate glycopeptide linkage involves 2-acetamido-2-deoxy-D-galactosyl (which does not occur in corneal keratan sulphate) residues linked to L-serine or L-threonine residues, but this may not be the only linkage type for this glycosaminoglycan.

Corneal keratan sulphate is apparently unique among the glycosaminoglycans in its glycopeptide linkage in that the glycopeptide structure somewhat represents that which occurs in many glycoproteins (129). The glycopeptide linkage has been characterised by the isolation from a partial-acid hydrolysate of N^4-(2-acetamido-2-deoxy-β-D-glucopyranosyl) hydrogen L-asparaginate, known sometimes as 2-acetamido-1-*N*-(L-aspart-4-oyl)-2-deoxy-β-D-glucopyranosylamine. It appears that neutral monosaccharide residues are adjacent to the 2-acetamido-2-deoxy-D-glucopyranosyl residues, and the linkage region conceivably has the structure (130).

Work on the structure of the glycopeptide linkages of chondroitin and hyaluronic acid has progressed least of all. That of chondroitin may be assumed to be identical with that for chondroitin 4-sulphate, etc., whereas that of hya-

R = rest of glycosaminoglycan chain

First disaccharide repeating unit

Linkage region

(127)

R = rest of glycosaminoglycan chain

First disaccharide repeating unit

Linkage region

(128)

luronic acid may involve D-glucose, L-arabinose, D-xylose and/or D-ribose and L-serine, but the various reports do not agree.

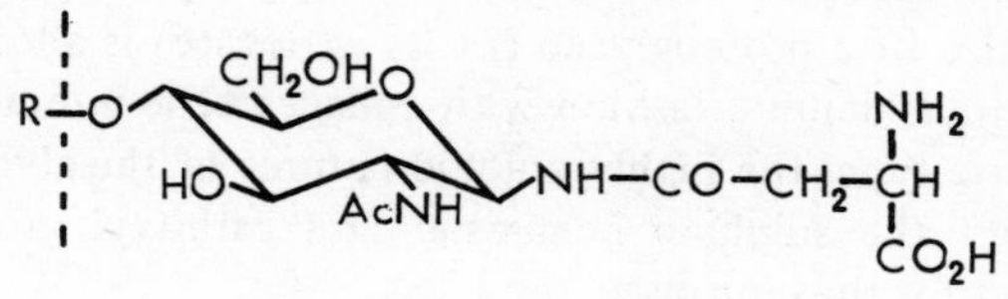

R–O–[2-acetamido-2-deoxy-glucopyranosyl: CH_2OH, HO, AcNH]–NH–CO–CH_2–CH(NH_2)(CO_2H)

(129)
R = glycosaminoglycan chain

$$\text{D-Gal}p\text{-}(1\rightarrow?)\text{-D-Man}p\text{-}(1\rightarrow?)\text{-D-Man}p\text{-}(1\rightarrow?)\text{-D-Glc}p\text{NAc-}(1\rightarrow4')\text{-L-Asn}$$

(L-Asn bonded to Protein above and Protein below)

(130)

Biological functions

One of the principal functions of connective tissues is to support and bind together the organs and bones – the functional parts of the body. Though the connective tissues assume different forms in different parts of the body, there is a fundamental similarity in the components. Considering the cells, the collagen fibres, cell membranes, the extracellular fibrils, the extracellular amorphous ground substance which surrounds the collagen and elastin fibres, cartilage and bone – all these very important materials contain proteoglycans, and therefore glycosaminoglycan chains, and thus these macromolecules must have a significant role. Since it is becoming increasingly apparent that connective tissues have many physiological functions in addition to being a supporting medium, they must allow transport of the many and varied 'moving' molecules. Therefore proteoglycans presumably have a profound influence on the processes of individual cells and tissues.

Considering the general location of greatest occurrence of glycosaminoglycans – connective tissue, it has been known for a long time that the tissue consists largely of collagen fibrils (see Ramachandran and Reddi, 1976, for the chemistry and biochemistry of collagen) and proteoglycans together with aqueous components. In the tissue, the protein chains of the proteoglycan chains and the collagen fibres lie side by side and the glycosaminoglycan chains of the proteoglycans interact with the collagen to give a non-covalent binding effect. The general representation of this situation (Fig. 9.9) as proposed some time ago still holds good. It must be realized, remembering the definition of a proteoglycan (many polysaccharide chains), that the size of a proteoglycan is very large; furthermore, some proteoglycans are able to form multiple aggregates of a very large size by specified interaction with a glycoprotein-like component. Thus, the overall size of a proteoglycan (or its aggregate) is adequate to permit it to interact in a multidentate fashion with collagen. The interactions are of an ionic or electrostatic type, the highly anionic natures of the glycosaminoglycan chains arising from the sulphate hemi-ester and carboxyl groups which are therefore important for this purpose.

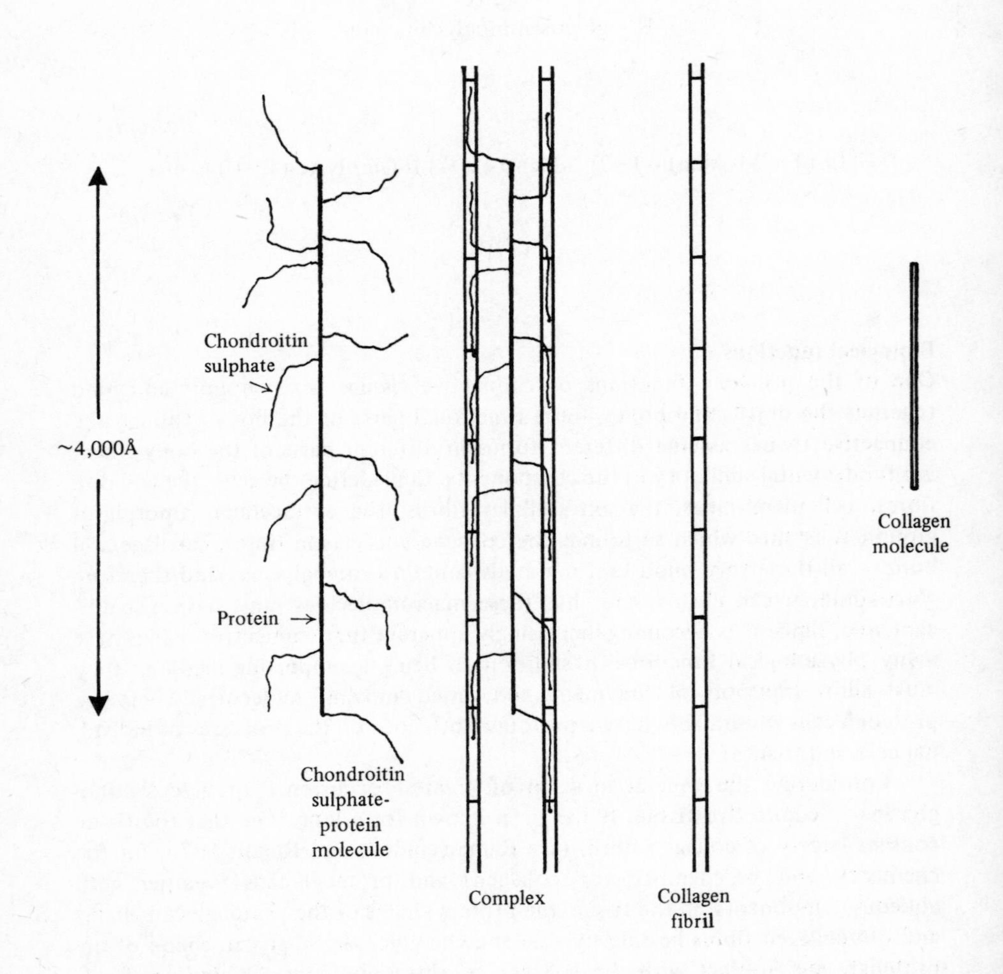

Fig. 9.9 – Schematic representation of interaction between collagen and chondroitin sulphate proteoglycan.

There are at least three binding sites for chondroitin 4-sulphate on a collagen molecule. The result of this interaction is to provide a three-dimensional network or matrix which resists disruption or separation (that is, holds the tissue together), which is suitable for the enclosure and location of an organ, which provides a buffer to external forces, but which is flexible and resilient. On account of the network, the matrix also permits the flow of important diffusable molecules and can be threaded through by arteries and veins. On this basis, it is clear that the presence of the glycosaminoglycan chains is of the utmost importance in maintaining the overall structure of tissues and the whole body and this is why they may be implicated in many of the processes of health, and, if they become faulty, in processes of disease.

In this picture of the proteoglycan and collagen fibre, it might be considered that the type of glycosaminoglycan is more important than the proteoglycan backbone, but it must be recalled that the protein dictates the spacing of the glycosaminoglycan chains and therefore is equally important. It has not yet been documented as to whether all the glycosaminoglycans are involved in this interactive way and it is unclear as to what differences there are between the function of, for example, chondroitin 4-sulphate and keratan sulphate in such a situation. Of course, the main proteoglycan chain axis to collagen fibre distance will vary since the acidic groups of the various glycosaminoglycan chains are of different strengths and exposed to different extents, not to mention any variation in the chain length of the glycosaminoglycan. Fibrillation is enhanced when connective tissue proteoglycan interacts with α-elastin, and glycosaminoglycans have been implicated in the regulation of collagen fibril diameters. Glycosaminoglycans and proteoglycans interact in fact with a whole variety of molecules *in vitro* and some of these interactions may represent functions *in vivo* (see Kennedy, 1979a).

Other justifiable suggestions for the functions of glycosaminoglycans and proteoglycans *in vivo* include wound healing, urine concentration in the kidney, storage and release of biogenic amines, maintenance of stable transport media, etc., contributions to elastogenesis, maintenance of the cornea in a transparent form, acting as cofactors in the formation of platelet clumping substance, acting as controlling agents for the hydration properties of cornea, acting as a molecular regulator to leukocyte lysosomal enzymes, maintenance of an orderly arrangement of fibrils of the corneal stroma, and metabolite control contributions to erythropoiesis.

In conclusion to the section, it is noteworthy that for the identification of the functions of carbohydrate and protein of many glycoproteins it has been possible to use their biological activity, which is frequently expressed at the nanogram or lower levels, as a sensing device. The search for specific roles of the carbohydrate and protein parts of proteoglycans is somewhat fraught with difficulty. Apart from heparin, no biological activity, in the accepted sense of the term, which can be monitored during diagnostic modification/manipulation of the macromolecule has been discovered. This of course does not mean that they do not possess such activities. Immunostudies may hold potential for sensing systems. Furthermore, there is the problem that many measurements can only be made outside of the natural environment and generally reflect only one of the multiple interactions and processes in which the macromolecule may be involved.

Involvement in diseases

Since proteoglycans occur so widely in the human body it is likely that the defects in many diseases will be related in some way to the proteoglycans. Furthermore, considering the large number of enzymes that must be involved in the biosynthesis and metabolism of the proteoglycans (see Chapter 5) and the interrelationship and control that exists between them, and between them and other tissue, etc., components, it is clear that an effect upon one of them may have far-reaching results in the reversible discontinuation of the maintenance of the production of and function of perfect tissue, etc. With this in mind, it is predictable that proteoglycan involvement is spread right across the disease types.

Whereas a considerable number of conditions has been reported to involve proteoglycan disorders in some way or other, the course of development of investigations has been greatly influenced, and dramatically limited to a narrow compass by various phenomena. First, much of the work reported is derived from studies carried out more than ten years ago when the proteoglycans were largely unrecognized as discrete intact molecules, attention therefore being focussed on the glycosaminoglycans. Secondly, attention has continued to be focussed on glycosaminoglycan rather than proteoglycan since the former is far more easily discernable in the presence of other tissue, etc., components. Thirdly, many of the clinical conditions originally examined are characterised by excessive urinary excretion of glycosaminoglycan peptides, that is, molecules from which most of the protein has been lost. Fourthly, there has been no significant systematic investigation of the involvement of the proteoglycans and/or their components – rather, discoveries of disorders have come from the recognition of an excess of material, often by histological methods. An excess may be many orders of magnitude different from the normal, whereas a decrease has a smaller maximum range and small differences go unnoticed more easily. Consequently, the diseases involving proteoglycan disorders reported to date are in fact almost exclusively characterised by an overproduction or accumulation of material. Thus the coverage of disorders by the literature relevant to this section is regrettably limited; in most instances only an end product of the disorder has been studied, usually on generally qualitative and quantitative bases without structural investigation. Since this end product has been determined as a glycosaminoglycan, it is only to be assumed that a proteoglycan disorder is involved. What is missing from an understanding of many of the disorders involving the glycosaminoglycan chain in some way is, in addition to structural information, the manner in which the whole proteoglycan is affected and at what stage(s) in the biosynthetic and degradative processes the primary disorder occurs. Of course, it may be that in some instances the proteoglycan disorder is a secondary effect. Most important, of course, is an understanding of the processes so that successful therapy can be devised.

A full review of the involvement and possible involvement of proteoglycans in various diseases is available (Kennedy, 1979a) but a brief discussion of genetic hyperglycosaminoglycanuria follows.

By far the greatest attention to the involvement of proteoglycans/glycosaminoglycans in disease has been given to a group of hereditary diseases which are characterized by the excessive urinary excretion of glycosaminoglycan material and have come to be known as the 'mucopolysaccharidoses'. The term 'mucopolysaccharidosis' is misleading, not only on account of the vague meaning of the word 'mucopolysaccharide', but also because it implies that glycosaminoglycans (acidic mucopolysaccharides) are the only storage material. This undoubtedly arises from the fact of the easily determined excessive glycosaminoglycan excretion, but it is apparent that the diseases involve complex storage disorders in which both glycosaminoglycan and other materials are affected and accumulate in tissues. It is difficult to give an expressive term to this group of diseases, but the term 'genetic hyperglycosaminoglycanuria' coined earlier (Kennedy, 1976) and selected for use here is acceptable although the urinary aspects are only some of the initial indications of far-reaching disorders. Nevertheless, there is now evidence that the primary defect in these hereditary disorders is faulty breakdown of the glycosaminoglycan(s); this therefore permits their distinction from other genetic diseases in which proteoglycan disorders and hyperglycosaminoglycanuria are secondary effects. However, it may emerge that there is also a disorder in proteoglycan protein metabolism in genetic hyperglycosaminoglycanuria since aminoacid differences as well as carbohydrate differences have been observed.

The classification (see later) of a disease as a case of genetic hyperglycosaminoglycanuria (or 'mucopolysaccharidosis') can be very arbitrary. Although some disease types are proven hereditary disorders, many diseases give rise to hyperglycosaminoglycanuria although not all of them are proven to be non-hereditary. Furthermore, there are undoubtedly many hereditary disorders which have not been investigated for (hyper)glycosaminoglycanuria. The term mucopolysaccharidosis has become something of a band wagon, with various workers seeking to add, to the classification of confirmed hereditary hyperglycosaminoglycanuria, other conditions characterised by hyperglycosaminoglycanuria but for which no genetic information is available. Variations within the classifications of individual cases has also prompted some to propose additional types. It must also be recognized that there may be several genetic disorders which exhibit a glycosaminoglycan disorder but which do not involve faulty glycosaminoglycan/proteoglycan shedding as a result of the primary enzyme defect.

In this chapter the term genetic hyperglycosaminoglycanuria is reserved for hereditary conditions where faulty proteoglycan metabolism is considered to be a primary defect. Nevertheless, whereas the meaning of this statement is immediately obvious and accepted, in so saying it is worthwhile questioning what is meant by the term primary defect. After all, whereas one may recognize that the primary defect is an enzyme deficiency (as in the case of genetic hyperglycosaminoglycanuria) or other alteration, such changes are really the result of changes in the amounts or structures of other molecules, the function of which impinges upon the molecule considered to be the primary defect. On this basis it is arguable that all primary defects can, since they involve enzymes, be traced back to protein disorders and thence back to nucleic acid sequences, etc. Furthermore, it is noticeable that anomalies are beginning to arise. For example, Sandhoff's disease is not normally considered as a case of genetic hyperglycosaminoglycanuria but it does generate a glycosaminoglycan disorder which is characterised by an accumulation of glycosaminoglycan in cultured fibroblasts and is due to a deficiency of a carbohydrate-directed enzyme.

As already indicated, the attention given to genetic hyperglycosaminoglycanuria is out of all proportion to the large range of clinical conditions known to involve proteoglycan disorders. One company has even gone to the extent of marketing urinary test papers for hyperglycosaminoglycanuria. The frequency of genetic hyperglycosaminoglycanuria is claimed to be relatively high amongst inherited diseases, although this attitude is to be questioned. The frequency of cases is of course related to the genetics; for example, the Hurler syndrome is autosomal recessive whereas Hunter syndrome is X-linked recessive, and this has a bearing on their relative frequency. The overall frequency in a Western population has been estimated at less than one per 10^5 live births. However, it is to be hoped that the consideration given these particular diseases is paving the way for successful and more rapid and direct investigation of other proteoglycan disorders, particularly with respect to identifying the primary defect. Regrettably, as yet there is little evidence of a trend to such an investigation of other diseases.

Considerable details are now known of the clinical features and the morphological, chemical and biological aspects of, and the glycosaminoglycans excreted by, the various types of genetic hyperglycosaminoglycanuria (see Table 9.4). Unfortunately, many of the conditions are manifested in childhood causing extensive deformity and mental retardation. Such characteristics of these diseases have been reviewed in detail (McKusick, 1972, and Kennedy, 1979a). The field is one in which new findings are frequently being reported.

In addition to examining the excreted end product of the disorders in genetic hyperglycosaminoglycanuria, the glycosaminoglycans, considerable attention has been given to a search for the primary defect. However, the examination of the likelihood that enzymic defects underlie the various types of genetic hyperglycosaminoglycanuria was hampered, since the enzymes in humans that normally degrade the glycosaminoglycans generally were largely, and perhaps remain, unidentified. It would seem that, in the light of certain evidence, the intracellular metabolism rather than the excretory mechanisms of the glycosaminoglycans is defective. There is now evidence for the defective degradation of glycosaminoglycans in genetic hyperglycosaminoglycanuria. Mutation, with respect to the enzymes involved in the biosynthesis and turnover of glycosaminoglycans, can result in deviation from normal control and may be responsible for these inherited disorders of connective tissue.

The six classified types of cases of genetic hyperglycosaminoglycanuria have been subclassified on the basis of clinical symptomatology, mode of genetic transmission, and nature of the glycosaminoglycan present in excess in the urine, and have been assigned type numbers (Table 9.4). Also given in the Table are relevant details of these diseases, together with details of some additional conditions which are classifiable in the same way. However, even acceptance of this somewhat arbitrary classification of types of hereditary glycosaminoglycanuria does not yield complete simplicity since hyperglycosaminoglycanuria is not always present. The best classification of these diseases would be based on the chemical aspects – the primary enzymic defect. Unfortunately, this is not yet possible since the specificities of even the enzymes which have been implicated still need to be investigated in further detail, and of course it is possible that other enzymes may be found to be involved.

Table 9.4 Classification of the better-establis

	Designation	Other names for condition	Clinical features
Type IH	Hurler syndrome	–	Early clouding of cornea grave manifestations, dea usually before age 10 ye
Type IS	Scheie syndrome	–	Stiff joints, cloudy corn aortic regurgitation, nor intelligence, ?normal life span
Type IH/S	Hurler-Scheie compound	–	Phenotype intermediate between Hurler and Sch
Type IIA	Hunter syndrome, severe	–	No clouding of cornea, milder course than in Ty IH but death usually bef age 15 years
Type IIB	Hunter syndrome, mild	–	Survival to 30s to 50s, f intelligence
Type IIIA	Sanfilippo syndrome A	Polydystrophic oligophrenia, Heparansulphaturia	Identical phenotype: mi somatic, severe central nervous system effects
Type IIIB	Sanfilippo syndrome B		
Type IIIC	Sanfilippo syndrome C		
Type IIID	Sanfilippo syndrome D		
Type IVA	Morquio-Ullrich syndrome	Morquio syndrome, Keratansulphaturia	Severe bone changes of distinctive type, cloudy cornea, aortic regurgitation
Type IVB			Mild symptoms, cloudy cornea, short stature
Type V	Vacant[b]		
Type VIA	Maroteaux-Lamy syndrome, classic form	Polydystrophic dwarfism	Severe osseous and corne change, normal intellect
Type VIB	Maroteaux-Lamy syndrome, mild form		
Type VII	β-D-Glucuronidase deficiency (more than one allelic form)	–	Hepatosplenomegaly, dysostosis multiplex, wh cell inclusions, mental retardation

a For details of other changes in levels of other glycosaminoglycans, see Kennedy (1979a).
b Due to modification of earlier classifications.
c The classification table given in McKusick (1972) records erroneously that dermatan sulfate is the principal excessive urinary glycosaminoglycan.

pes of genetic hyperglycosaminoglycanuria

Genetics	Principal excessive urinary glycosaminoglycans[a]	Enzyme deficient
omozygous for Hurler gene	Dermatan sulphate Heparan sulphate	α-L-Iduronidase (formerly called Hurler corrective factor)
omozygous for Scheie gene	Dermatan sulphate Heparan sulphate	α-L-Iduronidase
enetic compound of Hurler ıd Scheie genes	Dermatan sulphate Heparan sulphate	α-L-Iduronidase
emizygous for X-linked gene	Dermatan sulphate Heparan sulphate	L-Iduronic acid 2-sulphate sulphatase
emizygous for X-linked allele r mild form	Dermatan sulphate Heparan sulphate	L-Iduronic acid 2-sulphate sulphatase
omozygous for Sanfilippo gene	Heparan sulphate	Heparan sulphate sulphatase/ sulphamidase
omozygous for Sanfilippo gene (at different locus)	Heparan sulphate	α-*N*-Acetyl-D-glucosaminidase
	Heparan sulphate	D-Glucosamine acetyltransferase
	Heparan sulphate	*N*-Acetyl-D-glucosamine 6-sulphate sulphatase
omozygous for Morquio-Ullrich ne	Keratan sulphate Chondroitin 6-sulphate	*N*-Acetyl-D-galactosamine 6-sulphate sulphatase
	Keratan sulphate	β-D-Galactosidase
omozygous for M-L gene	Dermatan sulphate	*N*-Acetyl-D-galactosamine 4-sulphate sulphatase
omozygous for allele at M-L cus	Dermatan sulphate	
omozygous for mutant gene β-D-glucuronidase locus	Chondroitin 4-sulphate[c] Chondroitin 6-sulphate	β-D-Glucuronidase

CHAPTER 10

Glycolipids

Lipids, which are naturally occurring esters of long-chain fatty acids, rarely exist in a free state in an organism. They are more usually combined with protein or carbohydrate material, and as such they occur as part of the outer surface of cells. One class of lipids, known as glycolipids because of their content of a carbohydrate moiety, is a diffuse class of molecular species covering a wide range of structural types, which overlaps with other classes of lipids containing common structural features, making classification difficult and confusing. For example, some sulpholipids (sulphur-containing lipids) and phospholipids (phosphate-containing lipids) are also glycolipids, and different authors describe specific compounds in different classes, depending on the system used. For a full description of the many varied classes of lipid and the structural and biochemical properties involved the reader should refer to a general text on lipids (such as Gurr and James, 1980).

Glycolipids are widely distributed, but are usually minor components of the lipid mixture. They are associated with membrane proteinaceous material by noncovalent bonds, with the hydrophobic region buried in the outer membrane lipids and the carbohydrate region extending into the aqueous phase. The glycolipids can be removed from the membrane by use of neutral solvents, detergents, etc., in the same way as for other lipid material. Glycolipids are, therefore, a component of nearly all lipid extracts from tissue lipoproteins. The exact function of glycolipids is still the subject of much speculation. It has been suggested that glycolipids function as 'carriers' to transport carbohydrate moieties across cell membranes or as 'modifiers' which alter the physical properties of the membrane to provide the most suitable membrane for the purpose it exists. Glycolipids are known to be involved in the biosynthesis of glycoproteins and complex polysaccharides in that the lipid moiety acts as a carrier for the growing carbohydrate moiety which, on completion of its biosynthesis, is transferred to the protein (for glycoproteins) or carbohydrate (for polysaccharides) backbone (see Chapter 5). There is also good evidence to show that glycolipids exert a controlling mechanism in biosynthesis of proteoglycans (see Kennedy, 1979a) and in the inhibition of the biological activity of toxins and antiviral agents. It is known that a number of toxins, including cholera and tetanus toxins, bind to the carbohydrate moiety of glycolipids and that the specific action of such toxins can be inhibited by gangliosides, thereby providing a possible means of protection against the diseases caused by these toxins. Interferon interacts with gangliosides to provide or increase the antiviral activity of interferon. This has been demonstrated by the treatment with gangliosides of mouse cells which do not respond to interferon treatment, with the result that the cells become responsive (see Sharon, 1980).

The subsequent discussion on glycolipids uses divisions based on origin rather than chemical types. This reflects the structural complexities of glycolipids from various sources, with those from animal sources having much less diversity (mainly in the carbohydrate moiety) than those obtained from plant and microbial sources (with variations in all parts of the molecule).

For reviews of glycolipids and the background to work carried out prior to 1970 the reader should refer to McKibbin (1970), Kiss (1970), Cook and Stoddart (1973), Hakomori (1976), and Sweeley (1980), whilst up to date reviews of the occurrence, isolation, structures, chemistry and biochemistry of glycolipids from animal, plant, algal and microbial sources are available in an ongoing series of articles (see Various Authors, 1968 onwards). Nomenclature of glycolipids is governed by the IUPAC-IUB Rules for the Nomenclature of Lipids (Recommendation, 1976).

ANIMAL GLYCOLIPIDS

Most animal glycolipids are derivatives of long-chain bases related to sphingosine, the most common of which are shown in Table 10.1. Attached to these bases, *via* amide linkages, are long chain fatty acids (between 16 and 24 carbon atoms) which are usually completely or almost completely saturated (see Table 10.2). These simple sphingolipids are given the generic name ceramides (abbreviated to Cer for shorthand structural notation). Glycosphingolipids, in which carbohydrate moieties are attached *via* glycosidic linkages to the terminal hydroxyl group of the base, include neutral glycosylsphingoids (which contain no fatty acid) and glycosylceramides, acidic glycosylsphingolipids (gangliosides, which contain 5-acetamido-3,5-dideoxy-D-*glycero*-D-*galacto*-2-nonulopyranonic acid), and sulphoglycosylsphingolipids (non-recommended trivial name: sulphatides). The carbohydrate moiety usually contains between one and seven residues in a linear or branched structure, the most common components being D-glucose, D-galactose, L-fucose, 2-acetamido-2-deoxy-D-glucose and -D-galactose and 5-acetamido-3,5-dideoxy-D-*glycero*-D-*galacto*-2-nonulopyranonic acid, with D-galactose being the most common in plant glycolipids, D-glucose and D-mannose the most common in microbial glycolipids whilst 5-acetamido-3,5-dideoxy-D-*glycero*-D-*galacto*-2-nonulopyranonic acid is found mainly in gangliosides and globosides (see later in this chapter).

Table 10.1 Structure of the more common sphingoid bases[a]

Recommended name	Structure
Sphingosine (*trans*-4-sphingenine)	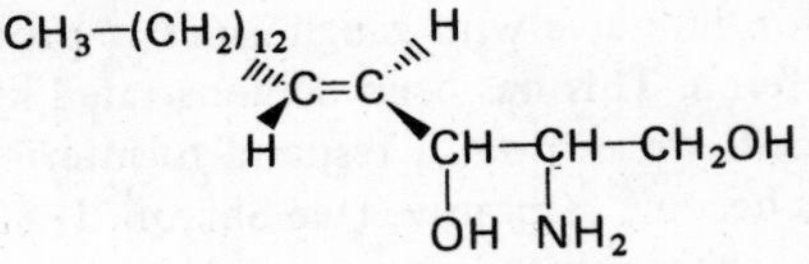 $CH_3{-}(CH_2)_{12}{-}CH{=}CH{-}CH(OH){-}CH(NH_2){-}CH_2OH$ (trans: H and H on opposite sides of C=C)
Sphinganine (dihydrosphingosine)	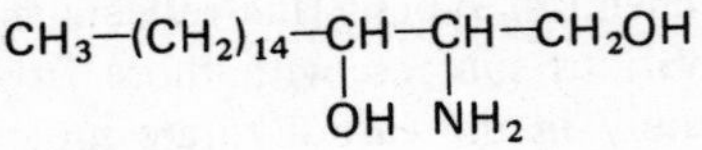 $CH_3{-}(CH_2)_{14}{-}CH(OH){-}CH(NH_2){-}CH_2OH$
4D-Hydroxysphinganine (phytosphingosine)	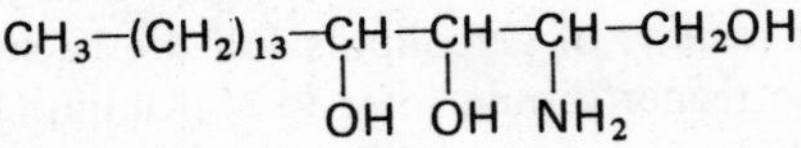 $CH_3{-}(CH_2)_{13}{-}CH(OH){-}CH(OH){-}CH(NH_2){-}CH_2OH$
4D-Hydroxy 8-sphingenine (dehydrophytosphingosine)	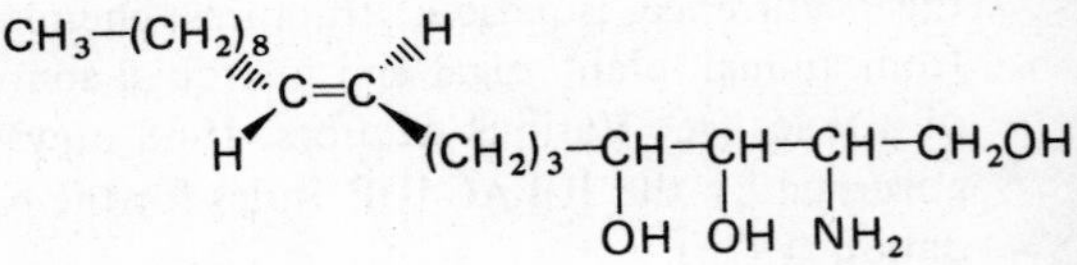 $CH_3{-}(CH_2)_8{-}CH{=}CH{-}(CH_2)_3{-}CH(OH){-}CH(OH){-}CH(NH_2){-}CH_2OH$ (trans: H and H on opposite sides of C=C)
Icosasphinganine[b]	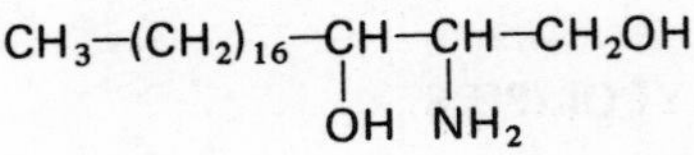 $CH_3{-}(CH_2)_{16}{-}CH(OH){-}CH(NH_2){-}CH_2OH$
Icosasphingosine[c]	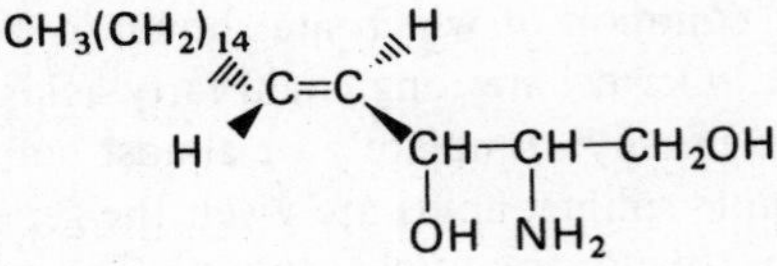 $CH_3(CH_2)_{14}{-}CH{=}CH{-}CH(OH){-}CH(NH_2){-}CH_2OH$ (trans: H and H on opposite sides of C=C)
Hexadecasphinganine	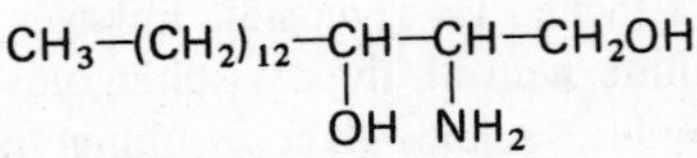 $CH_3{-}(CH_2)_{12}{-}CH(OH){-}CH(NH_2){-}CH_2OH$

a D-*Erythro*-configuration of sphingosine bases is implied, that is,

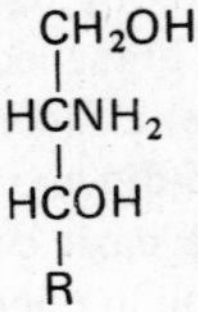

$$\begin{array}{c} CH_2OH \\ | \\ HCNH_2 \\ | \\ HCOH \\ | \\ R \end{array}$$

b Formerly eicosaphinganine

c Formerly eicosaphingosine

Table 10.2 Structure of the higher fatty acids

Trivial name	Structure $CH_3-[R]-COOH$	Systematic name	Numerical[a] symbol
Capric acid[b]	$-[CH_2]_8-$	Decanoic acid	10:0
Lauric acid	$-[CH_2]_{10}-$	Dodecanoic acid	12:0
Myristic acid	$-[CH_2]_{12}-$	Tetradecanoic acid	14:0
Palmitic acid	$-[CH_2]_{14}-$	Hexadecanoic acid	16:0
Palmitoleic acid	$-[CH_2]_5-CH=CH-[CH_2]_7-$	9-Hexadecenoic acid	16:1
Stearic acid	$-[CH_2]_{16}-$	Octadecanoic acid	18:0
Oleic acid	$-[CH_2]_7-CH=CH-[CH_2]_7-$	cis-9-Octadecenoic acid	18:1(9)
Vaccenic acid	$-[CH_2]_5-CH=CH-[CH_2]_9-$	11-Octadecenoic acid	18:1(11)
Linoleic acid	$-[CH_2]_3-[CH_2-CH=CH]_2-[CH_2]_7-$	*cis, cis*-9,12-Octadienoic acid	18:2(9,12)
(9,12,15)-Linoleic acid	$-[CH_2-CH=CH]_3-[CH_2]_7-$	9,12,15-Octadecatrienoic acid	18:3(9,12,15)
(6,9,12)-Linoleic acid	$-[CH_2]_3-[CH_2-CH=CH]_3-[CH_2]_4-$	6,9,12-Octadecatrienoic acid	18:3(6,9,12)
Eleostearic acid	$-[CH_2]_3-[CH=CH]_3-[CH_2]_7-$	9,11,13-Octadecatrienoic acid	18:3(9,11,13)
Arachidic acid	$-[CH_2]_{18}-$	Icosanoic acid[c]	20:0
	$-[CH_2]_6-[CH_2-CH=CH]_2-[CH_2]_6-$	8,11-Icosadienoic acid[c]	20:2(8,11)
	$-[CH_2]_6-[CH_2-CH=CH]_3-[CH_2]_3-$	5,8,11-Icosatrienoic acid[c]	20:3(5,8,11)
Arachidonic acid	$-[CH_2]_3-[CH_2-CH-CH=CH]_4-[CH_2]_3-$	5,8,11,14-Icosatetraenoic acid[c]	20:4(5,8,11,14)
Behenic acid	$-[CH_2]_{20}-$	Docosanoic acid	22:0
Lignoceric acid	$-[CH_2]_{22}-$	Tetracosanoic acid	24:0
Nervonic acid	$-[CH_2]_7-CH=CH-[CH_2]_{13}-$	*cis*-15-Tetracosenoic acid	24:1
Cerotic acid	$-[CH_2]_{24}-$	Hexacosanoic acid	26:0
Montanic acid	$-[CH_2]_{26}-$	Octacosanoic acid	28:0

a The notation giving the number of carbon atoms and of double bonds (separated by a colon) and the position of double bonds (in parentheses) can be used to described fatty acids, for example 16:0 for palmitic acid and 18:1(11) for 11-octadecenoic acid.
b Not recommended because of confusion with caproic (hexanoic) and caprylic (octanoic) acids.
c Formerly 'Eicosa-'.

Glycosylceramides (cerebrosides)

The simplest glycolipids which are found in most mammalian tissues, and particularly in high concentrations in the central nervous system, are the monoglycosylceramides or cerebrosides. Variations in the sugar, base and fatty acid components are possible, with sugar composition depending largely on the tissue source. Brain cerebrosides contain mainly D-galactosyl groups (131) whilst those from serum contain D-glucosyl groups (132), with fatty acids typically being behenic, lignoceric, nervonic or cerebronic (α-hydroxytetrasanoic) acid (see Table 10.1). A related D-galactosylsphingoid (psychosine, 133), which contains no fatty acid, and a sulphoglycosylceramide (D-galactopyranosyl 3-sulphate ceramide, 134) are as generally distributed as cerebrosides.

$CH_3-(CH_2)_{12}$ C = C H H CH–CH–CH_2–O–R OH NH CO $(CH_2)_n$ CH_3

n = 16–22

(131) –R = [HOH_2C, OH, OH, OH galactopyranosyl]

(132) –R = [HOH_2C, OH, OH, OH glucopyranosyl]

$CH_3-(CH_2)_{12}$ C=C H H CH–CH–CH_2–O– [HOH_2C, OH, OH, OH] OH NH_2

(133)

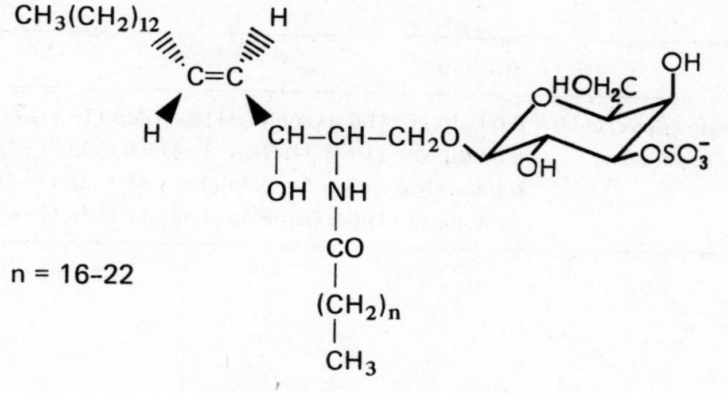

$CH_3(CH_2)_{12}$ H
C=C
H CH–CH–CH_2O — HOH_2C, OH, OH, OSO_3^-
OH NH
CO
n = 16–22
$(CH_2)_n$
CH_3

(134)

Diglycosylceramides are also widely distributed, with lactosyl ceramide (135) being the most common. It is a precursor for more-complex glycosyl-ceramides and gangliosides. A minor diglycosylceramide found in the kidney is digalactosylceramide which contains the disaccharide 4-*O*-β-D-galactopyranosyl-D-galactopyranose. Higher oligosaccharides are attached to ceramide to give tri- and tetra-glycosylceramides (see Table 10.3) which occur to various degrees throughout animal tissues. The oligosaccharide components often bear a striking resemblance to those of glycoproteins produced within the same tissue, particularly in the terminal part of the chain (not the core region). Thus the monosaccharide sequences of the soluble blood-group substance glycoproteins (see Chapter 9, p. 201 and Fig. 9.5) are identical to those of the glycolipids found on the surfaces of red blood cells. Such substances exhibit identical antigenic specificities because antibodies use these external sequences for recognition.

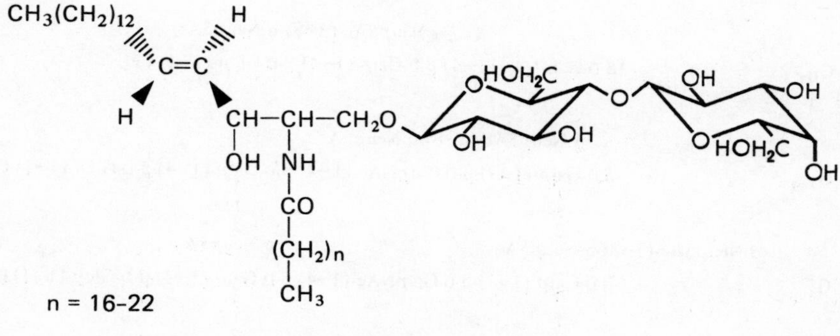

$CH_3(CH_2)_{12}$ H
C=C
H CH–CH–CH_2O — HOH_2C, OH, OH, O, OH, OH, HOH_2C, OH
OH NH
CO
$(CH_2)_n$
CH_3
n = 16–22

(135)

Table 10.3 Structures of some common tri- and tetra-glycosylceramides

Name	Structure
Di-D-galactosyl-D-glucosylceramide	β-D-Gal*p*-(1→4)-β-D-Gal*p*-(1→4)-β-D-Glc*p*-(1→1)-Cer
Aminoglycolipids (globosides)	β-D-Gal*p*NAc-(1→3)-β-D-Gal*p*-(1→4)-β-D-Gal*p*-(1→4)-β-D-Glc*p*-(1→1)-Cer
	α-D-Gal*p*NAc-(1→3)-β-D-Gal*p*-(1→4)-β-D-Gal*p*-(1→4)-β-D-Glc*p*-(1→1)-Cer
Ganglioside G_{A_1}	β-D-Gal*p*-(1→3)-β-D-Gal*p*NAc-(1→4)-β-D-Gal*p*-(1→4)-β-D-Glc*p*-(1→1)-Cer

Table 10.4 Structures of some gangliosides

Ganglioside	Structure
G_{M_1}	β-D-Gal*p*-(1→3)-β-D-Gal*p*NAc-(1→4)-β-D-Gal*p*-(1→4)-β-D-Glc*p*-(1→1)-Cer 3 (on β-D-Gal*p* linked to Glc*p*) ↑ 2 α-Neu*p*5Ac
G_{M_2}	β-D-Gal*p*NAc-(1→4)-β-D-Gal*p*-(1→4)-β-D-Glc*p*-(1→1)-Cer 3 (on β-D-Gal*p*) ↑ 2 α-Neu*p*5Ac
G_{M_3}	α-Neu*p*5Ac-(2→3)-β-D-Gal*p*-(1→4)-β-D-Glc*p*-(1→1)-Cer
$G_{D_{1a}}$	β-D-Gal*p*-(1→3)-β-D-Gal*p*NAc-(1→4)-β-D-Gal*p*-(1→4)-β-D-Glc*p*-(1→1)-Cer 3 (on terminal β-D-Gal*p*) ↑ 2 α-Neu*p*5Ac 3 (on β-D-Gal*p* linked to Glc*p*) ↑ 2 α-Neu*p*5Ac
$G_{D_{1b}}$	β-D-Gal*p*-(1→3)-β-D-Gal*p*NAc-(1→4)-β-D-Gal*p*-(1→4)-β-D-Glc*p*-(1→1)-Cer 3 (on β-D-Gal*p* linked to Glc*p*) ↑ 2 α-Neu*p*5Ac-(2→8)-α-Neu*p*5Ac
G_{D_2}	β-D-Gal*p*NAc-(1→4)-β-D-Gal*p*-(1→4)-β-D-Glc*p*-(1→1)-Cer 3 (on β-D-Gal*p*) ↑ 2 α-Neu*p*5Ac-(2→8)-α-Neu*p*5Ac
$G_{T_{1a}}$	β-D-Gal*p*-(1→3)-β-D-Gal*p*NAc-(1→4)-β-D-Gal*p*-(1→4)-β-D-Glc*p*-(1→1)-Cer 3 (on terminal β-D-Gal*p*) ↑ 2 α-Neu*p*5Ac-(2→8)-α-Neu*p*5Ac 3 (on β-D-Gal*p* linked to Glc*p*) ↑ 2 α-Neu*p*5Ac
$G_{T_{1b}}$	β-D-Gal*p*-(1→3)-β-D-Gal*p*NAc-(1→4)-β-D-Gal*p*-(1→4)-β-D-Glc*p*-(1→1)-Cer 3 (on terminal β-D-Gal*p*) ↑ 2 α-Neu*p*5Ac 3 (on β-D-Gal*p* linked to Glc*p*) ↑ 2 α-Neu*p*5Ac-(2→8)-α-Neu*p*5Ac

Gangliosides

This class of glycosphingolipids contains one or more residues of 5-acetamido-3,5-dideoxy-D-*glycero*-D-*galacto*-2-nonulopyranonic acid linked to a glycosyl-ceramide, (for example, $G_{D_{1a}}$ and $G_{D_{1b}}$ in Table 10.4) either as a single residue side chain or as a disaccharide residue side chain; these are referred to by the trivial names mono-, di-, tri-sialogangliosides, etc., depending on the number of residues of the nine carbon sugar. Because of their complex structures, and hence cumbersome chemical names, many shorthand notation systems have been employed, one of the most common being that due to Svennerholm (1963). They are all given the prefix G (for ganglioside) with subscripts A, M, D or T to denote the number of 5-acetamido-3,5-dideoxy-D-*glycero*-D-*galacto*-2-nonulopyranosyl-onic acid residues present; A (none), M (mono), D (di) and T (tri). To further differentiate the various gangliosides a number is assigned to each, which is usually based on the order in which the compounds separate on thin-layer chromatography. A system, devised by Weignandt (1973), is becoming more common due to its advantage that structures can be deduced from the shorthand notation once the symbols have been learnt, but, as yet, it has not superceded the Svennerholm system. The structures of some gangliosides, together with their Svennerholm notation are given in Table 10.4; the major bases present are sphingosine and icosasphingosine with some of the saturated analogues being present in minor amounts. Stearic acid is the predominant fatty acid, accounting for 85–95% of the total ganglioside fatty acid in human brain.

Gangliosides have been found in many tissues, commonly spleen, liver and kidney with the grey matter of brain being the major site of occurrence where the major gangliosides occur in the ratio G_{M_1}:$G_{D_{1a}}$:$G_{D_{1b}}$:$G_{T_{1a}}$ of 3:4:2:1. They also occur in erythrocytes, where, in common with gangliosides from spleen, the amino-group of 5-amino-3,5-dideoxy-D-*glycero*-D-*galacto*-2-nonulopyranonic acid is glycolated rather than acetylated.

Globosides

Closely related to the gangliosides are a number of glycolipids which contain an additional D-galactopyranosyl residue linked to the D-galactopyranosyl residue adjacent to the lipid moiety. These compounds, referred to by the trivial name of globosides, have very similar carbohydrate structures to the gangliosides in the regions further away from the lipid moiety (see Table 10.3).

Miscellaneous animal glycolipids

Trace quantities of glycosylglycerides (see plant glycolipids, p. 243) and phospholipids (see microbial glycolipids, p. 246) are found in the central nervous systems of several animals.

Table 10.5 Classification of the better established lipidoses

Designation	Other names	Clinical features	Principle glycolipid stored	Enzyme deficient
Fabry's disease	Angiokeratoma corporis diffusum Hereditary dystopic lipidosis	Pain, angiokeratoma, corneal opacities, skin lesions, renal failure.	D-Galactosyl-D-galactosyl-D-glucosyl ceramide	D-Galactosyl-D-galactosyl-D-glucosyl ceramidase
Farber's lipogranulomatosis	Ceramidase deficiency	Subcutaneous and joint swellings, dysphonia, dermatitis pyrexia, paralysis. Death by 2 years.	Ceramide	Ceramidase
Gaucher disease type I (juvenile form)	Glucosyl ceramide lipidosis (non-cerebral)	Hepatosplenomegaly, 'Gaucher' cells, variable neuropathy, encephalopathy.	D-Glucosyl ceramide	D-Glucosyl ceramidase
Gaucher disease type II (infantile form)	Glucosyl ceramide lipidosis (cerebral)	Hepatosplenomegaly, retroflexion of head, strabismus, dysphagia, choking spells, hypertonicity. Death by 1 year.	D-Glucosyl ceramide	D-Glucosyl ceramidase
Gaucher disease type III (juvenile and adult)	Glucosyl ceramide lipidosis (cerebral)	Hypersplenism, bone lesions, skin pigmentation, pingueculae occur.	D-Glucosyl ceramide	D-Glucosyl ceramidase
$G_{D_{1a}}$ Gangliosidosis		Progressive psychomotor deterioration, sensitivity to noise.	$G_{D_{1a}}$	
G_{M_1} Gangliosidosis type I	Neurovisceral gangliosidosis Pseudo-Hurler disease	Progressive psychomotor deterioration, hepatosplenomegaly, skeletal abnormalities, cherry-red macular spot. Death by 2 years.	G_{M_1}	β-D-Galactosidase
G_{M_1} Gangliosidosis type II	Juvenile type G_{M_1}	Diffuse angiokeratoma. Onset at 2 years with survival to 10 years.	G_{M_1}	β-D-Galactosidase (isoenzymes B and C)
G_{M_1} Gangliosidosis type III	Adult type G_{M_1}	Less severe than types I and II. Angiokeratoma, spondyloepiphyseal dysplasia.	G_{M_1}	β-D-Galactosidase
G_{M_2} Gangliosidosis type I	Tay-Sachs disease	Progressive psychomotor deterioration, dementia, blindness, cherry-red macular spot. Death by 3 years.	G_{M_2}	2-Acetamido-2-deoxy-D-galactosidase (isoenzyme A)

	(infantile type) Globoside storage disease	ation, early blindness, startle reaction, doll-like face, cherry-red macular spot. Death by 3 years.		galactosidase (isoenzymes A and B)
	Sandhoff's disease (juvenile type)	Progressive cerebellar ataxia, psychomotor retardation.	G_{M_2}	2-Acetamido-2-deoxy-D-galactosidase (isoenzymes A and B)
G_{M_2} Gangliosidosis type III	Juvenile type G_{M_2}	Ataxia, deterioration to decerebrate rigidity. Onset between 2 and 6 years with death by 5 to 15 years.	G_{M_2}	2-Acetamido-2-deoxy-D-galactosidase (isoenzyme A)
G_{M_2} Gangliosidosis type IV	Adult (chronic) type G_{M_2}	Slow progressive deterioration of gait and posture, mild ataxia, normal intelligence and vision.	G_{M_2}	2-Acetamido-2-deoxy-D-galactosidase (isoenzyme A)
G_{M_3} Gangliosidosis		Poor psychomotor development, gradual hepatosplenomegaly, stubby hands and feet, large inguinal herneas, unresponsive. Early death.	G_{M_3}	UDP-2-acetamido-2-deoxy-D-galactose : G_{M_3}-2-acetamido-2-deoxy-D-galactosyl transferase.
Krabbe's leucodystrophy	Globoid cell leucodystrophy	Progressive psychomotor deterioration, deafness, blindness.	D-Galactosyl ceramide	β-D-Galactosyl ceramidase
Lactosyl ceramidosis		Poor psychomotor development, hypotonia, optic atrophy, hepatosplenomegaly.	G_{A_3}	β-D-Galactosidase
Metachromatic leucodystrophy	Sulphatide lipidosis	Early onset. Psychomotor disturbances.	Sulphoglycosyl ceramides	Arylsulphatase (isoenzyme A)
	Adult type	Onset after 16 years. Schizophrenia, gallbladder nonfunction.	Sulphoglycosyl ceramides	Arylsulphatase (isoenzyme A)
	Late infantile type	Onset by 2 years with death by 5 years. Poor motor development, mental deterioration, hypotonia, muscle weakness.	Sulphoglycosyl ceramides	Arylsulphatase (isoenzyme A)
	Juvenile type	Onset between 4 and 10 years.	Sulphoglycosyl ceramides	Arylsulphatase (isoenzyme A)
Niemann-Pick disease	Sphingomyelinosis (types A–F)	Variable onset and severity. Hepatosplenomegaly, variable neuropathy, jaundice, cherry-red macular spot, corneal opacity.	Sphingomyelin	Sphingomyelin phosphodiesterase

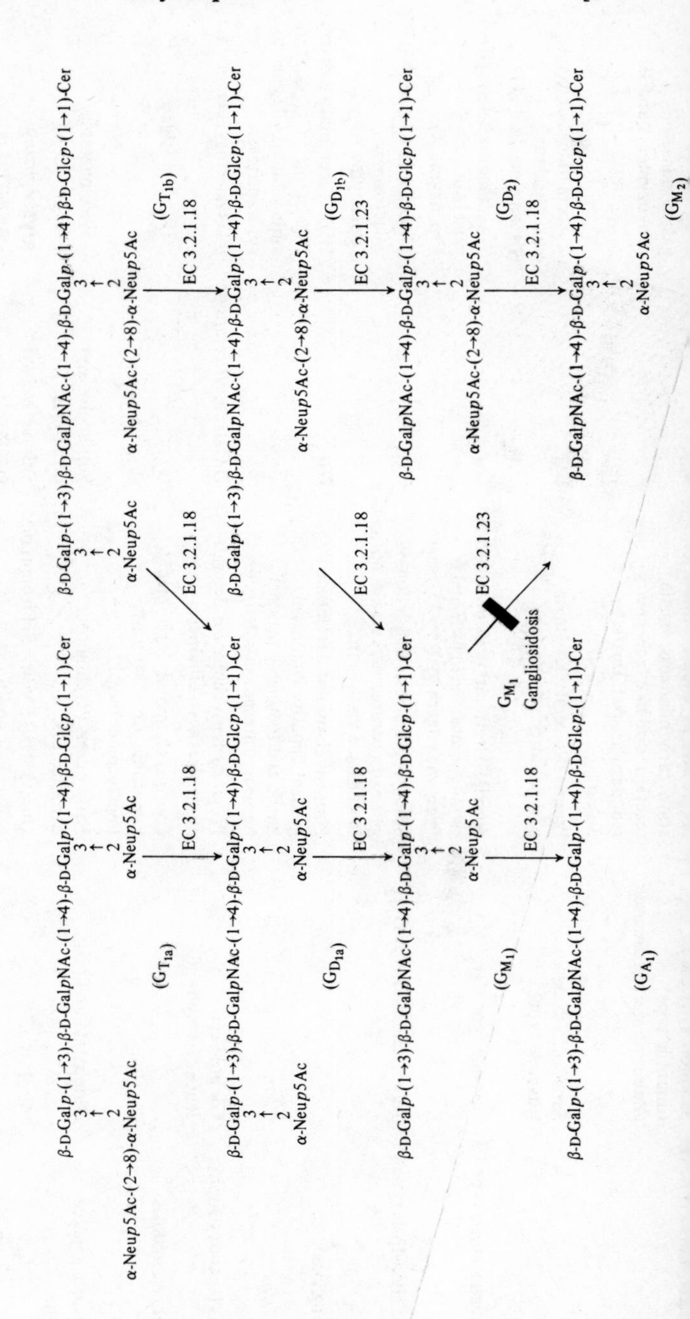
α-Neup5Ac-(2→8)-α-Neup5Ac
β-D-Galp-(1→3)-β-D-GalpNAc-(1→4)-β-D-Galp-(1→4)-β-D-Glcp-(1→1)-Cer
α-Neup5Ac
(G_T1a)
EC 3.2.1.18
β-D-Galp-(1→3)-β-D-GalpNAc-(1→4)-β-D-Galp-(1→4)-β-D-Glcp-(1→1)-Cer
α-Neup5Ac
α-Neup5Ac-(2→8)-α-Neup5Ac
(G_T1b)
EC 3.2.1.18
β-D-Galp-(1→3)-β-D-GalpNAc-(1→4)-β-D-Galp-(1→4)-β-D-Glcp-(1→1)-Cer
α-Neup5Ac
α-Neup5Ac
(G_D1a)
EC 3.2.1.18
β-D-Galp-(1→3)-β-D-GalpNAc-(1→4)-β-D-Galp-(1→4)-β-D-Glcp-(1→1)-Cer
α-Neup5Ac-(2→8)-α-Neup5Ac
(G_D1b)
EC 3.2.1.23
EC 3.2.1.18
β-D-Galp-(1→3)-β-D-GalpNAc-(1→4)-β-D-Galp-(1→4)-β-D-Glcp-(1→1)-Cer
α-Neup5Ac
(G_M1)
EC 3.2.1.18
EC 3.2.1.23
G_M1
Gangliosidosis
β-D-GalpNAc-(1→4)-β-D-Galp-(1→4)-β-D-Glcp-(1→1)-Cer
α-Neup5Ac-(2→8)-α-Neup5Ac
(G_D2)
EC 3.2.1.18
β-D-Galp-(1→3)-β-D-GalpNAc-(1→4)-β-D-Galp-(1→4)-β-D-Glcp-(1→1)-Cer
(G_A1)
β-D-GalpNAc-(1→4)-β-D-Galp-(1→4)-β-D-Glcp-(1→1)-Cer
α-Neup5Ac
(G_M2)

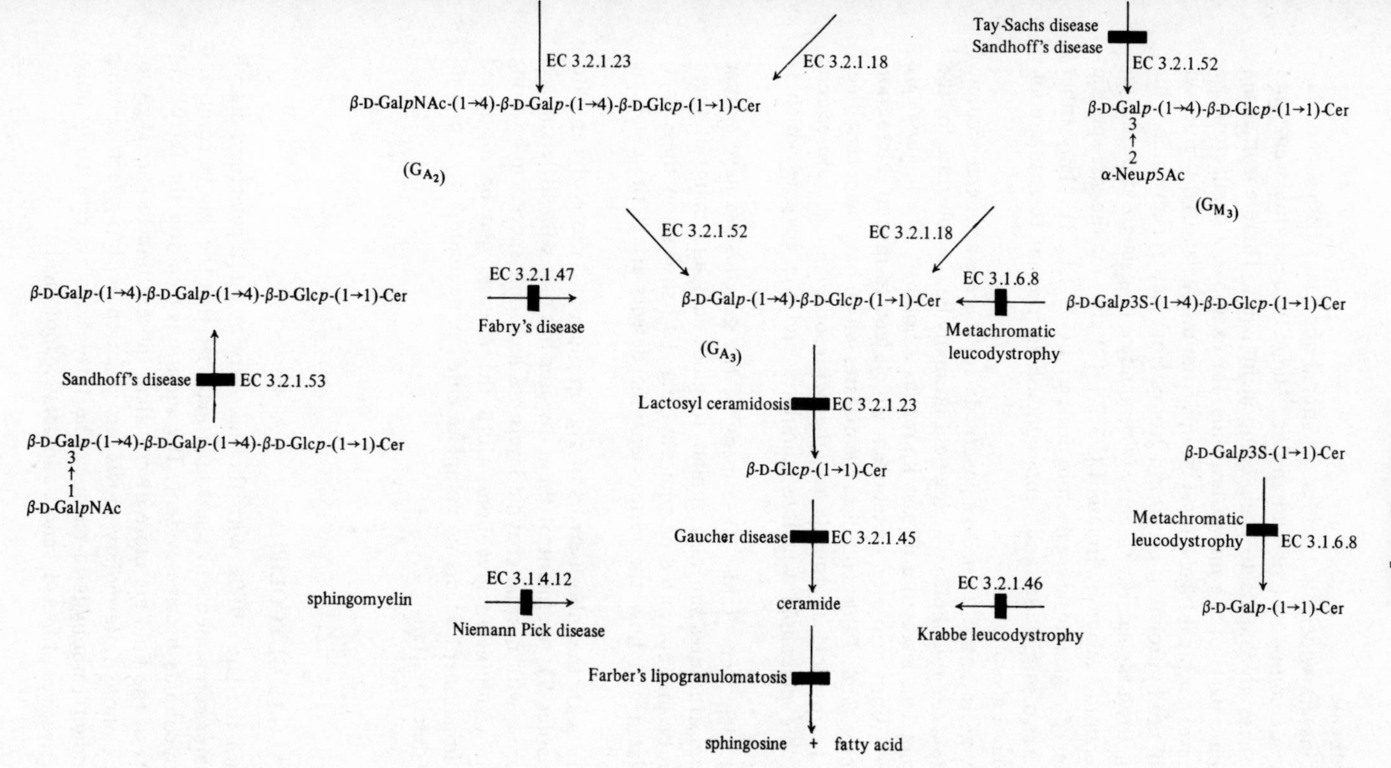

Fig. 10.1 – Glycolipid catabolic pathway showing the better established sites for blockages which give rise to the lipidoses.

Involvement in disease

Glycosphingolipids are widely distributed in animal and plant tissues and constitute a significant component of human diet. Many tissues contain enzymes which degrade these glycolipids to their component parts. These enzymes are specific for specific parts of the molecules, thus, for example, a D-galactosidase exists which cleaves terminal D-galactosyl groups from di-D-galactosyl-D-glucosyl ceramide but will not hydrolyse D-galactosyl groups from lactosyl ceramide. This enzyme is D-galactosyl-D-galactosyl-D-glucosylceramidase (D-galactosyl-D-galactosyl-D-glucosyl-ceramide galactohydrolase, EC 3.2.1.47) and not the less specific β-D-galactosidase (β-D-galactoside galactohydrolase, EC 3.2.1.23). Many other glycosidases and also sulphatases and neuraminidase are present to ensure complete degradation of glycolipids.

If any enzyme is deficient due to a genetic defect there is a general accumulation of that lipid (or glycolipid) which would normally be the substrate for the deficient enzyme. This leads to a disease known as a lipidosis. The lipidoses are a group of rare inborn errors of metabolism which lead to mental retardation and are frequently fatal. Table 10.5 lists the diseases which have been classified, on clinical and morphological grounds, together with some clinical features and, where known, the glycolipid which accumulates and the enzyme which is deficient.

As can be seen from Table 10.4, many of the gangliosides have similar structures; these structures are also similar to globosides and many of these compounds can be interconverted enzymically. Fig. 10.1 shows how these compounds are related and how the various enzyme defects affect the catabolic pathway.

In common with the inborn errors of glycoprotein and proteoglycan catabolism (see Chapter 9), treatment of these diseases has met with little success. Infusion of plasma with normal enzyme levels has had some success in lowering the amount of accumulated glycolipid. Until the diseases can be diagnosed before mental impairment occurs no complete cure can be made, only a prevention of further deterioration.

PLANT AND ALGAL GLYCOLIPIDS

It was not until the late 1950s, with the discovery of the D-galactosyl-diacylglycerols as major components in plant lipid material, that the unique complexity of plant glycolipids became evident. The variations found in the fatty acid and other parts of the lipid are more diverse than those found in animal glycolipids, but the carbohydrate moiety shows less variation, with D-galactose being the most abundant carbohydrate residue. The fatty acids are essentially polyunsaturated derivatives of hexadecanoic and octadecanoic acids.

Glycosylglycerides

The major glycerolipids of chloroplasts are mono-D-galactosyl diacylglycerol (136) and the sulpholipid, sulphoquinovosyl diacylglycerol (137) which contains a carbon–sulphur bond at C-6. The accepted lipid nomenclature for these compounds are 1,2-diacyl-[β-D-galactopyranosyl (1′→3)] -*sn*-glycerol and 1,2-diacyl-[6-sulpho-α-D-quinovopyranosyl-(1′→3)] -*sn*-glycerol but in carbohydrate-accepted nomenclature these compounds are β-D-galactopyranosyl-(1→1′)-2′,3′-diacyl-D-glycerol and 6-deoxy-α-D-glucopyranosyl 6-*C*-sulphate-(1→1′)-2′,3′-diacyl-D-glycerol respectively, (D-quinovose ≡ 6-deoxy-D-glucose). The mono-D-galactosyl diacyl glycerols from *Chlorella vulgaris* mainly contains the fatty acids octadecenoic acid and octadecadienoic acid when grown in the dark but up to 20% octadecatrienoic acid when grown in the light, whilst those from spinach chloroplast have 25% hexadecatrienoic and 75% octadecatrienoic acids and from *Euglena gracilis* contain hexadecatetrenoic acid. Sulphoquinovosyl diacylglycerols contain more of the saturated fatty acids (mainly palmitic acid) than are found in D-galactolipids, with that from spinach leaf containing 27% palmitic, 39% octadecadienoic and 28% octadecatrienoic acids.

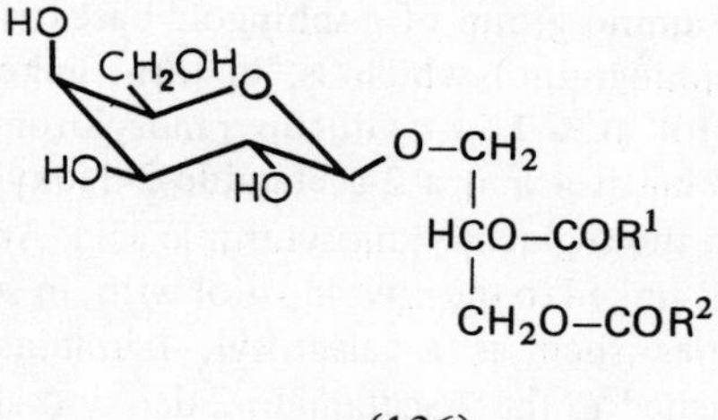

(136)

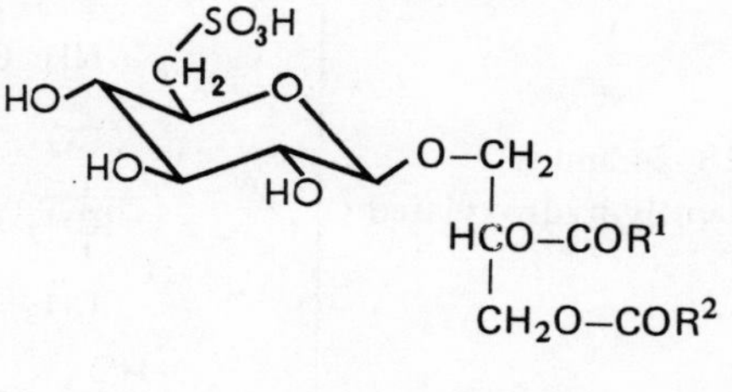

(137)

Another common glycosylglyceride found in higher plants and algae is digalactosyl diacylglycerol (138) which contains the same high proportions of polyunsaturated fatty acids (mainly octadecatrienoic acid) as the mono-D-galactosyl derivative.

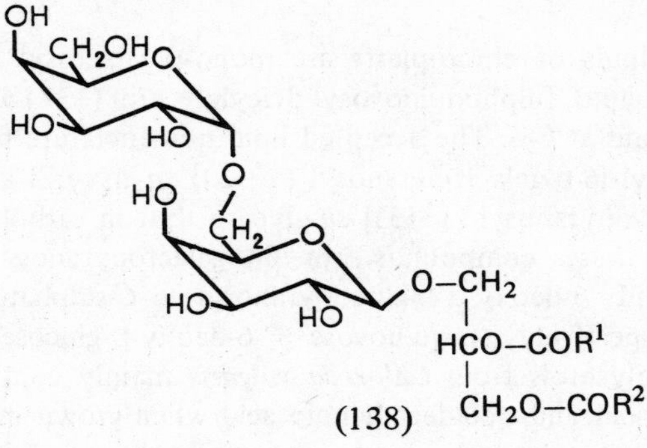

(138)

Miscellaneous glycolipids

A number of glycolipids occur as minor components in plants. These include D-glucocerebrosides (see animal glycolipids, p. 233), phospholipids (see microbial glycolipids, p. 246), phytoglycolipids, and sterol glycosides. In the group of phosphosphingolipids, isolated from corn, soya bean, wheatgerm, flax, cotton and sunflower seeds, known as phytoglycolipids (139), the fatty acid residue is linked to the amino group of a sphingoid base (usually phytosphingosine or dehydrophytosphingosine) which is, in turn, linked *via* phosphate to a residue of *myo*-inositol at C-1. α-D-Glucopyranosyluronic acid residues are attached at C-6 of the inositol and a 2-acetamido-2-deoxy-α-D-glucopyranosyl group is (1→4)-linked to the D-glucopyranosyluronic acid. An α-D-mannopyranosyl group is usually (1→2)-linked to the *myo*-inositol with, in some cases, additional carbohydrate moieties such as D-galactosyl, D-arabinosyl, and L-fucosyl residues and groups attached to the 2-acetamido-2-deoxy-D-glucosyl group.

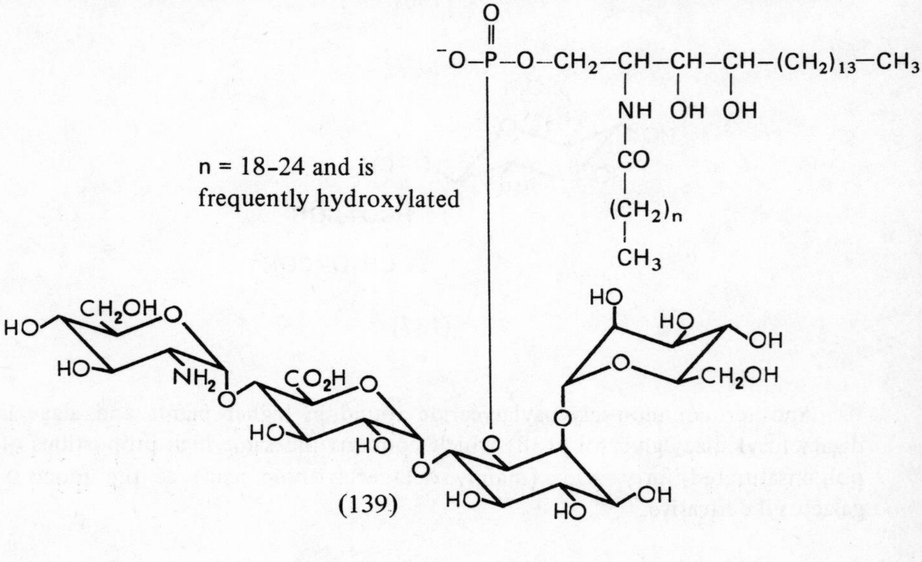

n = 18–24 and is frequently hydroxylated

(139)

A number of sterol and acylsterol glycosides (140) have been isolated in trace amounts from plants. In these glycolipids the sterol is usually β-sitosterol but cholesterol has now been identified in many higher plants and algae (particularly red algae). The carbohydrate is, in some cases, acylated at C-6 with fatty acids (particularly hexadecanoic and octadecanoic acids and their unsaturated derivatives), as shown in (140).

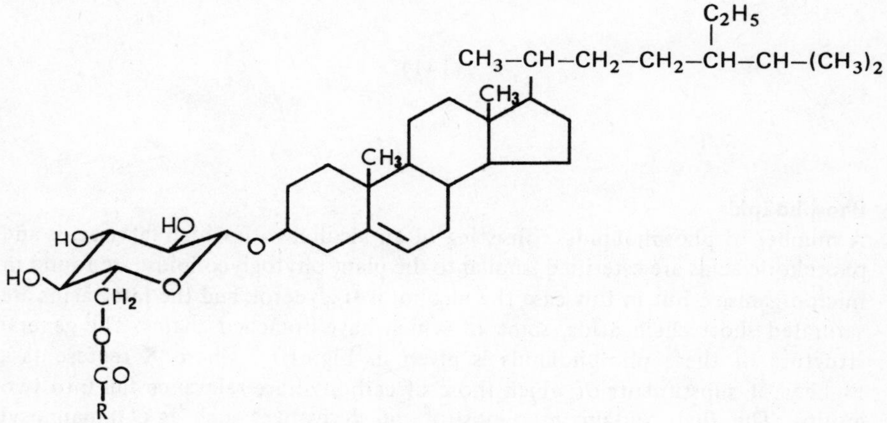

(140)

MICROBIAL GLYCOLIPIDS

The majority of bacterial carbohydrate–lipid materials are essentially high molecular weight polymers, known as lipopolysaccharides, distinguishable from glycolipids on account of their water solubility. The structures of these materials have been discussed in Chapter 8. The occurrence of sphingolipids and cerebrosides is rare, with the major component being glycosylglycerides, esters of carbohydrates and glycosides of hydroxy fatty acids and hydroxylated hydrocarbons containing terminal phenolic groups.

Glycosylglycerides

The glycosylglycerides found in bacteria, particularly the micrococci, are similar to those found in plants with the exceptions that D-mannose and D-glucose are more predominant than D-galactose and a large proportion of the fatty acids have branched chains. A di-D-mannosyl diacylglycerol (141) is most abundant although a mono-D-mannosyl derivative occurs. Glycosylglycerides isolated from *Pneumococci* consist of, *inter alia*, D-galactosyl-D-glucosyl-diacylglycerols whilst those from *Streptococci* and *Mycoplasma* contain mono- and di-D-glucosyl diacylglycerols.

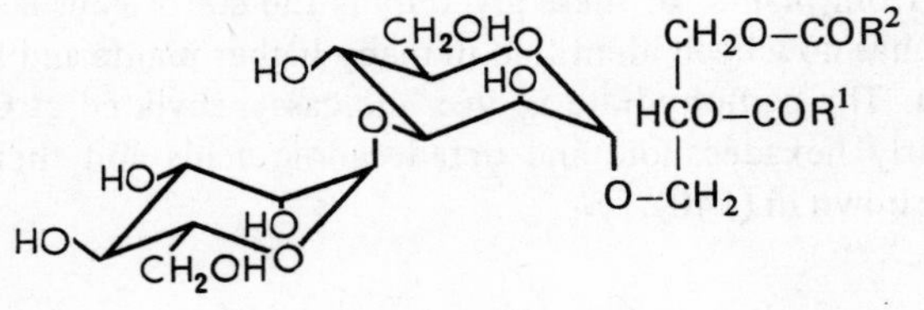

(141)

Phospholipids

A number of phospholipids, consisting of an alcohol with which fatty acids and phosphoric acids are esterified, similar to the plant phytoglycolipids, are found in microorganisms but in this case the alcohol is D-glycerol, and the fatty acids are saturated short chain acids, some of which have branched chains. The general structure of these phospholipids is given in Fig. 10.2 where X represents a number of substituents of which those of carbohydrate relevance fall into two groups. The first contains *myo*-inositol and derivatives such as *O*-D-mannosyl inositols. Thus structures such as (142), the so-called phosphatidyl inositol mannosides from *Micrococcus phlei* and *M. tuberculosis*, are obtained.

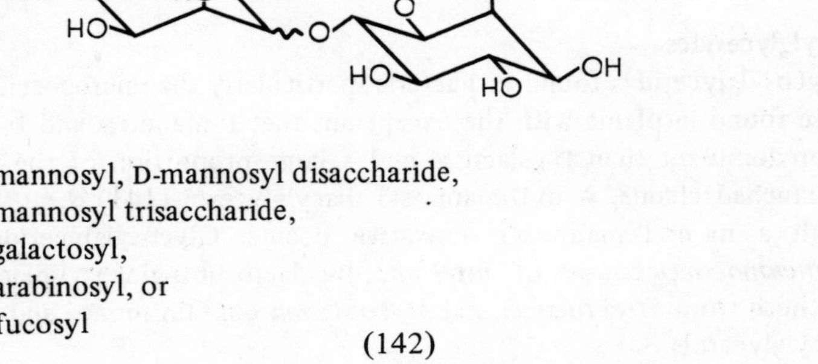

R = D-mannosyl, D-mannosyl disaccharide,
D-mannosyl trisaccharide,
D-galactosyl,
D-arabinosyl, or
L-fucosyl

(142)

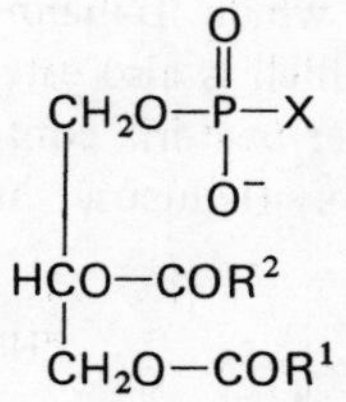

$$
\begin{array}{l}
\qquad\quad O \\
\qquad\quad \| \\
CH_2O-P-X \\
\;|\qquad\;\; | \\
\;|\qquad\;\; O^- \\
HCO-COR^2 \\
\;| \\
CH_2O-COR^1
\end{array}
$$

Fig. 10.2 – General representation of the glycerolphospholipids. X = organic bases, aminoacids, alcohols, carbohydrates, etc.

The second group of phospholipids, which are related to the above, have glycerol in place of inositol, and some of this group have 2-acetamido-2-deoxy-D-glucosyl residues linked to C-2 or C-3 of free glycerol residues (143).

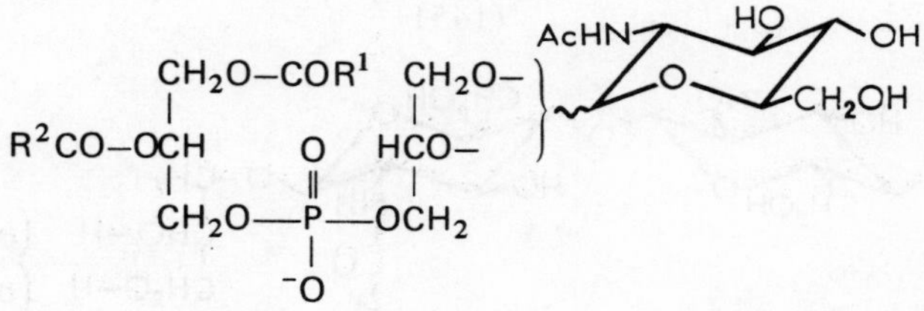

$$
\begin{array}{l}
CH_2O-COR^1 \quad CH_2O- \\
\;| \\
R^2CO-OCH \qquad O \quad HCO- \\
\;| \\
CH_2O-P-OCH_2 \\
\qquad\quad | \\
\qquad\; ^-O
\end{array}
$$

(2-acetamido-2-deoxy-D-glucosyl residue: AcHN, HO, OH, O, CH_2OH)

(143)

Carbohydrate esters

These constitute a whole family of complex glycolipids which are mainly confined to the *Mycobacteria*. Cord factor, so-called because it is found in the waxy capsular material of virulent strains of tubercule bacilli where it causes the bacteria to string together in a long cord or chain, is an ester of α,α-trehalose, it being esterified with two molecules of the mycolic acid fatty acids, such as the 60 carbon atom acid as found in cord factor from *Mycobacterium smegmatis* shown in (144).

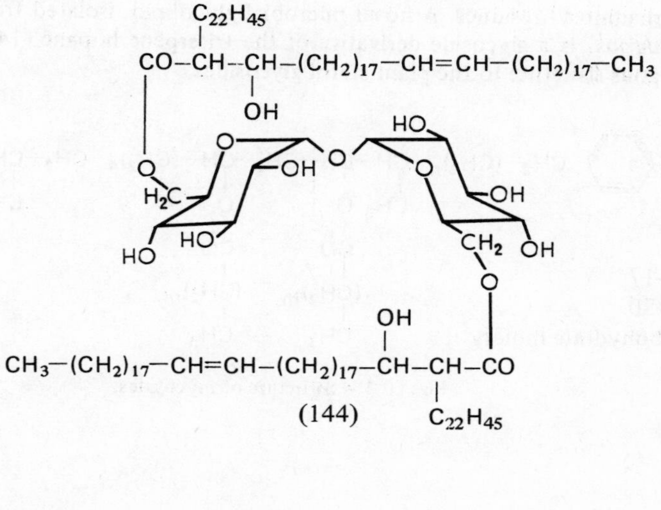

$$
CO-CH(C_{22}H_{45})-CH(OH)-(CH_2)_{17}-CH{=}CH-(CH_2)_{17}-CH_3
$$

(trehalose: O, H_2C, HO, HO, OH, O, HO, O, OH, CH_2, OH)

$$
CH_3-(CH_2)_{17}-CH{=}CH-(CH_2)_{17}-CH(OH)-CH(C_{22}H_{45})-CO
$$

(144)

Esters of *myo*-inositol in which D-mannosyl residues are glycosidically linked to C-2 of *myo*-inositol which is also esterified at both C-1 and C-6 (145) are found in *Mycobacteria*. Other bacteria contain esters of D-glucose and other sugars including 2-amino-2-deoxy-D-glucose, as in the glycolipid (146) from *Bacillus acidocaldarius*.

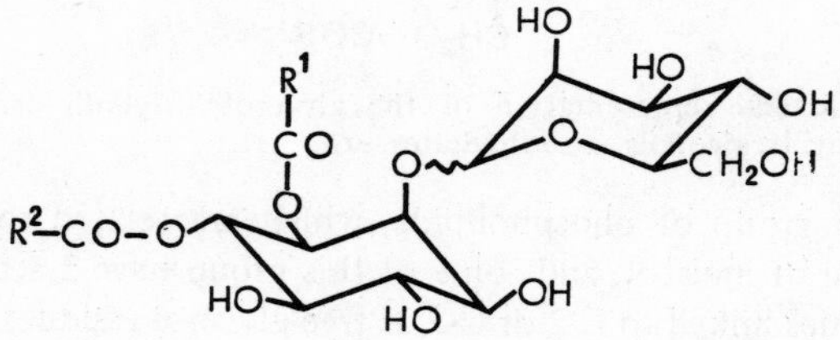

(145)

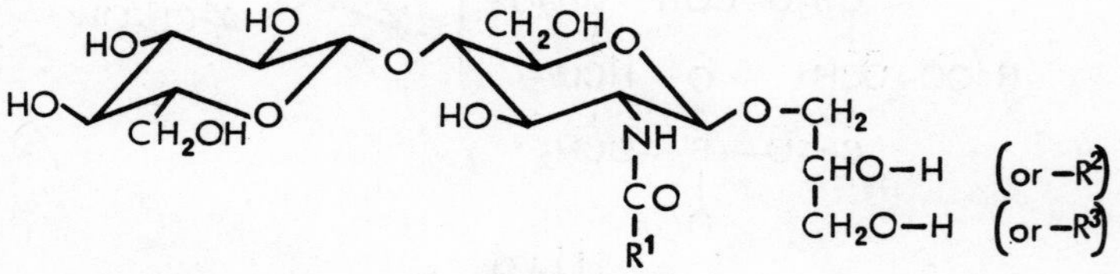

(146)

Carbohydrate glycosides

The major glycosides are the mycosides from *Mycobacteria* which contain a long chain, highly branched, hydroxylated hydrocarbon terminated by a phenol group to which the carbohydrate moiety is glycosidically linked (see Fig. 10.3). The carbohydrate is usually a di- or tri-saccharide and frequently contains 6-deoxy-L-talosyl, 3-*O*-methyl-6-deoxy-L-talosyl (which may also be acetylated at C-2 and C-4), 2-*O*-methyl-L-fucosyl, 2-*O*-methyl-L-rhamnosyl or 2,4-di-*O*-methyl rhamnosyl residues. A novel microbial glycolipid, isolated from *Bacillus acidocaldarius*, is a glycoside derivative of the triterpene hopane (147) and has an analogous structure to the plant sterol glycosides.

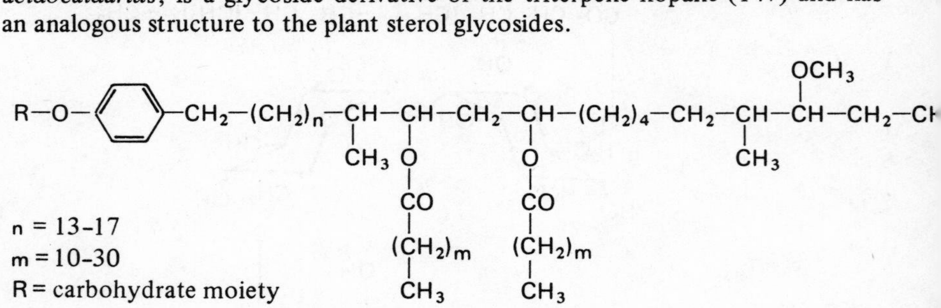

n = 13–17
m = 10–30
R = carbohydrate moiety

Fig. 10.3 – Structure of mycosides.

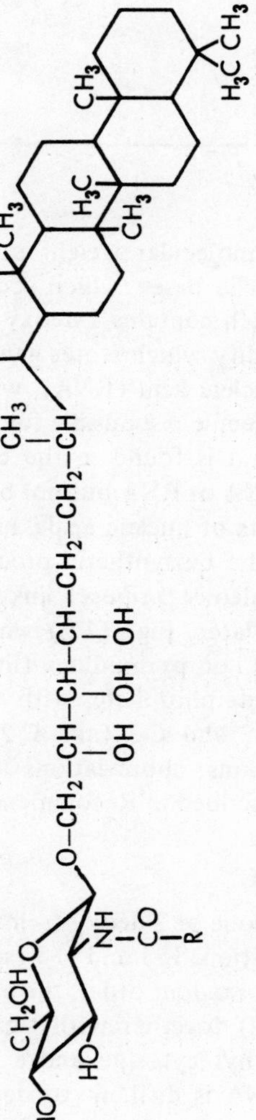

(147)

CHAPTER 11

Nucleic acids

Nucleic acids are high molecular weight polymers, containing carbohydrate, phosphate and heterocyclic bases, which occur in every living cell. Deoxyribonucleic acid (DNA), which contains 2-deoxy-D-ribose as the sole carbohydrate, is the molecule of heredity which stores genetic information in the nucleus of the cell, whereas ribonucleic acid (RNA), which contains D-ribose as the sole carbohydrate, is the molecule responsible for the transfer of genetic information into protein structure and is found in the cytoplasm outside the cell nucleus. Viruses contain either DNA or RNA but not both nucleic acids.

The component parts of nucleic acids, namely nucleosides and nucleotides, also occur as part of the biosynthetic process. Nucleosides consist of carbohydrate (D-ribose or 2-deoxy-D-ribose) linked *via* β-D-glycosidic bonds to the heterocyclic bases (see later, Fig. 11.2) which are either purines (linked *via* position 3 of the base), or pyrimidines (linked *via* position 1 of the base). Nucleotides are nucleoside phosphates with the phosphate linked to the carbohydrate at positions C-5 and C-3 (and C-2 in D-ribose). The nomenclature, conformational descriptions, abbreviations and symbols for nucleic acids and their constituents are described in Recommendations (1970 and 1981b).

PRIMARY STRUCTURE

DNA consists of a backbone of 2-deoxy-D-ribofuranosyl residues linked by phosphodiester bonds at positions C-3 and C-5 (see Fig. 11.1). To this backbone are linked, in an apparently random order, the bases adenine, guanine, cytosine and thymine (see Fig. 11.2) together with small or trace amounts of 5-methyl cytosine, 5-hydroxymethyl cytosine and a number of *N*-methyl purines. The molecular weight of DNA is difficult to determine due to degradation, which can be brought about even by simply pouring a solution from one vessel to another, but values up to 10^9 have been reported. The variable part of DNA is its sequence of bases which is the key to the biological role of DNA (see p. 259); however the bases are present as complementary base pairs in which the ratio of adenine:thymine and guanine:cytosine equal 1:1 but the amounts of adenine and guanine are not normally equal.

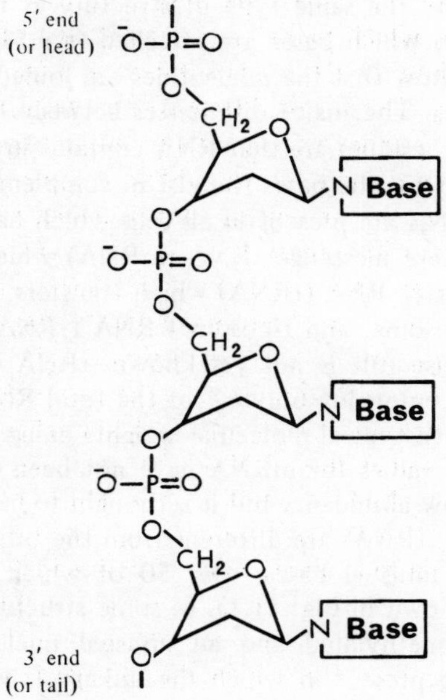
5′ end
(or head)
CH2
Base
CH2
Base
CH2
Base
3′ end
(or tail)

Fig. 11.1 – Repeating structures of DNA chains.

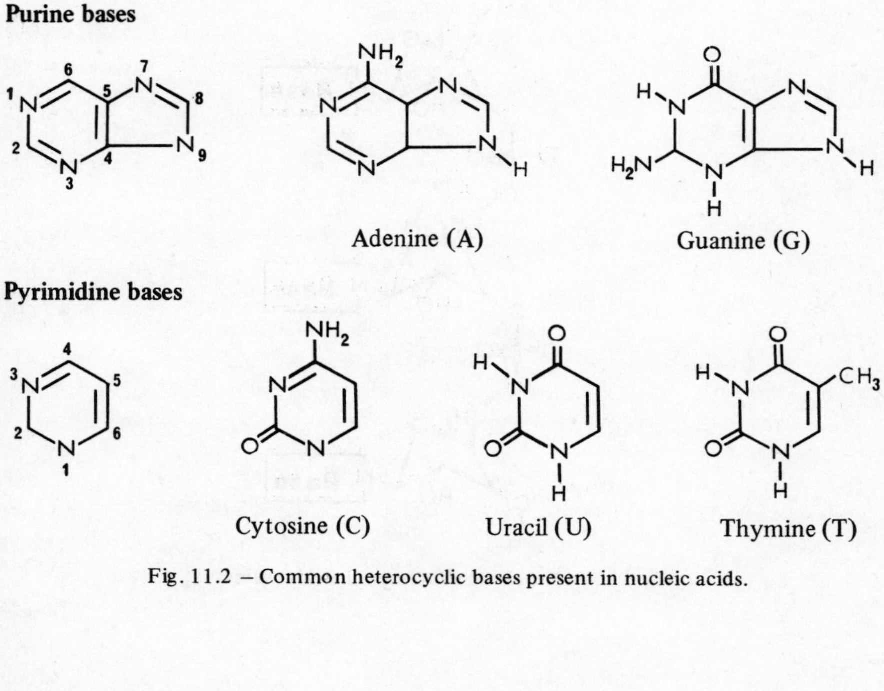
Purine bases
Adenine (A)
Guanine (G)
Pyrimidine bases
Cytosine (C)
Uracil (U)
Thymine (T)

Fig. 11.2 – Common heterocyclic bases present in nucleic acids.

RNA has essentially the same type of structure as DNA, namely a sugar-phosphate backbone to which bases are attached (see Fig. 11.3). No evidence has been obtained to show that the nucleotides are joined by other than 3′→5′ phosphodiester linkages. The major differences between DNA and RNA (apart from the carbohydrate residue) are that RNA contains uracil, not thymine and there is no requirement for the bases to exist in complementary base ratios. At least three kinds of RNA are present in all cells which have nuclei and/or synthesise protein. These are messenger RNA (mRNA) which is the template for protein synthesis, transfer RNA (tRNA) which transfers aminoacids in an activated form to the ribosome, and ribosomal RNA (rRNA) which is present in ribosomes but its precise role is not yet known. rRNA is the most abundant and mRNA the least, comprising only 5% of the total RNA. The sizes of these RNAs are different, with typical molecular weights being $\approx 10^6$ for rRNA, and $\approx 10^4$ for tRNA. The values for mRNA have not been determined due to its transient nature and low abundance but it is thought to have a molecular weight in the region of $\approx 10^5$. tRNAs are different from the other RNAs in that they contain a number of unusual bases, over 50 of which have been identified (some of which are shown in Fig. 11.1). In some structures C-2 of the D-ribofuranosyl residues is methylated and an unusual nucleoside, pseudouridine (Ψ, 148), can also be present in which the linkage is *via* the 5 and not the 1 position of uracil.

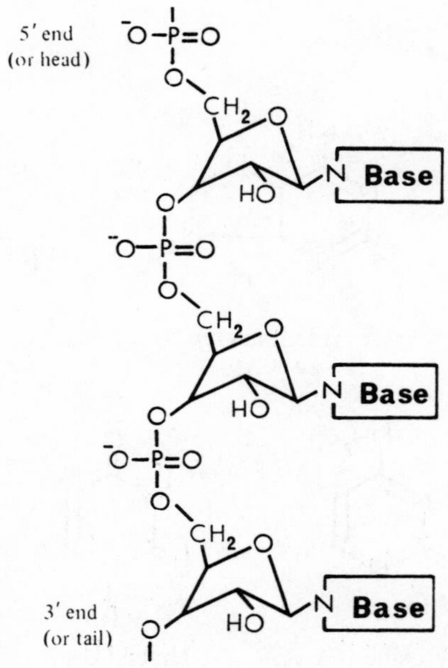

Fig. 11.3 – Repeating structures of RNA chains.

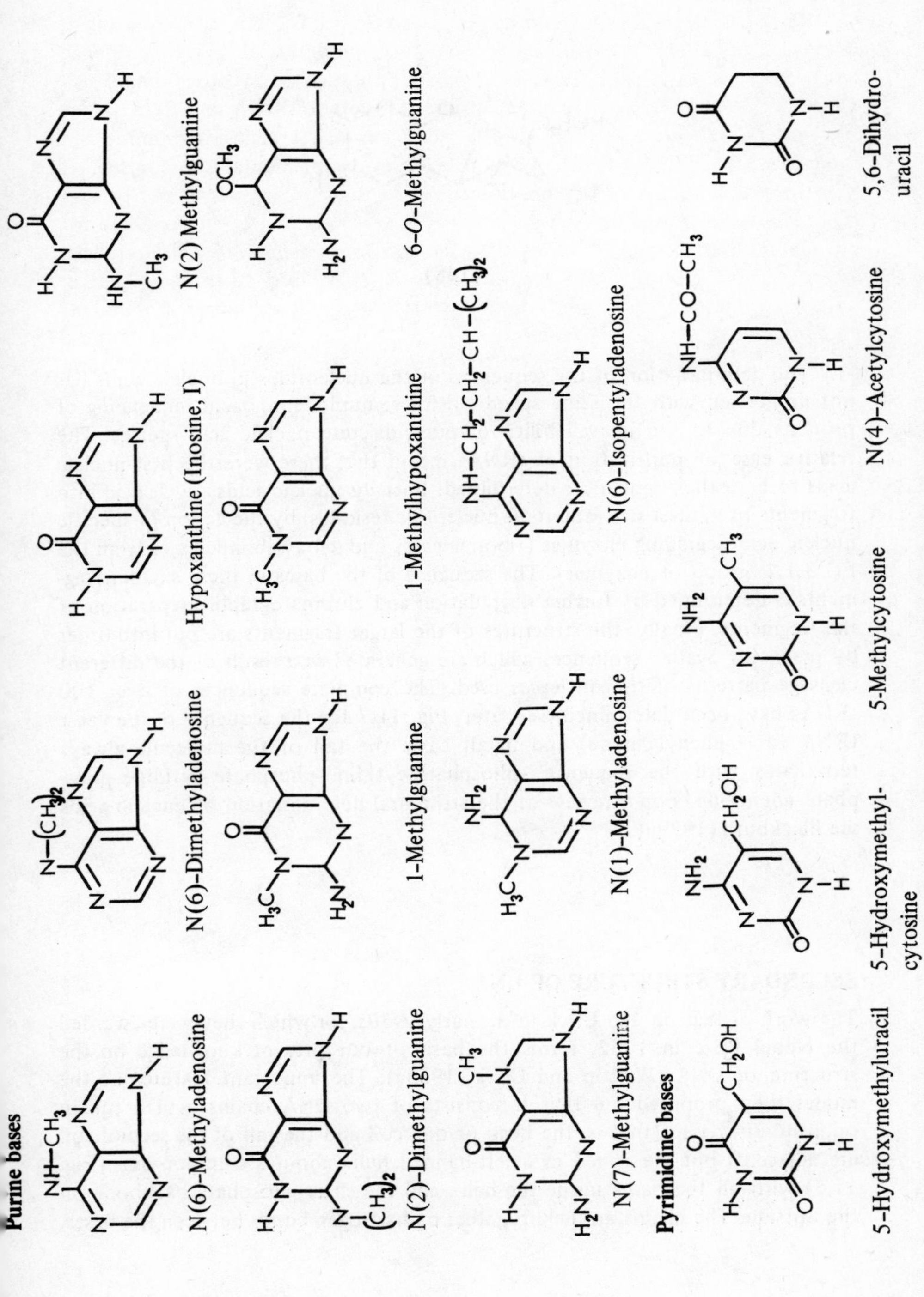

Purine bases

N(6)-Methyladenosine

N(6)-Dimethyladenosine

Hypoxanthine (Inosine, I)

N(2) Methylguanine

N(2)-Dimethylguanine

1-Methylguanine

1-Methylhypoxanthine

6-*O*-Methylguanine

N(7)-Methylguanine

N(1)-Methyladenosine

N(6)-Isopentyladenosine

Pyrimidine bases

5-Hydroxymethyluracil

5-Hydroxymethyl-cytosine

5-Methylcytosine

N(4)-Acetylcytosine

5,6-Dihydro-uracil

Fig. 11.4 – Some unusual bases found in tRNA.

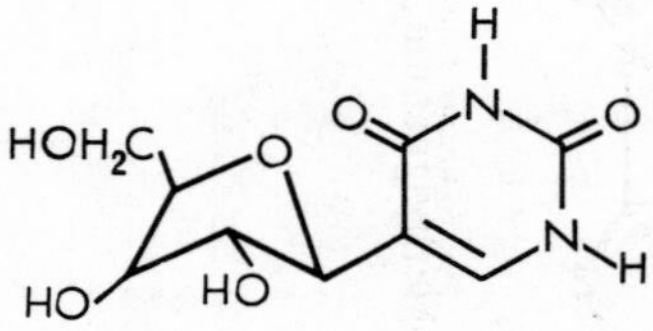

(148)

The determination of the sequences of the nucleotides in nucleic acids has not progressed with the same speed as, for example, aminoacid sequencing of proteins due to the unavailability of pure, discrete nucleic acid species. The relative ease of purification of tRNAs meant that these were the first nucleic acids to have their sequences determined. Initially nucleic acids are cleaved into fragments of modest size (about 50 nucleotide residues) by the action of specific nucleic acid degrading enzymes (ribonucleases and deoxyribonucleases from the EC 3.1.4. group of enzymes). The sequence of the bases in these smaller fragments is determined by further degradation and chromatographic separation of the fragments. Finally, the structures of the larger fragments are put into order by means of overlap sequences which are generated as a result of the different cleavage patterns of the nucleases used. The complete sequences of over 100 tRNAs have been determined (see later, Fig. 11.7 for the sequence of the yeast tRNA for L-phenylalanine) and in all cases the tail of the molecule always terminates with the sequence -phosphate–cytidine-phosphate-cytidine-phosphate–adenosine. For a review of the structural determination of nucleic acids see Blackburn (1979a).

SECONDARY STRUCTURE OF DNA

The work of Watson and Crick in the early 1950s, for which they were awarded the Nobel Prize in 1962, forms the basis of our present knowledge on the structure of DNA (Watson and Crick, 1953a). The important features of the model they proposed are that it consists of two DNA chains, which run in opposite directions (that is, the head of one coil and the tail of the second coil are adjacent) but are coiled in a left-handed helix about a common axis (Fig. 11.5) with all the bases inside the helix and the sugar-phosphate backbone on the outside. The chains are held together by hydrogen bonds between the bases, with adenine always paired with thymine and guanine always paired with cytosine (Fig. 11.6) which explains the requirement for complementary base pair ratios. There are ten residues per complete turn of the helix. There has been much refinement of the original double-helix model and various modifications have been proposed including a side-by-side model in which the coils turn into a right-handed double helix for 1.7 nm and then revert back to a left-handed double helix for a further 1.7 nm. For a review of the history of the development of models for the structure of DNA and the recent modifications, see Blackburn, 1979b.

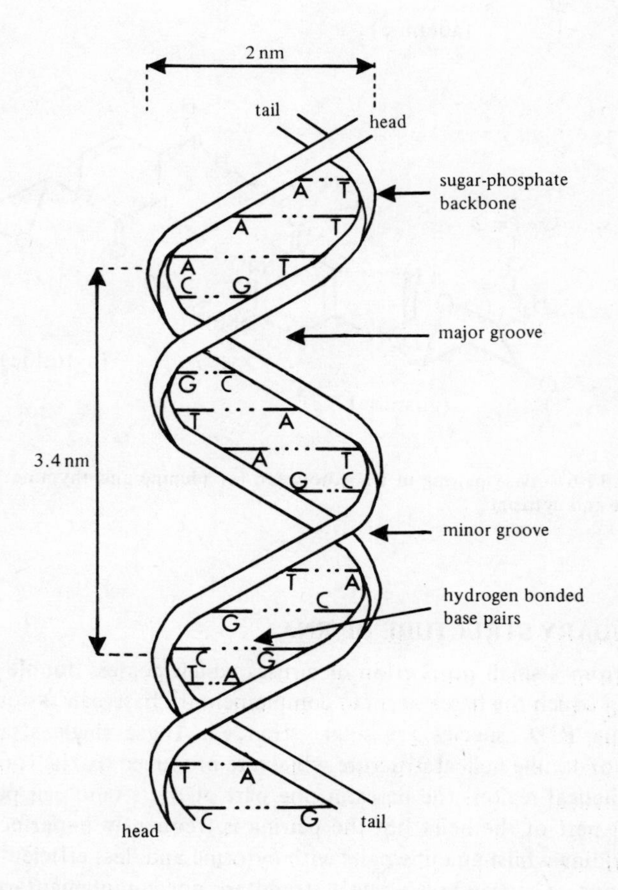

Fig. 11.5 – Double helical DNA showing hydrogen bonds between base pairs (A = adenine, C = cytosine, G = guanine, and T = thymine).

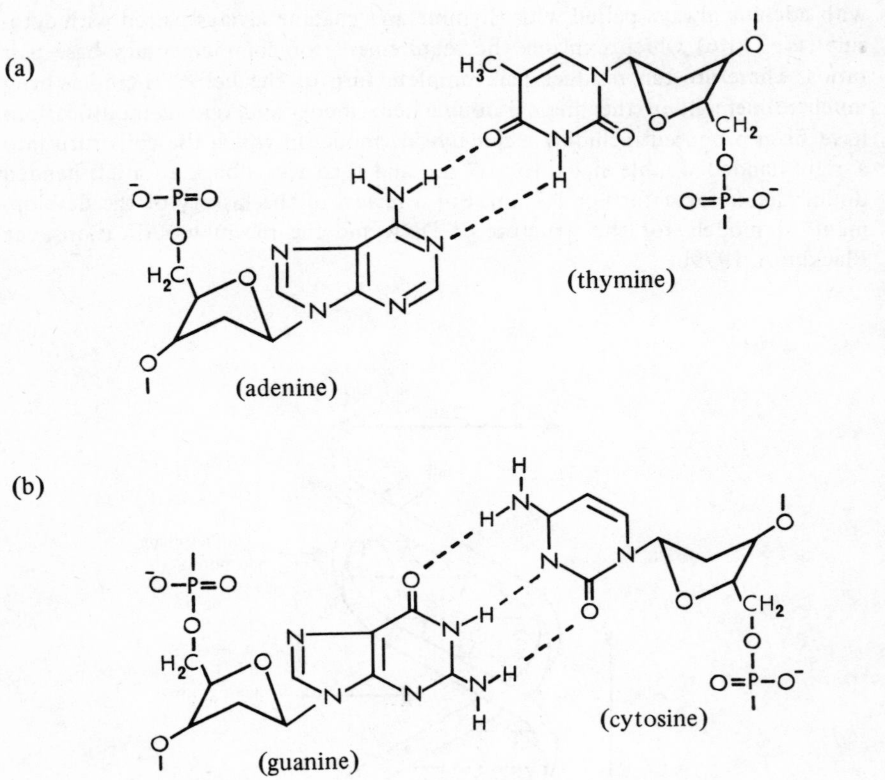

(a)

(thymine)

(adenine)

(b)

(cytosine)

(guanine)

Fig. 11.6 – Base pairing in DNA between (a) adenine and thymine, and (b) guanine and cytosine.

SECONDARY STRUCTURE OF RNA

Apart from a small proportion of viruses which possess double stranded helical RNA, in which the bases occur in complementary base pair ratios, most naturally occurring RNA species are single stranded. These single strands do contain regions of double-helical structure which are produced by the formation of loops. In the helical regions the bases in one part of the strand can pair with bases in another part of the helix but the pairing is frequently imperfect. Adenine pairs with uridine whilst guanine pairs with cytosine and, less efficiently, with uridine. Where one or more bases along a strand are not complementary with the other part of the strand then they may be looped out to facilitate pairing of the other bases. This results in about 50% of the bases present being paired and a common conformation of the nucleic acids has been shown to have a cloverleaf structure as typified by the yeast L-phenylalanine tRNA (Fig. 11.7). It is conventional to show this structure as a flat leaf but it must be remembered that each loop is in fact a helix. Further examples of the secondary structure of RNA are contained in the review by Blackburn (1979b).

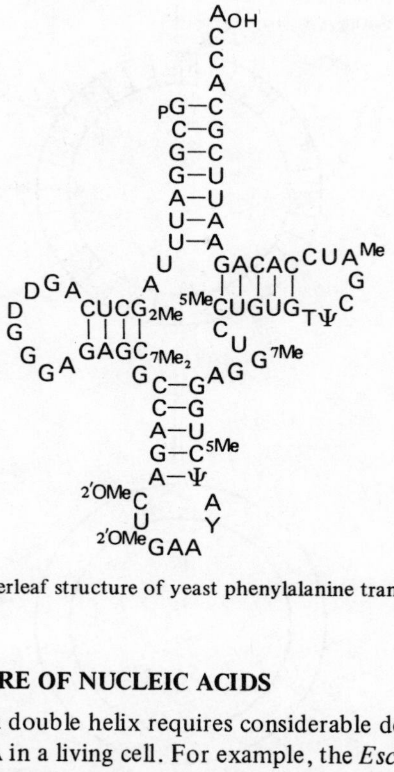

Fig. 11.7 – The cloverleaf structure of yeast phenylalanine transfer ribonucleic acid ($tRNA^{Phe}$).

TERTIARY STRUCTURE OF NUCLEIC ACIDS

The image of DNA as a double helix requires considerable development to bring it closer to that of DNA in a living cell. For example, the *Escherichia coli* bacterium is 2 μm long but the DNA chromosome extracted from it is 1.1 mm long and some DNA from bacteriophage particles expand in aqueous solution to occupy a volume 15 times the size of the head cavity of the phage. Thus the tertiary structure of DNA is obviously a very important aspect of its structure.

Electron microscopy has shown that some intact DNA molecules are circular. This structure is brought about by the helix having unequal length strands at each end such that, for example, both heads are longer than both tails by between four and ten bases, which have a complementary structure (see Fig. 11.8). These ends are held together by the complementary bases and the breaks in the strands converted into normal nucleotide linkages by a 'repair' enzyme (a ligase) to give a closed loop. This circular DNA can itself be twisted to give a more compact structure known as supertwisted DNA.

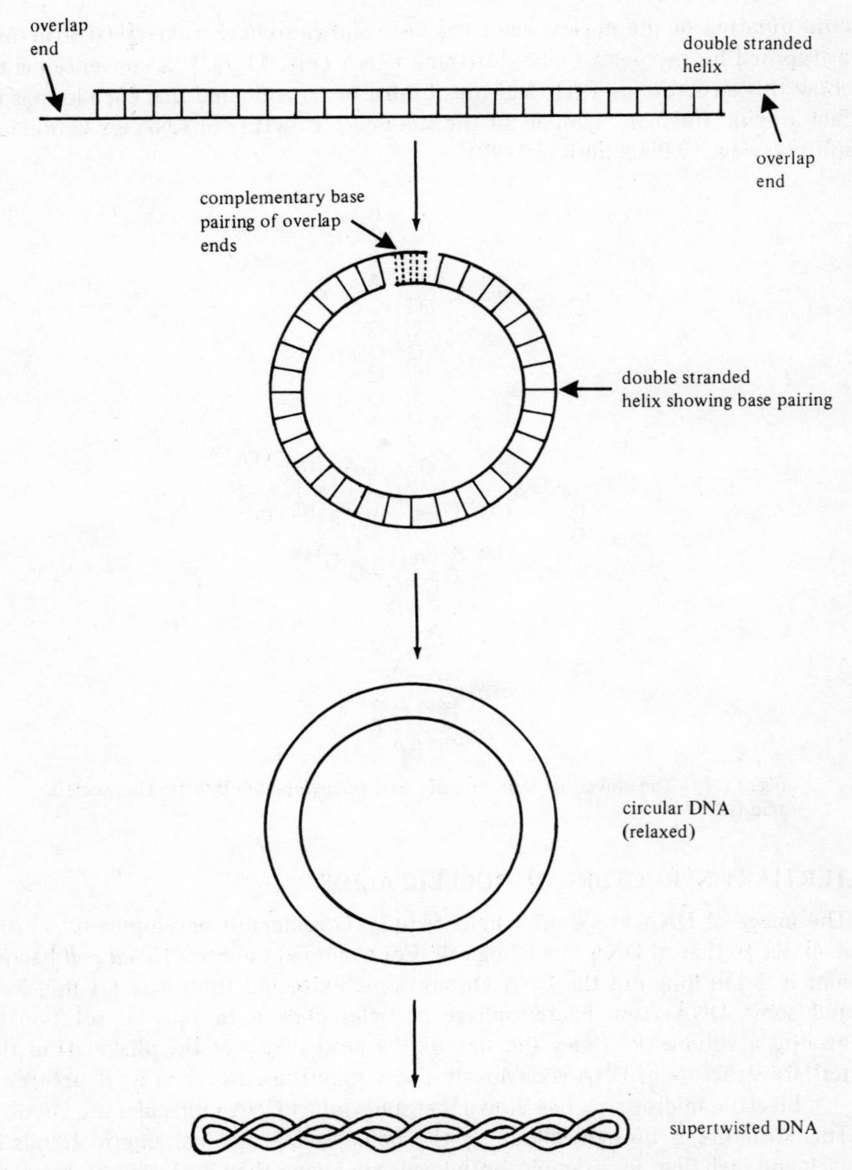

Fig. 11.8 – Schematic representation of the formation of tertiary structures of DNA.

Electron microscopy has also shown that RNA molecules have compact tertiary structures which are held together by hydrogen bonding between bases. These hydrogen bonds are not the same as those between the usual base pairs which hold the helices of the secondary structure together, with extensive use being made of the C-2 hydroxyl group of the D-ribofuranosyl residues as donors or acceptors of hydrogen bonds. The cloverleaf structure of tRNA is folded such that an 'L'-shaped conformation is obtained with the exclusion of water molecules with the anticodon region at one extremity and the tail of the molecule at the other. A more extensive discussion of the tertiary structure of nucleic acids can be found in Blackburn (1979b).

FUNCTION OF NUCLEIC ACIDS

The double helical model of DNA was used by Watson and Crick (1953b) to explain the mechanism for the replication of DNA. Each strand of a 'parent' DNA molecule acts as a template for a 'daughter' molecule which itself acts as a template for its other companion strands (Fig. 11.9), all of which contain an exact copy of the original due to the specific pairing of the bases. The unwinding of the DNA molecules, at the replicating forks, is promoted by binding numerous protein molecules to the parental DNA molecules. It has been proposed that complementary RNA (cRNA) is involved in this replication (see review by Blackburn, 1979c).

The base sequences in DNA hold the heredity data and control the sequence of aminoacids in protein synthesis. Two separate stages take place in this control. Firstly, the information is transcribed, in the cell nucleus from the DNA to mRNA by the DNA acting as a template for the formation of the mRNA, with biosynthesis taking place in a $5' \rightarrow 3'$ direction (Fig. 11.10). Only one strand of the DNA is used unless the DNA is damaged or if the strands are separated, when both strands act as templates. Secondly, the translation of the mRNA information into protein sequence takes place in the cytoplasm.

For successful translation of the information in the mRNA into protein sequence, the genetic code, a sequence of three bases (or codons) in the mRNA which codes for a given aminoacid, is followed, in the ribosome by specific tRNA–aminoacid complexes which have the complementary three bases in the anticodon region (see Fig. 11.7). The ribosome moves along the mRNA molecule binding an aminoacid from the correct tRNA-aminoacids, which have bound to the mRNA, to the forming protein chain through the action of the various ribosomal enzymes (see Fig. 11.11). *N*-Formylmethionine always acts as the first residue since this represents the starting code (AUG or GUG) for transmission whilst the codons UAA, UGA and UAG terminate the protein synthesis. Table 11.1 shows the possible codons and the aminoacids for which they code. After synthesis the protein can be modified by removal of a terminal residue or terminal sequence, formation of disulphide bonds or conversion of residues to others by, for example, hydroxylation. For a fuller explanation of protein synthesis and the genetic code the reader should consult Stryer (1981).

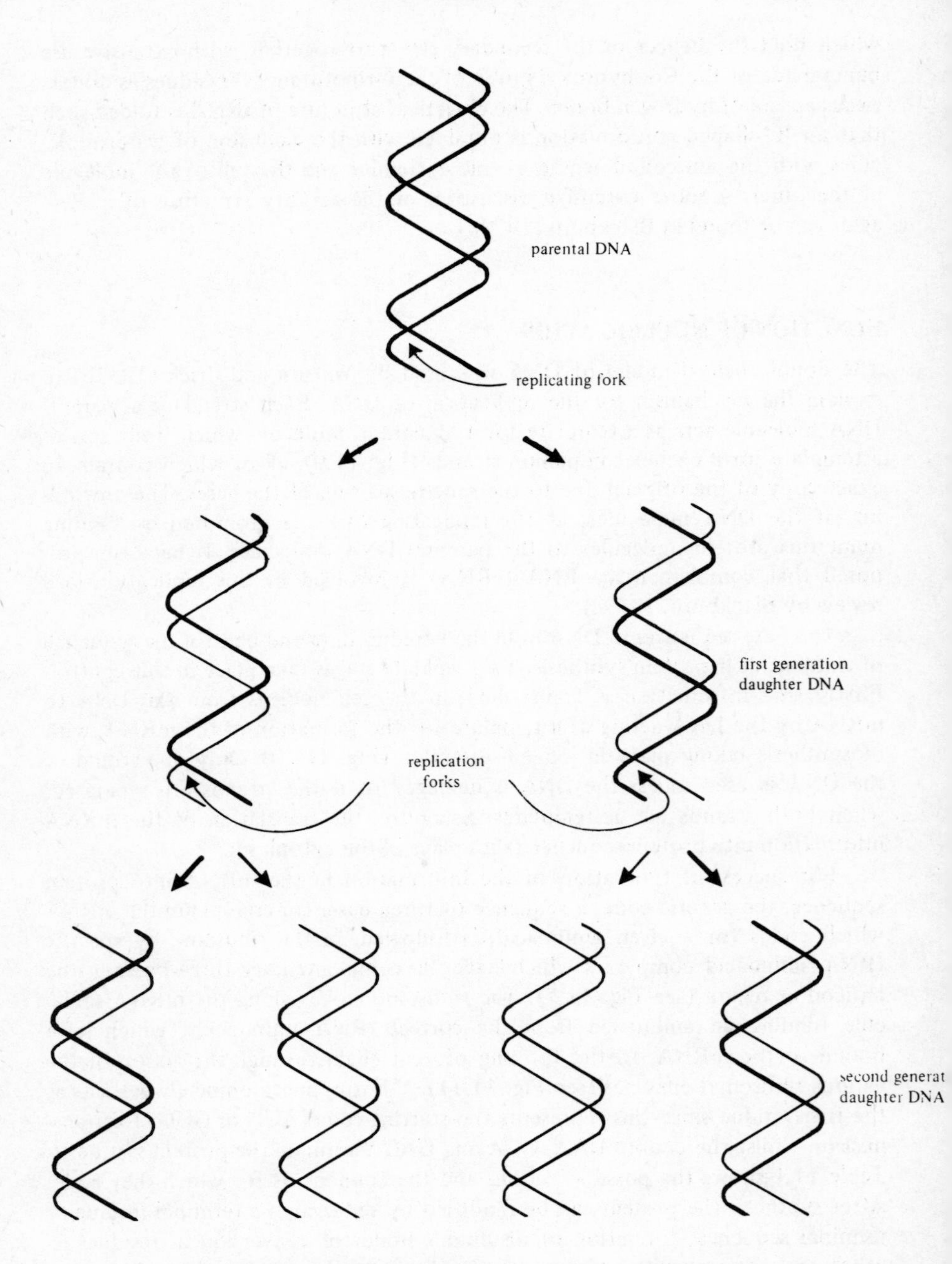

Fig. 11.9 – Replication of DNA (parental DNA strands shown in heavier print).

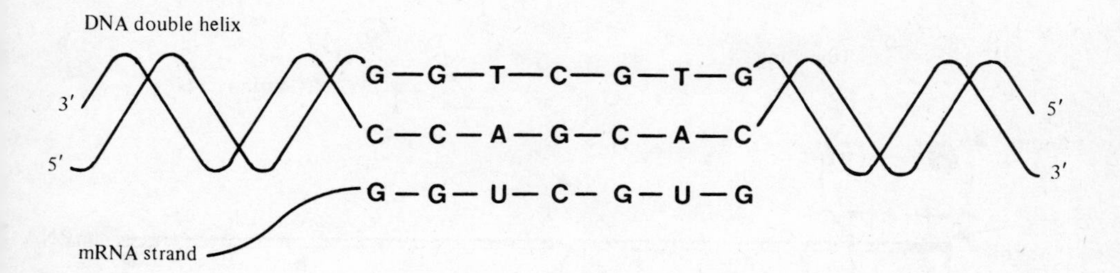

Fig. 11.10 – Biosynthesis of mRNA.

Table 11.1 The genetic code

First position (5′-end)	Second position				Third position (3′-end)
	U	C	A	G	
U	Phe[a]	Ser	Tyr	Cys	U
	Phe	Ser	Tyr	Cys	C
	Leu	Ser	Stop[b]	Stop[b]	A
	Leu	Ser	Stop[b]	Trp	G
C	Leu	Pro	His	Arg	U
	Leu	Pro	His	Arg	C
	Leu	Pro	Gln	Arg	A
	Leu	Pro	Gln	Arg	G
A	Ile	Thr	Asn	Ser	U
	Ile	Thr	Asn	Ser	C
	Ile	Thr	Lys	Arg	A
	Met	Thr	Lys	Arg	G
G	Val	Ala	Asp	Gly	U
	Val	Ala	Asp	Gly	C
	Val	Ala	Glu	Gly	A
	Val	Ala	Glu	Gly	G

a   All L-aminoacids except glycine

b   Termination signal, does not code for any aminoacid

a) initiation

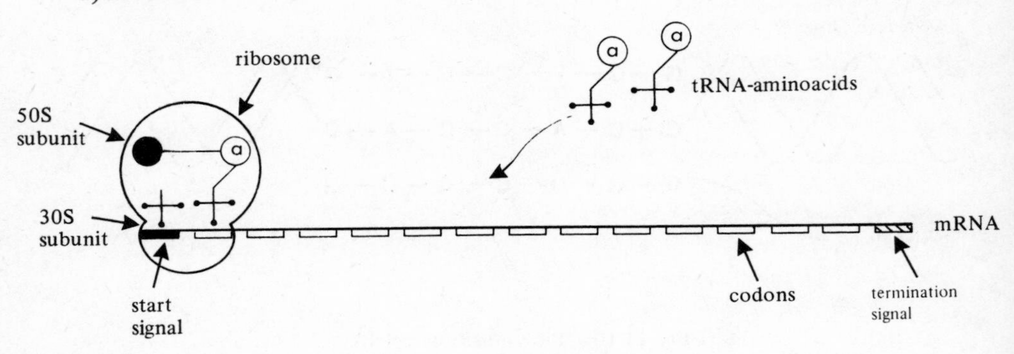

b) elongation

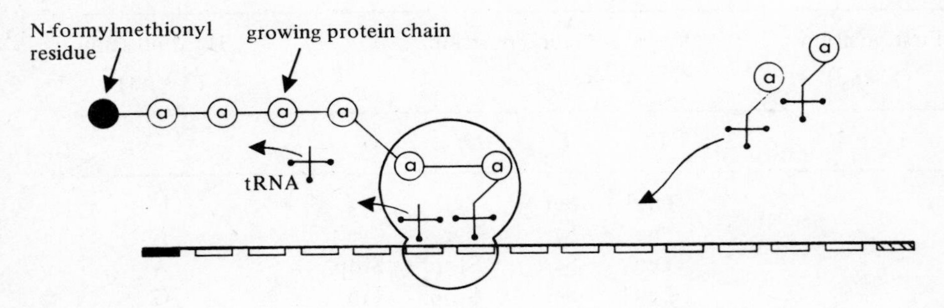

c) termination

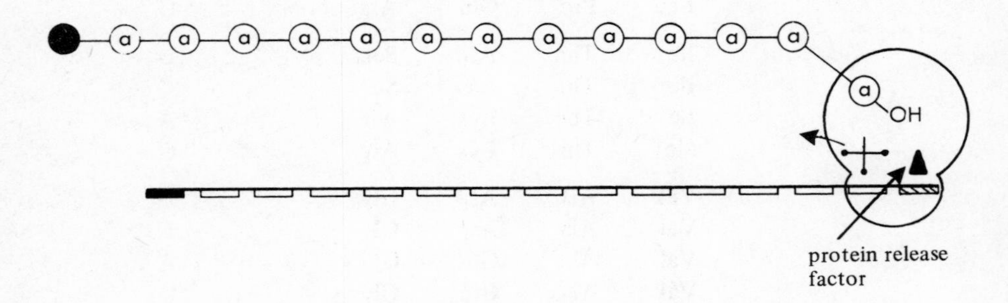

Fig. 11.11 – Transmission of the nucleic acid sequence in mRNA into protein sequence *via* the action of ribosomes and various tRNA aminoacids.

GENETIC ENGINEERING

Two of the fundamental objectives of genetic engineering are the correction of genetic defects and the augmentation of sound genes by others to improve the biosynthetic processes. To date, much work has been carried out in preparing synthetic genes to carry out a nonnatural but important process. This requires the insertion of the required genetic information into the DNA molecule of a chosen species (normally plasmids of *Escherichia coli* which are small, circular chromosomal elements found in the bacterial cytoplasm and which replicate with the bacterium).

A major success in this field is the synthesis of human insulin by *E. coli*. The first stage of the process is to prepare a complementary DNA (cDNA) to the mRNA for the synthesis of proinsulin (a precursor of active insulin). This is then inserted into the plasmid by cutting the plasmid with a specific DNA ligase and joining the DNA molecule to it to reform the circular shape in a manner similar to the formation of circular DNA discussed earlier (p. 257), using base pairing of unequal ends to hold the parts of the molecule together and 'repairing' the cut with another DNA ligase. The closed, circular nucleic acid containing the insulin gene is then incubated to increase the number of plasmids containing the insulin gene and allowed to react in a medium from which human insulin can ultimately be extracted.

Other examples in which genetic engineering of DNA molecules has led to the production of biologically important materials, which cannot be prepared with the required activity or purity by normal chemical methods, by bacterial fermentation include the production of peptide hormones (for example, somatostatin) and proteins (for example, interferon, the antiviral agent). One particularly important aspect is that a useful function from a pathological organism can be transferred into a nonpathological organism, thereby avoiding the use of a pathological organism. For further examples and discussion of the more practical aspects of genetic engineering including the application to industrial processes, see the reviews by Atherton *et al.* (1979), Wu (1979), Zadrazil and Sponar (1980), Gilbert and Villa-Komaroff (1980), and Setlow and Hollaender (1979 and 1980).

CHAPTER 12

Antibiotics

An antibiotic is a compound, usually elaborated by microorganisms, which inhibits the growth of other microorganisms and animal and plant tumours. Synthetic antibiotics, having structures similar to the naturally occurring materials, are produced by biosynthetic reactions, arising from natural organisms, or by chemical reactions (see Rinehart and Suami, 1980). Those elaborated by *Streptomyces* species are frequently given names ending in 'mycin' whilst those from other sources use the ending 'micin'. The mechanism of action often involves interference with DNA, RNA or protein synthesis. There are many varied types of compound which possess antibiotic properties (see Sammes, 1977 onwards) and as early as 1944, when streptomycin, an aminoglycoside antibiotic, which is used in the treatment of tuberculosis, was isolated, it was discovered that naturally-occurring antibiotics contained carbohydrate moieties. In the last 20 years the number of known carbohydrate-containing antibiotics has increased rapidly and the discovery of many unusual and unique amino-, deoxy- and branched-chain monosaccharides (see Fig. 12.1) has stimulated increased interest in the distribution of these unusual carbohydrates in nature. The number of known naturally occurring aminosugars reflects this interest with four being known in 1950, 20 in 1960, 50 in 1970 and to date over 80 are known.

The carbohydrate-containing antibiotics can be grouped into several broad classes, in some of which the compounds are largely or completely carbohydrate whilst in others the carbohydrate is only a small part of the molecule. For extensive coverage of the structures and action of carbohydrate-containing antibiotics the reader is referred to the numerous reviews and books which are available. Reviews of the earlier work on carbohydrate-containing antibiotics are available (for example, Henessian and Haskell, 1970, Umezawa, 1976, Reden and Dürckheimer, 1979, and Horton and Wander, 1980a), whilst the structure, isolation, properties and mode of action are reviewed in two ongoing series (Various Authors, 1968 onwards, and Sammes, 1977 onwards). In this chapter carbohydrates are referred to by their trivial names where their systematic name is given in Fig. 12.1.

Aldoses

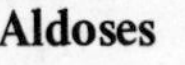

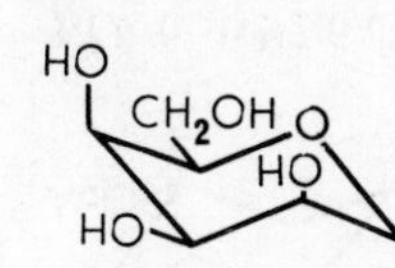

β-D-Talopyranose

β-D-Psicofuranose

2,6-Di-*O*-methyl-β-D-mannopyranose (Curamicose)

3-Deoxy-D-*glycero*-β-D-*glycero*-pentafuranose

4-*C*-Fluoro-5-sulphamido-β-D-ribofuranose

Deoxysugars

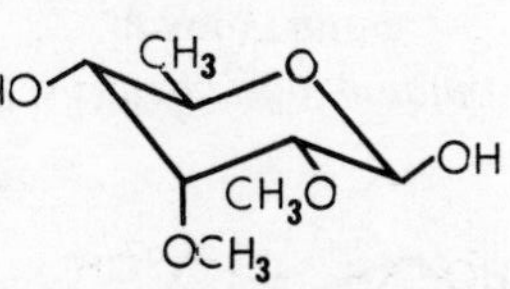

6-Deoxy-2,3-di-*O*-methyl-β-D-allopyranose (Mycinose)

6-Deoxy-4-*O*-methyl-β-D-galactopyranose (Curacose)

6-Deoxy-2,4-di-*O*-methyl-β-D-galactopyranose (Labilose)

6-Deoxy-β-D-*arabino*-5-hexulofuranose (Hygromycin A)

Fig. 12.1 – Structures of some of the less common monosaccharides found in antibiotics (***continued next page***)

Deoxysugars (*continued*)

2,6-Dideoxy-β-D-*arabino*-hexopyranose (Olivose)

2,6-Dideoxy-β-L-*lyxo*-hexopyranose (L-Oliose)

2,6-Dideoxy-β-D-*ribo*-hexopyranose (Digitoxose)

2,6-Dideoxy-β-D-*xylo*-hexopyranose (Boivinose)

6-Deoxy-β-D-*erythro*-2,5-hexodiulofuranose

2,6-Dideoxy-3-*O*-methyl-β-L-*arabino*-hexopyranose (L-Oleandrose)

2,6-Dideoxy-3-*O*-methyl-β-D-*ribo*-hexopyranose (Cymarose, Variose)

2,6-Dideoxy-4-*O*-methyl-β-D-*lyxo*-hexopyranose (Olivomose)

3-*O*-Acetyl-2,6-dideoxy-β-D-*lyxo*-hexopyranose (Acetyl-D-oliose)

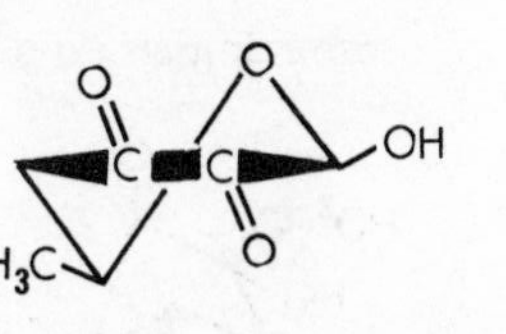

4,6-Dideoxy-2,3-hexodiulopyranose (Actinospectose)

2,3,6-Trideoxy-β-D-*erythro*-hexopyranose (Amicetose)

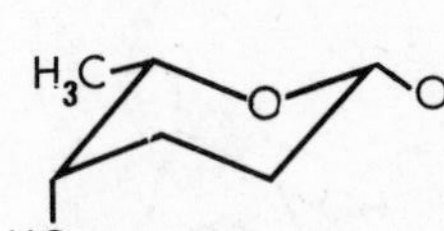

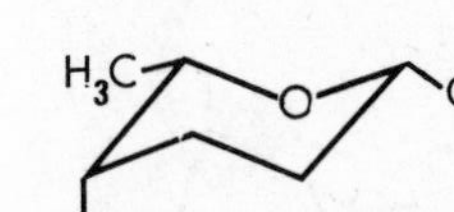

2,3,6-Trideoxy-β-L-*threo*-hexopyranose (Rhodinose)

Aminosugars

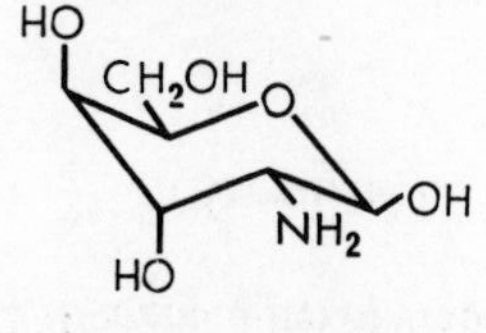

2-Amino-2-deoxy-β-D-gulopyranose

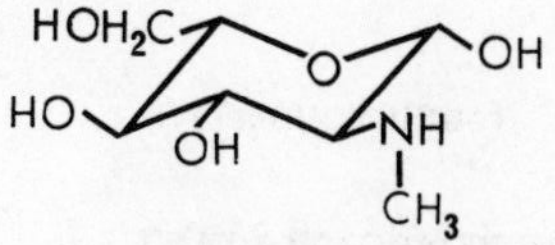

2-Deoxy-2-*N*-methyl-amino-β-L-glucopyranose

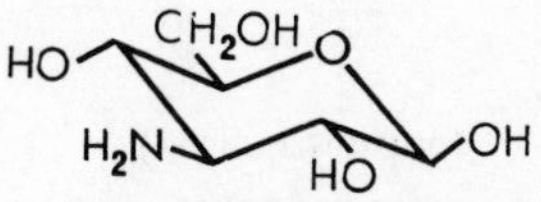

3-Amino-3-deoxy-β-D-glucopyranose (Kanosamine)

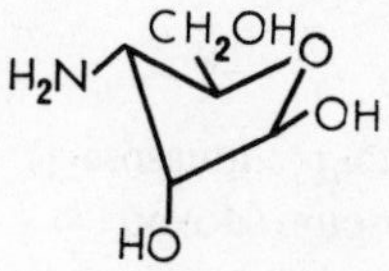

3-Amino-3-deoxy-β-D-ribofuranose

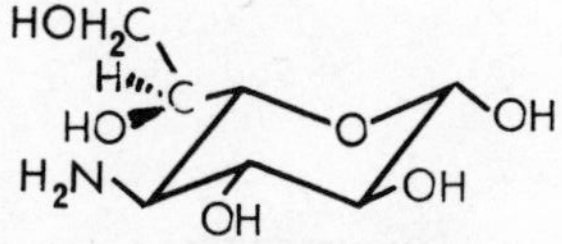

4-Amino-4-deoxy-β-L-*glycero*-L-*gluco*-heptopyranose

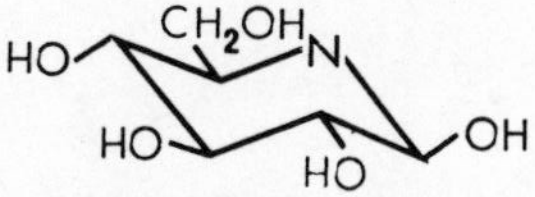

5-Amino-5-deoxy-β-D-glucopyranose (Nojirimycin)

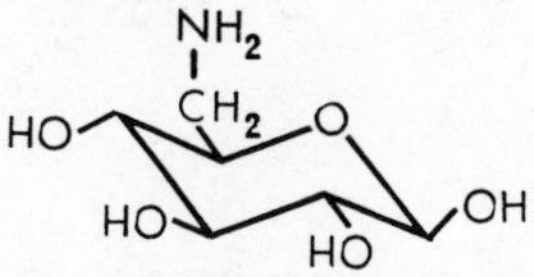

6-Amino-6-deoxy-β-D-glucopyranose

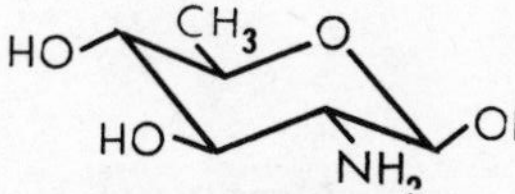

2-Amino-2,6-dideoxy-β-D-glucopyranose (Quinovosamine)

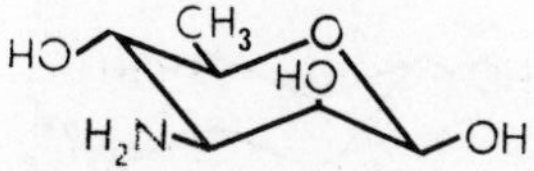

3-Amino-3,6-dideoxy-β-D-mannopyranose (Mycosamine)

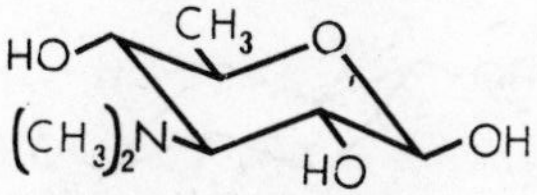

3,6-Dideoxy-3-*N,N*-dimethylamino-β-D-glucopyranose (Mycaminose)

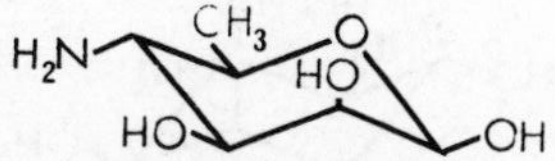

4-Amino-4,6-dideoxy-β-D-mannopyranose (Perosamine)

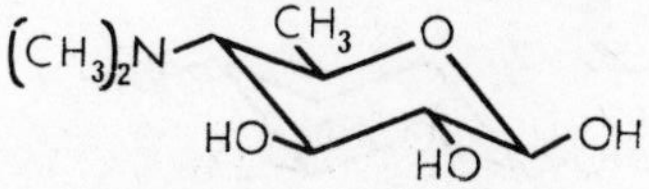

4,6-Dideoxy-4-*N,N*-dimethylamino-β-D-glucopyranose (Amosamine)

Fig. 12.1 – Structures of some of the less common monosaccharides found in antibiotics (*continued next page*)

Aminosugars (*continued*)

6-Amino-6,8-dideoxy-β-D-*erythro*-D-*galacto*-octopyranose
(Lincosamine)

2,6-Diamino-2,6-dideoxy-β-D-glucopyranose

(Neosamine C)

2,6-Diamino-2,6-dideoxy-α-L-idopyranose

(Paromose)

3-Amino-2,3,6-trideoxy-β-L-*ribo*-hexopyranose

(Ristosamine)

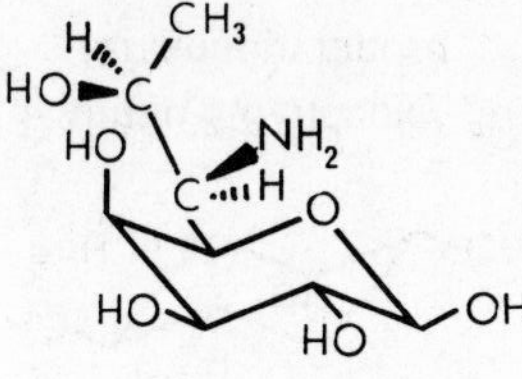

3-Amino-2,3,6-trideoxy-α-D-*arabino*-hexopyranose

(Acosamine)

3-Amino-2,3,6-trideoxy-β-L-*lyxo*-hexopyranose

(Daunosamine)

3-*N*,*N*-Dimethylamino-2,3,6-trideoxy-β-D-*xylo*-hexopyranose
(Angolosamine)

3-*N*,*N*-Dimethylamino-3,4,6-trideoxy-β-D-*xylo*-hexopyranose
(Desosamine, Picrocin)

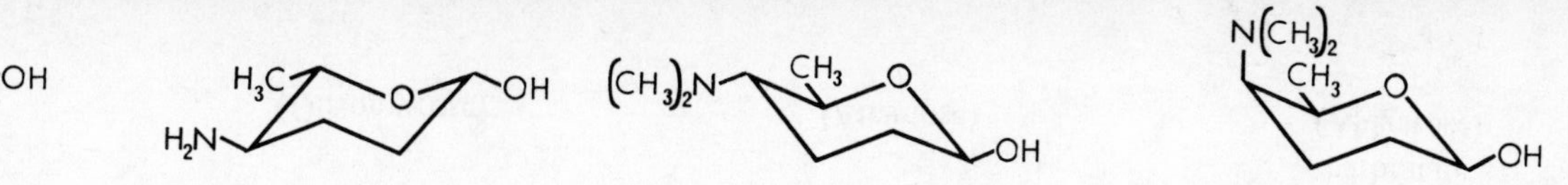

3-*N,N*-Dimethylamino-2,3,6-trideoxy-β-L-*lyxo*-hexopyranose (Rhodosamine)

4-Amino-2,3,4,6-tetradeoxy-β-L-*erythro*-hexopyranose (Tolyposamine)

4-*N,N*-Dimethylamino-2,3,4,6-tetradeoxy-β-D-*erythro*-hexopyranose (Forosamine)

4-*N,N*-Dimethylamino-2,3,4,6-tetradeoxy-β-D-*threo*-hexopyranose (Ossamine)

2,4-Diamino-2,3,4,6-tetradeoxy-β-D-*arabino*-hexopyranose (Kasugamine)

Aminouronic acids

4-Amino-4-deoxy-α-D-*erythro*-2-hexenopyranuronic acid

4-Amino-4-deoxy-β-D-glucopyranuronic acid

5-Amino-5-deoxy-β-D-allofuranuronic acid

Branched-chain monosaccharides

6-Deoxy-3-*C*-methyl-β-D-mannopyranose (Evalose)

6-Deoxy-3-*C*-methyl-α-L-talopyranose (Vinelose)

6-Deoxy-5-*C*-methyl-4-*O*-methyl-α-L-*lyxo*-hexopyranose (Noviose)

5-Deoxy-3-*C*-oxomethyl-α-L-lyxofuranose (Streptose)

Fig. 12.1 – Structures of some of the less common monosaccharides found in antibiotics (*continued next page*)

Branched-chain monosaccharides (*continued*)

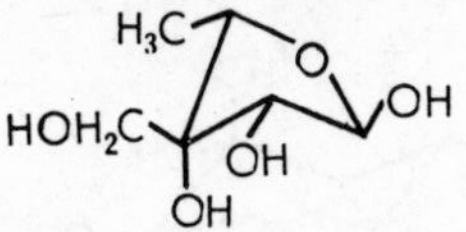

5-Deoxy-3-*C*-hydroxy-methyl-α-L-lyxofuranose (Dihydrostreptose)	6-Deoxy-5-*C*-methyl-4-*O*-methyl-3-*O*-(5-methyl-2-pyrrolyl)-β-L-*lyxo*-hexopyranose (Coumerose)	2,6-Dideoxy-3-*C*-methyl-3-*O*-methyl-β-L-*ribo*-hexopyranose (Cladinose)	2,6-Dideoxy-3-*C*-methyl-β-L-*ribo*-hexopyranose (L-Mycarose)
2,6-Dideoxy-3-*C*-methyl-β-D-*arabino*-hexopyranose (Evermicose)	4-*O*-Acetyl-2,6-dideoxy-3-*C*-methyl-β-L-*arabino*-hexopyranose (Olivomycose)	2,6-Dideoxy-3-*C*-methyl-3-*O*-methyl-α-L-*xylo*-hexopyranose (Arcanose)	4,6-Dideoxy-3-*C*-[(S)-1-hydroxyethyl]-β-D-*ribo*-hexopyranose 3,3′-cyclic carbonate (Aldgarose)

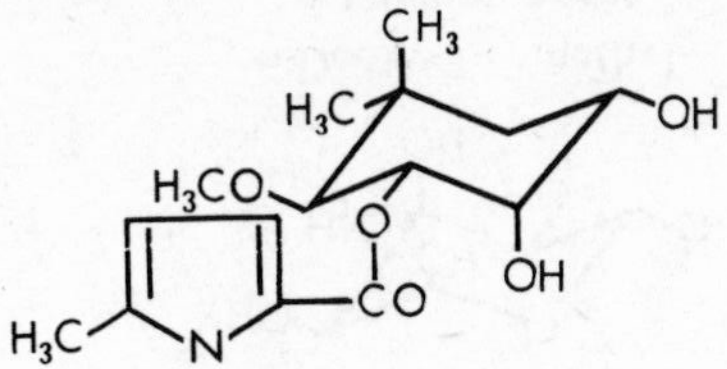

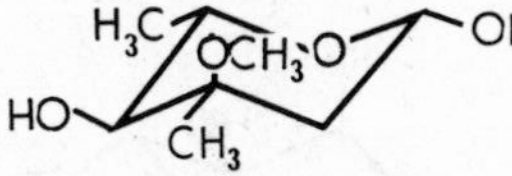

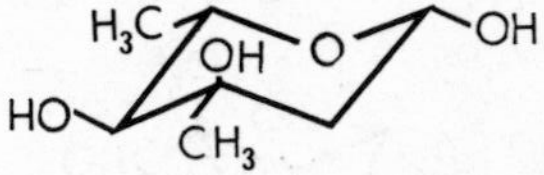

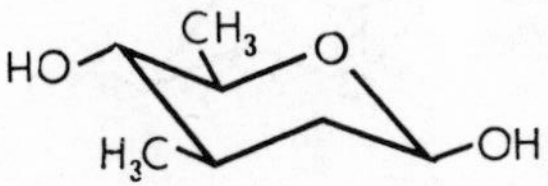

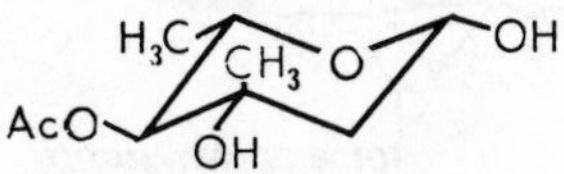

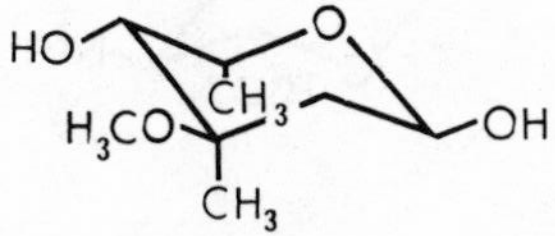

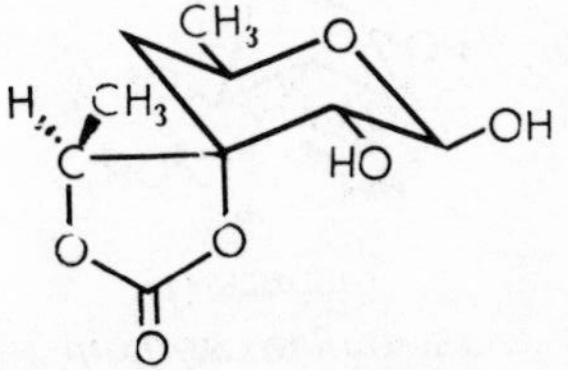

4-*C*-(1-Hydroxyethyl)-2,6-dideoxy-β-L-*lyxo*-hexopyranose

3-Amino-2,3,6-trideoxy-3-*C*-methyl-β-L-*lyxo*-hexopyranose (Vancosamine)

4-*O*-Methyl-3-*C*-methyl-3-nitro-2,3,6-trideoxy-β-L-*arabino*-hexopyranose (Evernitrose)

2,3,6-Trideoxy-4-*C*-(1-oxo-2-hydroxyethyl)-β-L-*threo*-hexopyranose (L-Pillarose)

Cyclitol derivatives

1-Deoxy-3-*O*-carbomoyl-1-guanidino-*scyllo*-inositol

(Bluensidine)

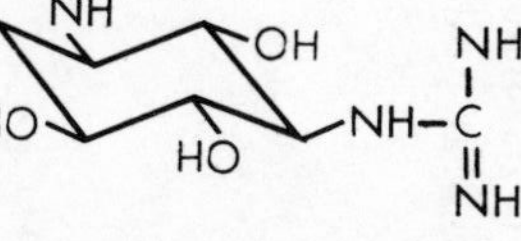

1,3-Dideoxy-1,3-diguanidino-*scyllo*-inositol

(Streptidine)

1,3-Di-(*N*-methylamino)-*myo*-inositol

(Actinamine)

1D-(1,3,5/4,6)-1,3-Diamino-1,2,3-trideoxy-4,5,6-cyclohexanetriol

(2-Deoxystreptamine)

1D-(1,3,5/4,6)-1-Amino-1,2,3-trideoxy-3-*N*-methylamino-4,5,6-cyclohexanetriol (Hyosamine)

1D-1-Amino-1-deoxy-*neo*-inositol

Fig. 12.1 – Structures of some of the less common monosacchardies found in antibiotics (shown in one anomeric form only). See Figs. 2.8 and 8.9 for other monosaccharide structures.

NUCLEOSIDE ANTIBIOTICS

Only a limited number of nucleosides can be accepted by nature for inclusion in the biosynthetic processes, and therefore, most nucleoside antibiotics contain either a 'normal' sugar or base whilst the corresponding base or sugar has a structure which is not normally encountered in the common nucleosides which make up DNA or RNA (see Chapter 11).

Purine derivatives

Puromycin (149) was the first member of the nucleoside antibiotic group to be discovered – in the early 1950s. It contains the unusual aminosugar, 3-amino-3-deoxy-D-ribose. A family of related adenine-containing nucleosides have been isolated and include nucleocidin (49) which contains a sulphuric acid derivative of a fluorosugar, septacidin (150) which, on acid hydrolysis, gives the unusual aminosugar 4-amino-4-deoxy-L-*glycero*-L-*gluco*-heptose, glycine and isopalmitic acid, cordycepin (151), which contains 3-deoxy-D-*glycero*-D-*glycero*-pentose, and the angustmycins A (decoyinine, 152) and C (psicofuranine, 153) which contain the ketosugar psicose or the related diketosugar 6-deoxy-D-*erythro*-2,5-hexodiulose. Examples of antibiotics containing 'non-natural' bases include formycin (154) and tubercidin and its analogues toyocamycin and sangivamycin (155) in which the nitrogen atom at position 7 is replaced by a substituted carbon atom.

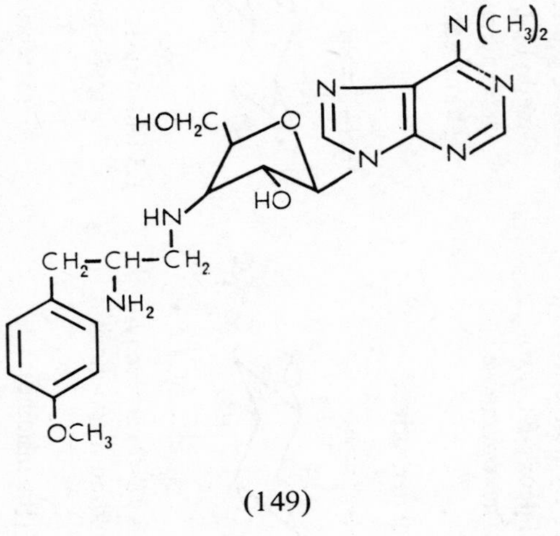

$N(CH_3)_2$ N N N N HOH_2C O HO HN $CH_2-CH-CH_2$ NH_2 OCH_3

(149)

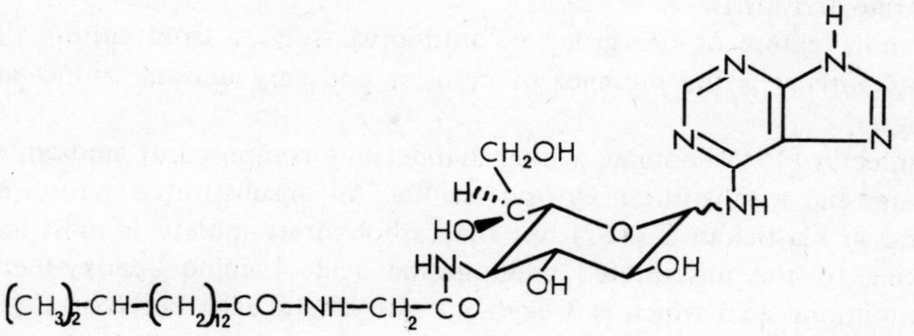

(150)

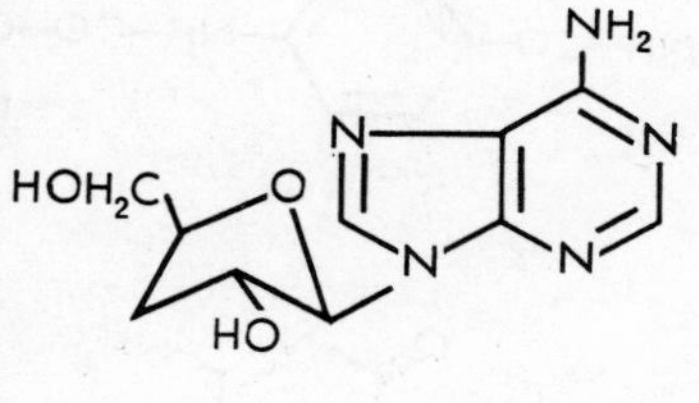

(151)

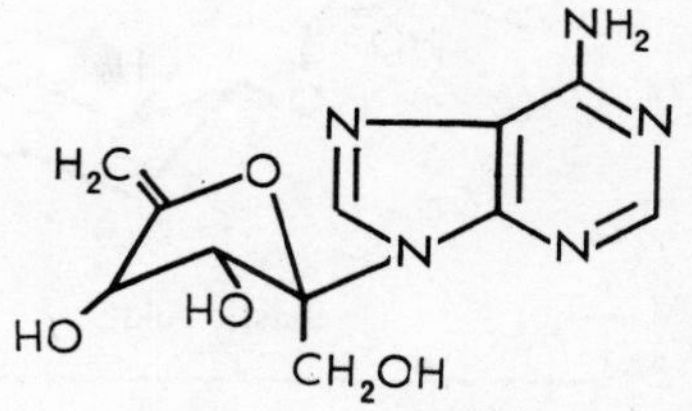

(152)

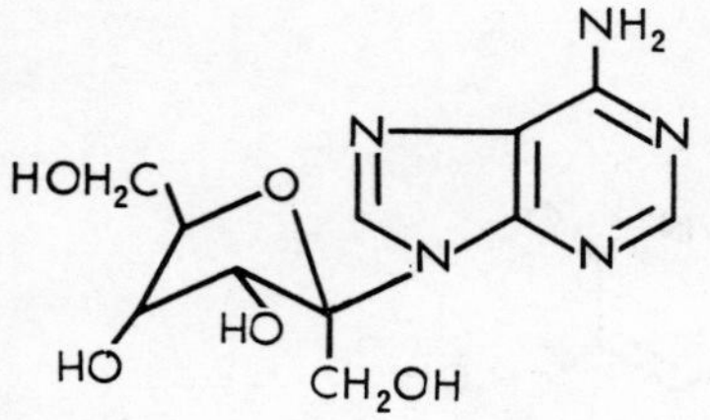

(153)

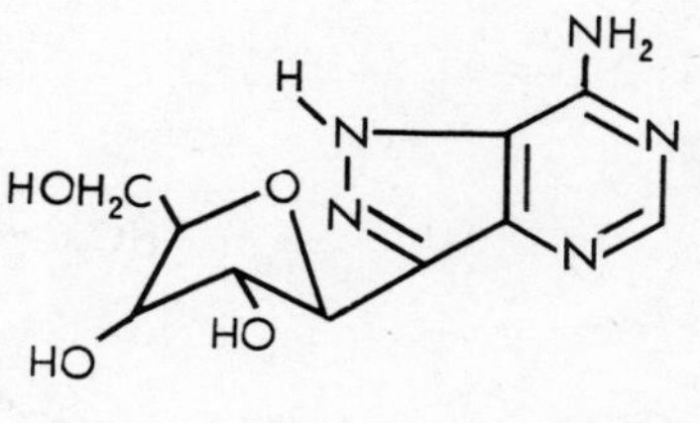

(154)

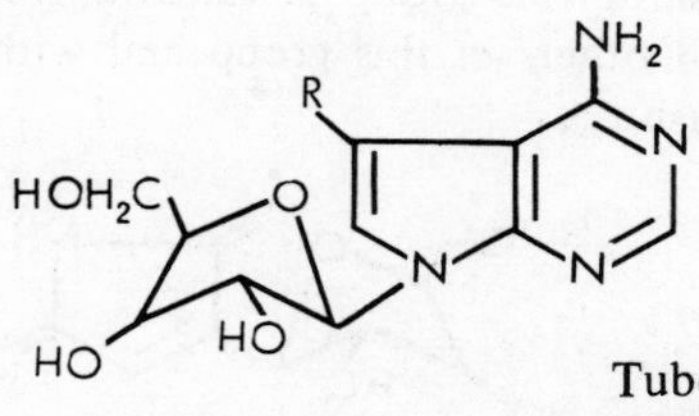

(155)

Tubercidin: R = H
Toyocamycin: R = CN
Sangivamycin: R = $CONH_2$

Pyrimidine derivatives

A common feature of this group of antibiotics isolated from culture filtrates of *Streptomyces* is the presence of cytosine and very unusual amino- and deoxy-sugars.

Amicetin (156) contains a disaccharide unit composed of amosamine and amicetose and a substituted cytosyl residue. An unsubstituted cytosyl residue is found in blasticidin S (157) but the carbohydrate moiety is most unusual, consisting of the unsaturated aminouronic acid, 4-amino-4-deoxy-D-*erythro*-2-hexenouronic acid which is *N*-acylated with blastic acid. Polyoxin C (158) is an example of an antibiotic containing 5-hydroxymethyluracil. It also contains 5-amino-5-deoxy-D-allofuranuronic acid.

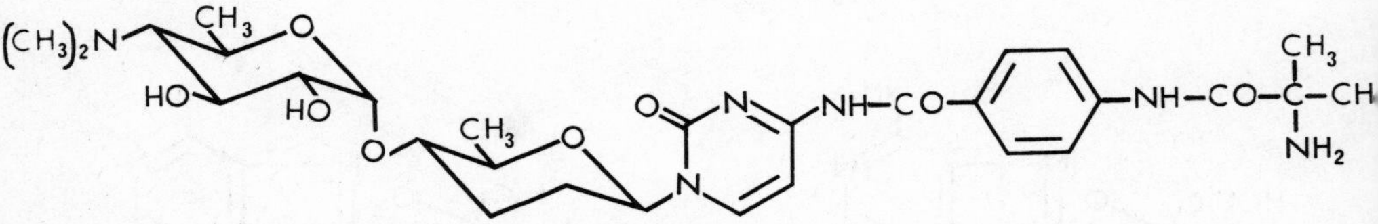

(156)

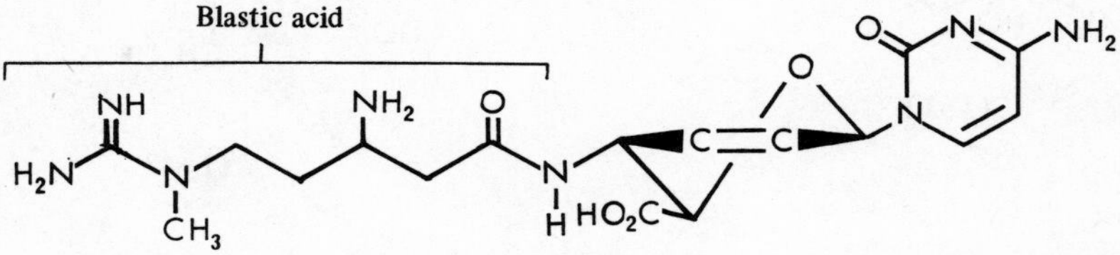

Blastic acid

(157)

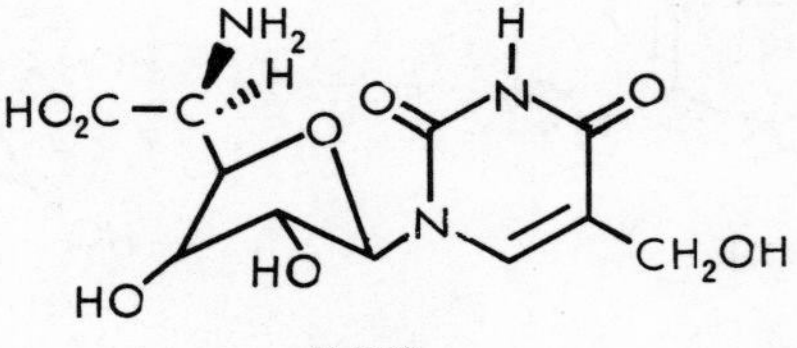

(158)

A number of nucleosides, which contain non-natural pyrimidine bases, have been identified. Typical of this group is showdomycin (159), a *C*-glycoside which, in common with others in this group and with the non-natural purine nucleosides, contains D-ribose.

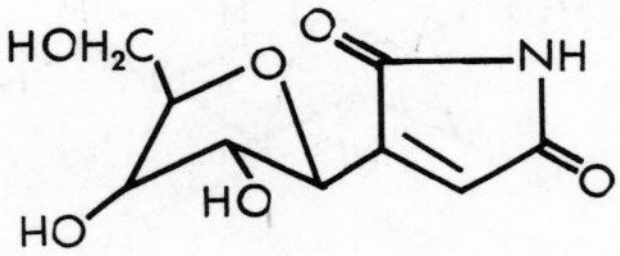

(159)

ANTIBIOTICS CONTAINING AROMATIC GROUPS

This group of antibiotics have characteristic aromatic systems to which one to three carbohydrate residues are attached, either singly or as oligosaccharide units. Typical examples of this group include those frequently referred to as the anthracycline antibiotics because of the tetracyclic aromatic system, such as daunomycin (160) which contains daunosamine, and the cinerubins, pyrromycins and rhodomycins, a family of red coloured antibiotics which contain *inter alia* the *N,N*-dimethyl analogue of daunosamine, rhodosamine (see Fig. 12.1). Pyrromycin has the structure (161) whilst other members of the rhodomycins contain the tetracyclic structure (162) and one or two carbohydrate residues in addition to rhodosamine. Rhodomycin II contains 2,6-dideoxy-L-*lyxo*-hexose, whilst rhodomycin III contains, in addition, the trideoxysugar rhodinose.

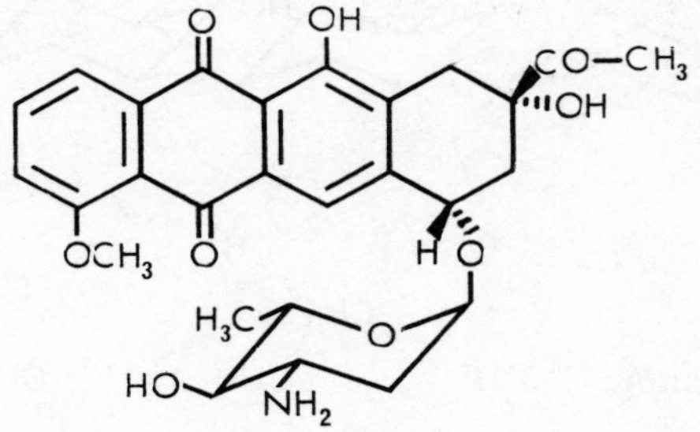

(160)

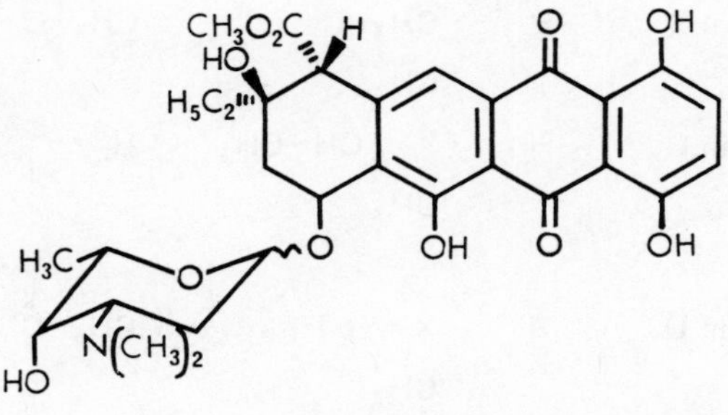

(161)

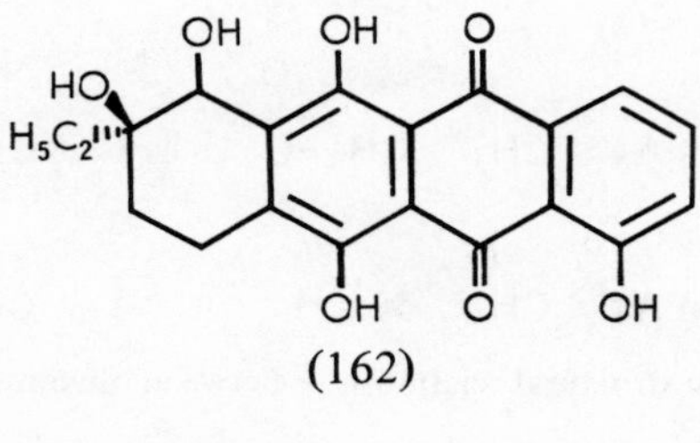

(162)

Another family of antibiotics in this group which contains a characteristic aromatic system is that of the chromomycin antibiotics which include both olivomycins and chromomycins, the general structures and relationships being shown in Fig. 12.2.

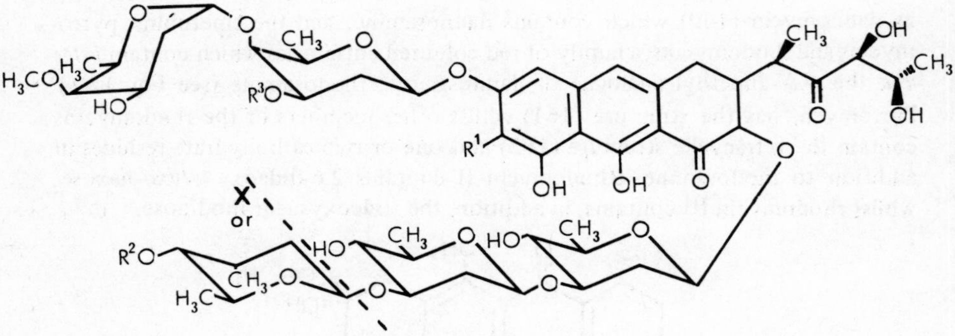

	R^1	R^2	R^3
Olivomycin A	H	$(CH_3)_2CH{-}CH_2$	$CH_3{-}C(=O)$
Olivomycin B	H	$CH_3{-}C(=O)$	$CH_3{-}C(=O)$
Olivomycin C	H	$(CH_3)_2CH{-}CH_2$	H
Olivomycin D	H	X = H	$CH_3{-}C(=O)$
Chromomycin A_2	CH_3	$(CH_3)_2CH{-}CH_2$	$CH_3{-}C(=O)$
Chromomycin A_3	CH_3	$CH_3{-}C(=O)$	$CH_3{-}C(=O)$
Chromomycin A_4	CH_3	X = H	$CH_3{-}C(=O)$

Fig. 12.2 – The structural relationships between olivomycins and chromomycins.

MACROLIDE ANTIBIOTICS

The sugar moiety of macrolide antibiotics forms only a small but important part of the molecule, the main component being a large lactone ring. This ring can contain a conjugated polyene system or a polyfunctional, almost macrocyclic ring.

Polyene derivatives

These substances, which are elaborated by *Actinomyces* and usually have pronounced activity in inhibiting the growth of fungi, exhibit complex ultraviolet spectra on account of the conjugated polyene system. The carbohydrate-containing macrolide antibiotics in this group belong to two general types, those with conjugated tetraene and those with conjugated heptaene systems. Typical of the first type, in which all the double bonds have the *trans* arrangement, is rimocidin (163) which contains a single residue of mycosamine, a common feature of many other polyene macrolides. The conjugated heptaenes, which have stronger antifungal activities than the tetraenes, do not always have the all-*trans* arrangement found in the tetraenes, but their general insolubility and instability have impeded complete characterisation. Antibiotic 67-121-C is the first example, reported in 1977, of a disaccharide component in a heptaene antibiotic; it contains 4-*O*-β-D-mannopyranosylmycosamine.

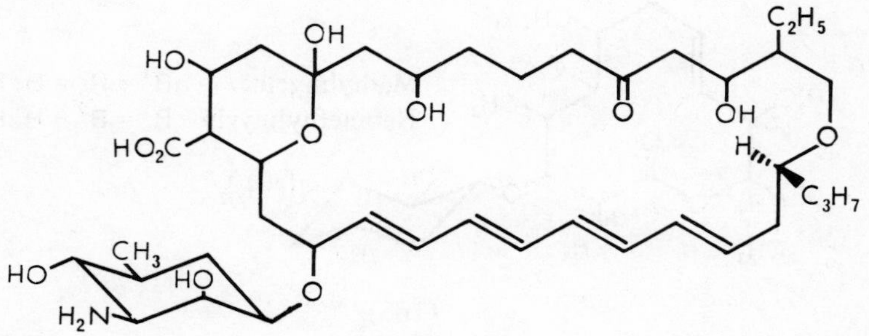

(163)

Macrocyclic-lactone derivatives (macrolides)

This extensive group of antibiotics which is active against Gram-positive organisms contain a characteristic large lactone ring which has a large number of substituents on it. The differences in biological action are a result of subtle changes in the size of the macrolide ring, the number and type of functional groups present and the nature and position of attachment of the carbohydrate moiety. There are essentially two types of macrocyclic-lactone antibiotics, those which contain aminosugars and those which contain deoxysugars.

Picromycin (164) was the first macrolide antibiotic to be discovered, and, together with the related narbomycin and the pair of antibiotics, methylmycin and neomethylmycin (165), contains the aminosugar desosamine, but the member of this group most widely used medicinally is erythromycin (166). This antibiotic contains the branched-chain monosaccharide cladinose, in addition to desosamine.

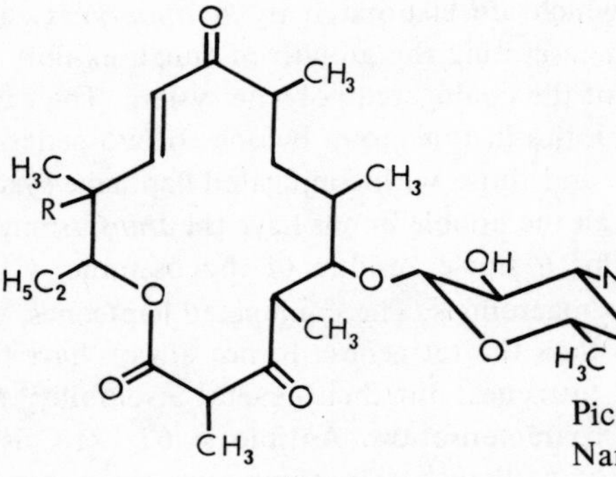

Picromycin: R = OH
Narbomycin: R = H

(164)

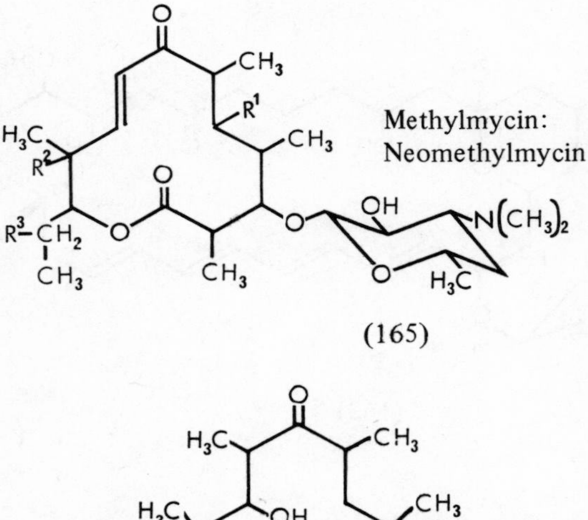

Methylmycin: $R^1 = R^3 = H$; $R^2 = OH$
Neomethylmycin: $R^1 = R^2 = H$; $R^3 = OH$

(165)

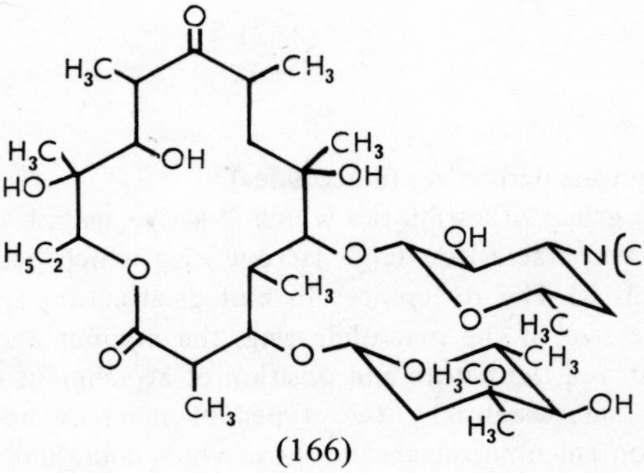

(166)

The spiramycin complex (foromacidins) which contains three components (167) consists of a single forosaminyl residue and a disaccharide unit of mycarose and mycaminose.

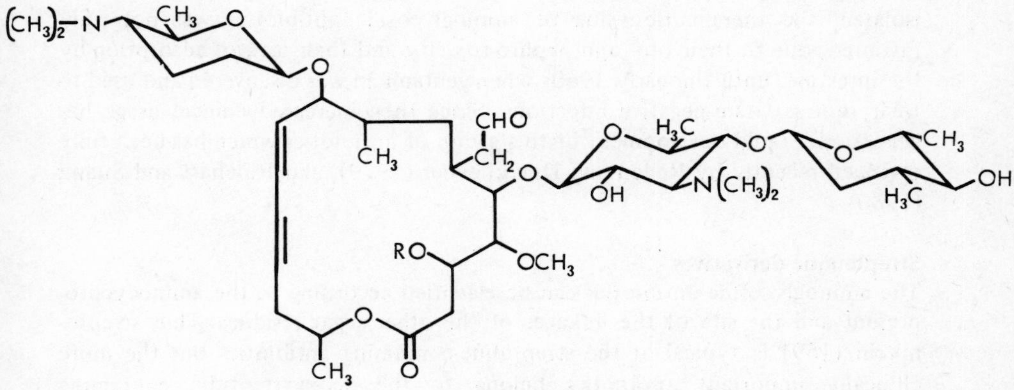

(167)

An example of those macrocyclic-lactone antibiotics which only contain deoxysugars as the carbohydrate moiety is provided by lankamycin (168) which contains the deoxysugars, chalcose (4,6-dideoxy-3-*O*-methyl-D-*xylo*-hexose) and the branched-chain monosaccharide, arcanose.

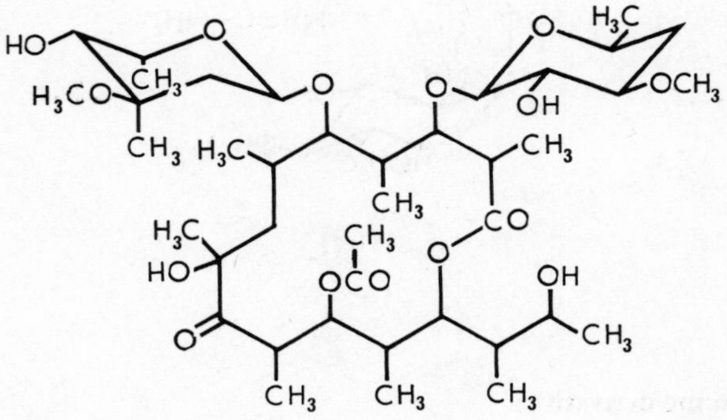

(168)

AMINOGLYCOSIDE ANTIBIOTICS

This class of antibiotics, referred to by some authors as aminocyclitol antibiotics, is perhaps the most interesting to carbohydrate chemists and biochemists in that these antibiotics consist completely of carbohydrate material. The fundamental structural feature of this group is an aminocyclitol residue (hence the alternative name) to which is attached one or more aminosugars and occasionally other carbohydrate residues. They are stable, colourless, basic compounds which are readily soluble in water but only slightly soluble or insoluble in organic solvents. Although streptomycin was the first carbohydrate-containing antibiotic to be isolated, the therapeutic value of aminoglycoside antibiotics was not fully favoured, due to their oto- and nephro-toxicity and their lack of adsorption by the intestine, until the early 1960s when gentamicin was discovered and used to treat serious Gram-negative infections. Since then increased clinical usage has generated a rapid development of this group of antibiotics which has been fully reviewed recently by Reden and Dürckheimer (1979), and Rinehart and Suami (1980).

Streptamine derivatives

The aminoglycoside antibiotics can be classified according to the aminocyclitol present and the site of the linkages of the other sugar residues. Thus streptomycin (169) is typical of the streptidine-containing antibiotics but the more clinically important antibiotics belong to the deoxystreptidine-containing antibiotics which can be subdivided into three groups.

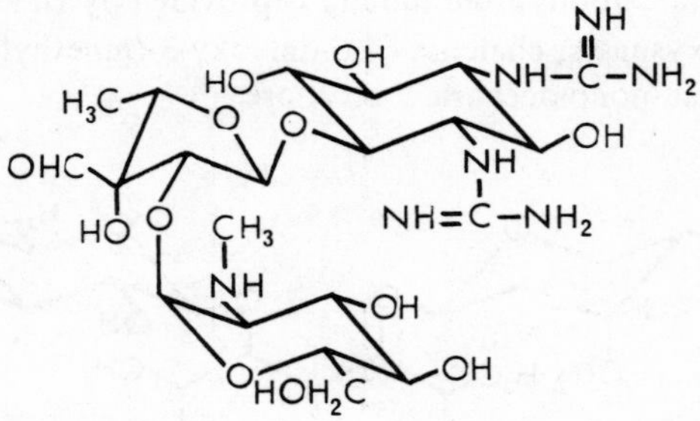

(169)

Deoxystreptamine derivatives

The 4-substituted deoxystreptamine antibiotics, which have no real clinical importance, include neamine (neomycin A, 170). Neomycins B and C (see Fig. 12.3) are 4,5-disubstituted deoxystreptamine antibiotics and therefore contain additional carbohydrate residues attached to the 5 position of the deoxystreptamine residue of neamine. Another related group of 4,5-disubstitued deoxystreptamine antibiotics is the paromomycins (see Fig. 12.3) which contain 2-amino-2-deoxy-D-glucose rather than 2,6-diamino-2,6-dideoxy-D-glucose which is found in neomycins.

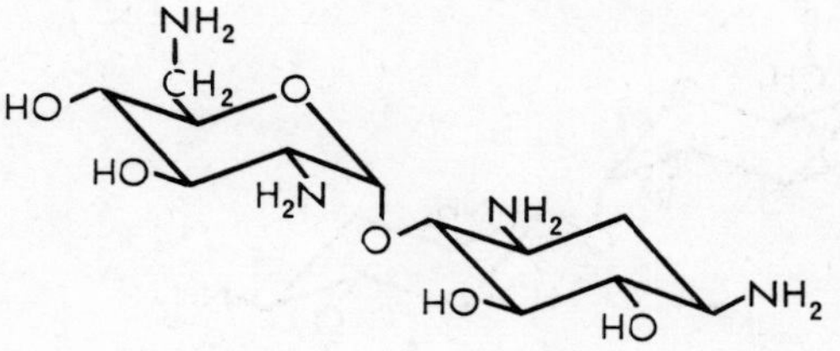

(170)

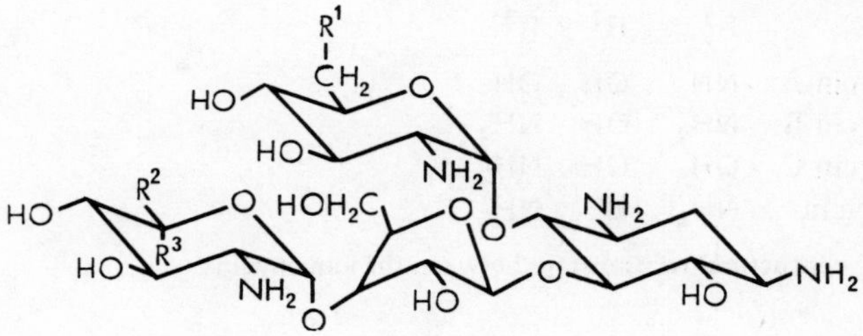

	R^1	R^2	R^3
Neomycin B	NH_2	H	CH_2NH_2
Neomycin C	NH_2	CH_2NH_2	H
Paromomycin I	OH	H	CH_2NH_2
Paromomycin II	OH	CH_2NH_2	H

Fig. 12.3 – The structural relationships between neomycins and paromomycins.

The third group of deoxystreptamine antibiotics is the 4,6-disubstituted deoxystreptamine derivatives, which can be readily distinguished from the 4,5-disubstituted derivatives on account of their larger optical rotations ($[\alpha]_D$ = 120–160° compared to 4–80°). These are probably the most important of the three groups, consisting of the kanamycins and the gentamicins. These related antibiotic types differ from each other in the residue attached to the 6-position of deoxystreptamine, the kanamycins containing kanosamine (see Fig. 12.4), whilst the gentamicins contain 3-deoxy-3-*N*-methylaminopentoses, some of which contain 4-*C*-methyl groups. Fig. 12.5 lists only five gentamicins of the many which are naturally occurring; others contain different substituents (usually hydroxyl, amino or methyl groups or hydrogen atoms) at the eight positions indicated. Sisomicin (171) is related to the gentamicins but contains an unsaturated aminosugar attached to the 4-position of deoxystreptamine.

	R^1	R^2	R^3
Kanamycin A	NH_2	OH	OH
Kanamycin B	NH_2	OH	NH_2
Kanamycin C	OH	OH	NH_2
Tobramycin	NH_2	H	NH_2

Fig. 12.4 – Structural relationships between the kanamycins.

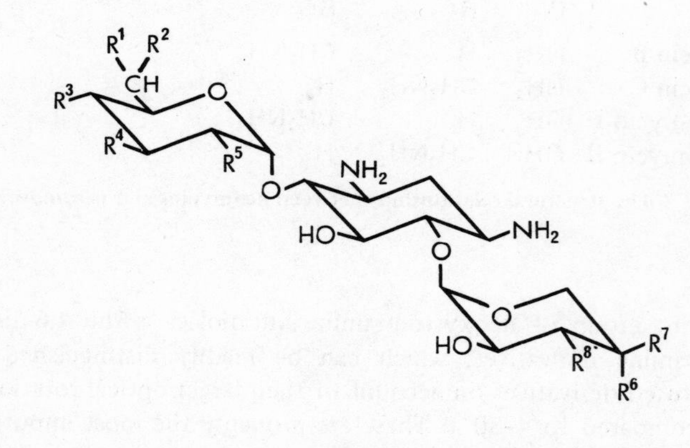

	R^1	R^2	R^3	R^4	R^5	R^6	R^7	R^8
Gentamicin C_1	CH_3	$NHCH_3$	H	H	NH_2	OH	CH_3	$NHCH_3$
Gentamicin C_2	CH_3	NH_2	H	H	NH_2	OH	CH_3	$NHCH_3$
Gentamicin C_{1a}	H	NH_2	H	H	NH_2	OH	CH_3	$NHCH_3$
Gentamicin B	H	NH_2	OH	OH	OH	OH	CH_3	$NHCH_3$
Gentamicin A	H	OH	OH	OH	NH_2	H	OH	$NHCH_3$

Fig. 12.5 – Structural relationships between the gentamicins.

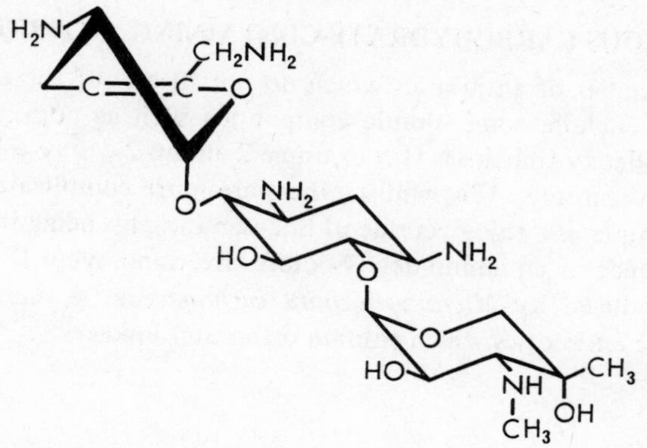

(171)

Miscellaneous aminoglycoside antibiotics

There are a number of aminoglycoside antibiotics which do not contain either of the above cyclitol derivatives. These include the simple 1D-*chiro*-inositol derivative kasugamycin (172), which contains kasugamine as its amidine derivative, and apramycin (173) which contains an octadiose that exists as a rigid bicyclic system.

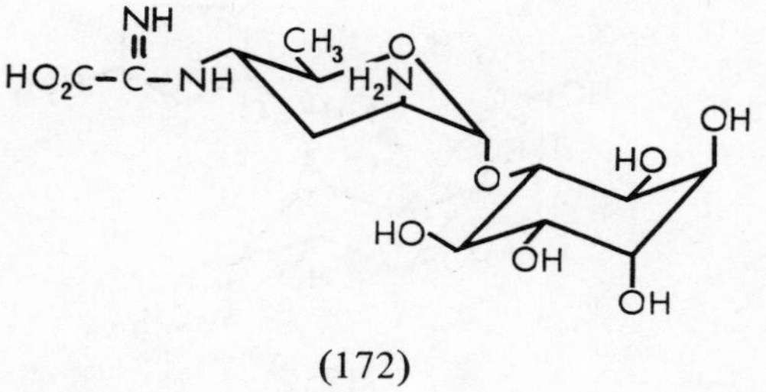

(172)

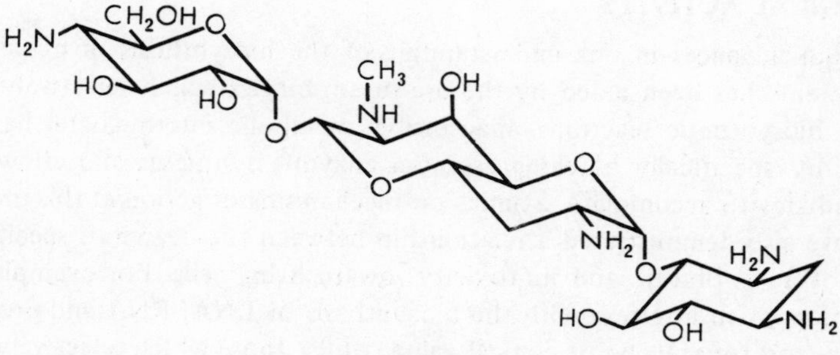

(173)

MISCELLANEOUS CARBOHYDRATE-CONTAINING ANTIBIOTICS

There are a number of antibiotics which do not fit any of the above classes or groups. These include some simple compounds such as nojirimycin (43) and 2/4-amino-2/4-deoxy-trehaloses (for example 2-amino-2-deoxy-α-D-glucopyranosyl-α-D-glucopyranoside (174) whilst others are more complicated. Lincomycin (175), for example, is a thioglycoside of lincosamine, this being the first example of the occurrence of an amino-deoxy-octose. Everninomycin D (176), an octasaccharide produced by *Micromonospora carbonaceae*, is one of a group of oligosaccharide antibiotics which contain orthoester linkages.

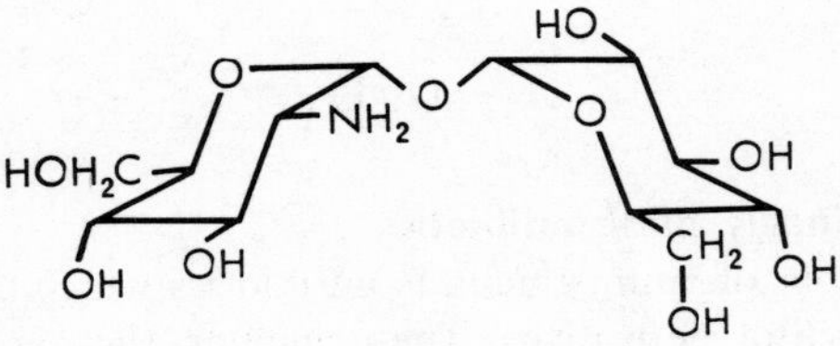

(174)

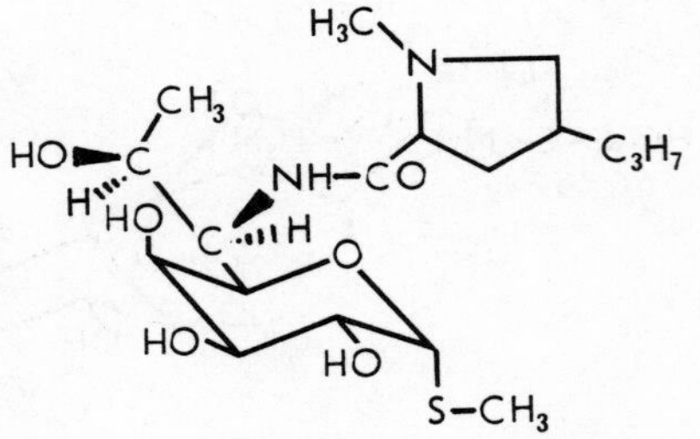

(175)

BIOLOGICAL ACTIVITY

The major advances in our understanding of the biosynthesis of nucleic acids and proteins has been aided by the use of antibiotics which selectively inhibit specific biosynthetic reactions. Inaccessible metabolic intermediates have been studied by specifically blocking essential enzymic syntheses and allowing the intermediates to accumulate. Studies on mechanisms of action at this molecular level have also demonstrated a relationship between the degree of specificity of action of an antibiotic and its toxicity toward living cells. For example, those antibiotics which interfere with the biosynthesis of DNA, RNA and protein are normally too toxic to be of clinical value, whilst those which selectively inhibit protein synthesis are frequently nontoxic and clinically useful.

(176)

The nucleoside antibiotics interfere with the synthesis of DNA and RNA by virtue of their close structural similarity to normal nucleosides. They are therefore highly toxic to living cells. Cordycepin (151) is incorporated into RNA by the normal synthetic processes, but since the 3′-hydroxyl group is lacking, no further nucleotides can be added and chain termination occurs. Puromycin (149) bears close structural resemblance to the aminoacid-bearing end of tRNA and inhibits protein synthesis by competing with aminoacyl tRNA in the ribosome for the activated carboxyl group on the growing peptide chain.

The antibiotics containing aromatic groups have different modes of action. The anthracycline antibiotics such as daunomycin (160) bind strongly to DNA and inhibit the synthesis of RNA through steric interference with the RNA polymerase. Daunomycin binds to both strands of DNA by formation of hydrogen bonds with both the amino groups of daunosamine and hydroxyl groups of the aromatic ring system. The aromatic moiety of the antibiotic fits in between the base pairs whilst the carbohydrate moiety projects into the minor grooves of DNA (see Fig. 11.5) and thereby inhibits the RNA polymerase. The chromomycin antibiotics (Fig. 12.2) also bind to DNA but do not intercalate with the base pairs. They inhibit DNA replication.

The exact mode of action of the macrolide antibiotics has not been elucidated fully but they are thought to render cell membranes permeable to small metabolites, with the result that subsequent metabolic processes are hindered by the loss of these metabolites. Erythromycin (166) is known to inhibit cell-free protein synthesis by interacting with the 50S ribosomal subunit (see Fig. 11.11), thereby inhibiting the synthesis of proteins induced by mRNA.

The aminoglycoside antibiotics appear to have modes of action similar to those of the macrolide antibiotics since they inhibit protein synthesis whilst allowing uninhibited synthesis of DNA or RNA. Streptomycin (169) binds irreversibly to the 30S ribosomal subunit when no mRNA is bound and thereby alters the function but not the formation of the mRNA, tRNA and ribosome complex. In altering the function of the ternary complex, streptomycin inhibits the incorporation of L-phenylalanine, stimulated by polyuridine, into protein, but stimulates the incorporation of other aminoacids (L-isoleucine, L-serine and L-leucine) for which polyuridine does not normally act as the code. The neomycins (Fig. 12.3), kanamycins (Fig. 12.4) and gentamicins (Fig. 12.5) also disturb the fidelity of reading polynucleotides, supposedly by alteration of the triplet code and configuration of the 30S ribosome subunit. From studies on the many structural variations, which have been isolated or synthesised, it has become evident that the number and position of the amino groups are of importance for the efficacy of these antibiotics.

Lincomycin (175), classified under the miscellaneous group of antibiotics, interferes with the protein synthesis in Gram-positive organisms by inhibiting the binding of aminoacyl tRNA to the 50S ribosomal subunit.

Much of the interest in the area of carbohydrate-containing antibiotics is in the production, either by biosynthesis or chemical methods (or a combination of the two), of novel derivatives of antibiotics to alter specific properties of the compound (such as solubility, stability, toxicity or specificity). The reviews by Reden and Dürckheimer (1979) and Rinehart and Suami (1980) and the ongoing series of reviews (Various Authors, 1968 onwards, and Sammes, 1977 onwards) provide up to date descriptions of the work which has been carried out in this area.

CHAPTER 13

Synthetic derivatives of polysaccharides

This chapter describes a number of derivatives of carbohydrates, which are more frequently prepared from polysaccharides than from monosaccharides, etc., and applications of such derivatives. Derivatives of carbohydrates which can be prepared from all types of carbohydrates, such as ethers, esters, etc., have already been described (see Chapter 3) and therefore compounds such as cellulose acetate and rayon are not described in this chapter. The applications of a number of industrially important derivatives is discussed in Chapter 14.

DYE DERIVATIVES

The dyeing of polysaccharides has been studied as an art for centuries for the production of coloured cotton fabrics which are cellulose-based. Traditional methods of dyeing cotton depended on the formation of insoluble dye molecules on the fibres, or of hydrogen bonds between the dye molecules and the fibres, but in the strict sense these cannot be considered as dye derivatives. However, two classes of dyes react with polysaccharides to produce covalently bound derivatives.

The di- and tri-azine dyes are formed by the attachment of chromogenic groups to, for example, 2,4,6-trichloro-*sym*-triazine. The triazinyl group then reacts with primary hydroxyl groups (although secondary hydroxyl groups are not inert) in the polysaccharide to give a dyed polysaccharide (177). Dyes in this group include the Procion® dyes (dichloro-*sym*-triazinyl) and Cibacron® Blue 3G-A (a monochloro-*sym*-triazinyl) and a number of dyed polysaccharides are commercially available including Amylochrome® (Roche) and Blue Dextran® 2000 (Pharmacia). A related group of dyes, the Reactone® dyes which contain trichloropyrimidinyl groups, react in an analogous way to the trichloro-*sym*-triazinyl dyes, but the extent of their subsequent reaction is somewhat less.

The Remazol® type of reactive dyes combine with polysaccharide hydroxyl groups (preferentially the primary hydroxyl groups), by way of unsubstituted intermediates, to give polysaccharide ethers (178).

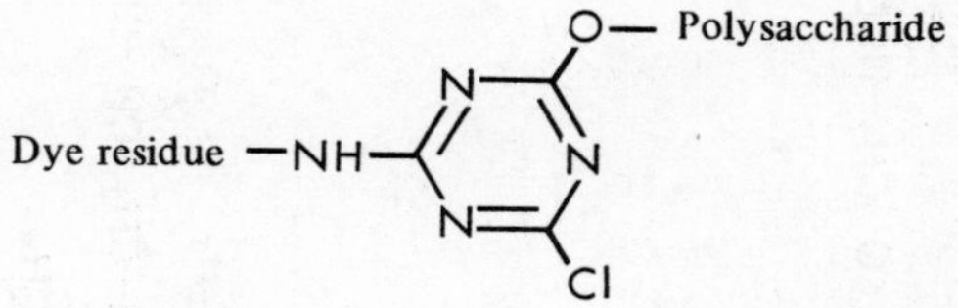

O— Polysaccharide

Dye residue —NH— [triazine ring: N, N, N] — Cl

(177)

Dye residue—NH—SO_2—CH_2—CH_2—O—Polysaccharide

(178)

ALKYL ETHERS

The reaction of monochlorocarboxylic acids with polysaccharides produce a range of carboxyalkyl ethers, essentially with the secondary hydroxyl groups, the most common being carboxymethyl ethers (179) which are produced by the action of monochloroacetic acid (Scheme 13.1). Derivatives of carboxyalkyl ethers which are important synthetically include the hydrazide (180) which, for carboxymethyl cellulose, is commercially available, the azide (181) and the isocyanate (182), which can be prepared from the parent carboxyalkyl derivative as shown in Scheme 13.1.

Reaction of polysaccharide with an aminoalkyl chloride, such as 2-diethylaminoethyl chloride (Scheme 13.2), produces an ether of which *O*-(2-diethylaminoethyl)-polysaccharides (183) are the most common and are commercially available as derivatives of dextran (DEAE-Sephadex®) and agarose. However, other aminoalkyl and alkylaminoalkyl groups may be introduced by using the appropriate activated amine to give 2-aminoethyl, 2-ethylaminoethyl and 2-triethylammoniumethyl derivatives and other derivatives containing different alkyl substituents. The amino or substituted amino group can be further derivatised using, for example, epichlorohydrin, to give di- (184) or mono- (185) chlorohydrins from 2-aminoethyl or 2-ethylaminoethyl derivatives. With 2-diethylaminoethyl derivatives the epichlorohydrin acts as an alkyl halide to produce a quaternary ammonium salt containing an epoxy group (Scheme 13.2).

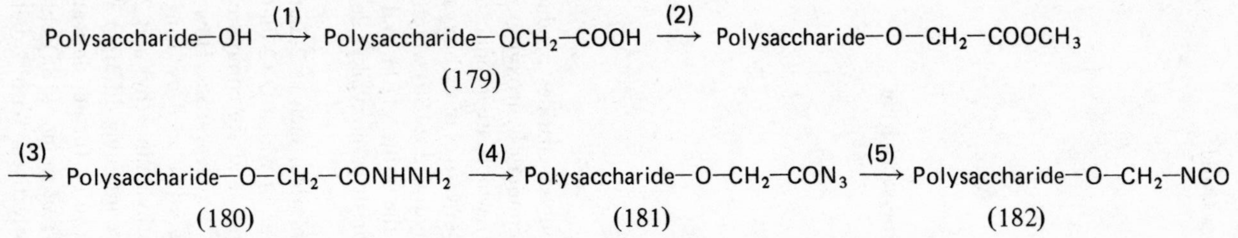

$$\text{Polysaccharide—OH} \xrightarrow{(1)} \underset{(179)}{\text{Polysaccharide—OCH}_2\text{—COOH}} \xrightarrow{(2)} \text{Polysaccharide—O—CH}_2\text{—COOCH}_3$$

$$\xrightarrow{(3)} \underset{(180)}{\text{Polysaccharide—O—CH}_2\text{—CONHNH}_2} \xrightarrow{(4)} \underset{(181)}{\text{Polysaccharide—O—CH}_2\text{—CON}_3} \xrightarrow{(5)} \underset{(182)}{\text{Polysaccharide—O—CH}_2\text{—NCO}}$$

Reagents: (1) $ClCH_2COOH$, ^-OH; (2) CH_3OH, H^+; (3) NH_2NH_2; (4) $NaNO_2$, H^+; (5) H^+

Scheme 13.1

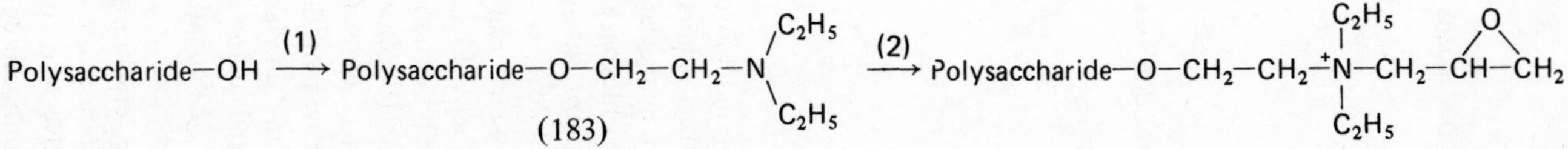

$$\text{Polysaccharide—OH} \xrightarrow{(1)} \underset{(183)}{\text{Polysaccharide—O—CH}_2\text{—CH}_2\text{—N(C}_2\text{H}_5)_2} \xrightarrow{(2)} \text{Polysaccharide—O—CH}_2\text{—CH}_2\text{—}\overset{+}{\text{N}}(\text{C}_2\text{H}_5)_2\text{—CH}_2\text{—}\underset{\text{O}}{\text{CH—CH}_2}$$

Reagents: (1) $(C_2H_5)_2NCH_2CH_2Cl$, ^-OH; (2) $\overset{O}{CH_2—CH}—CH_2—Cl$, H^+

Scheme 13.2

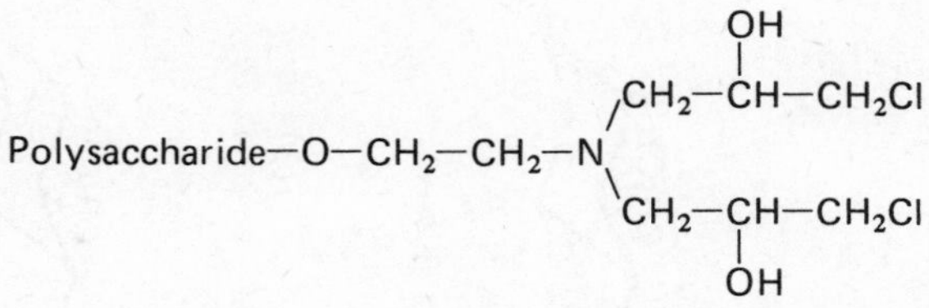

$$\text{Polysaccharide}-O-CH_2-CH_2-N\begin{cases} CH_2-CH(OH)-CH_2Cl \\ CH_2-CH(OH)-CH_2Cl \end{cases}$$

(184)

$$\text{Polysaccharide}-O-CH_2CH_2-N\begin{cases} C_2H_5 \\ CH_2-CH(OH)-CH_2Cl \end{cases}$$

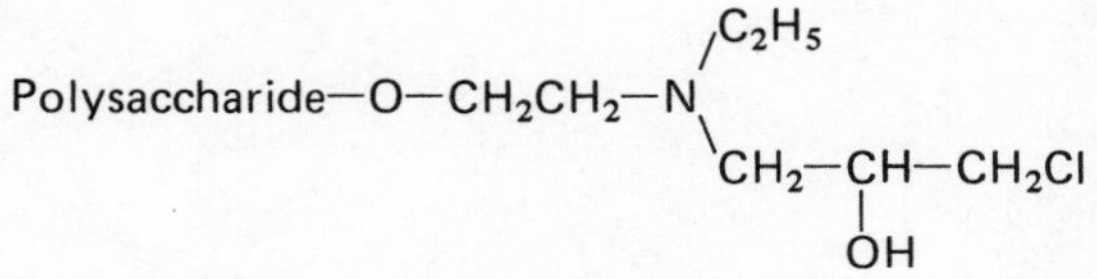

(185)

CYCLIC CARBONATES

Although cyclic and acylic carbonates of monosaccharides have been known for some time, little interest was shown in such derivatives of polysaccharides until the early 1970s, when the conditions for the production of cyclic derivatives, with minimum acylic substitution, were developed. The reaction of ethylchloroformate in anhydrous organic solvents produces cyclic carbonate derivatives (186) of a number of polysaccharides, including cellulose, nigeran, xylan, inulin and cyclomalto-oligosaccharides. In inulin strained *trans*-4,6- (187) and -1,3-carbonate (188) groups can be formed with the D-fructofuranosyl residues. Simple compounds, such as aminoacids, have been shown to react with these derivatives through nucleophilic attacks (Scheme 13.3). A stable covalent bond is produced *via* opening of the carbonate ring in either of two ways to give two possible products (189 or 190).

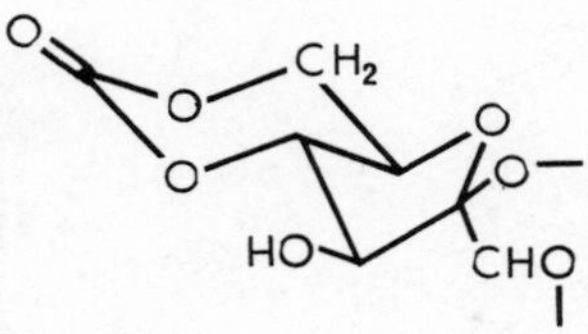

(187)

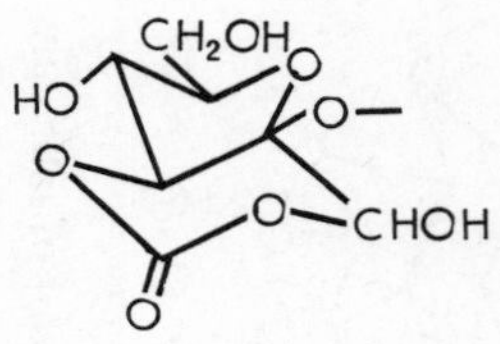

(188)

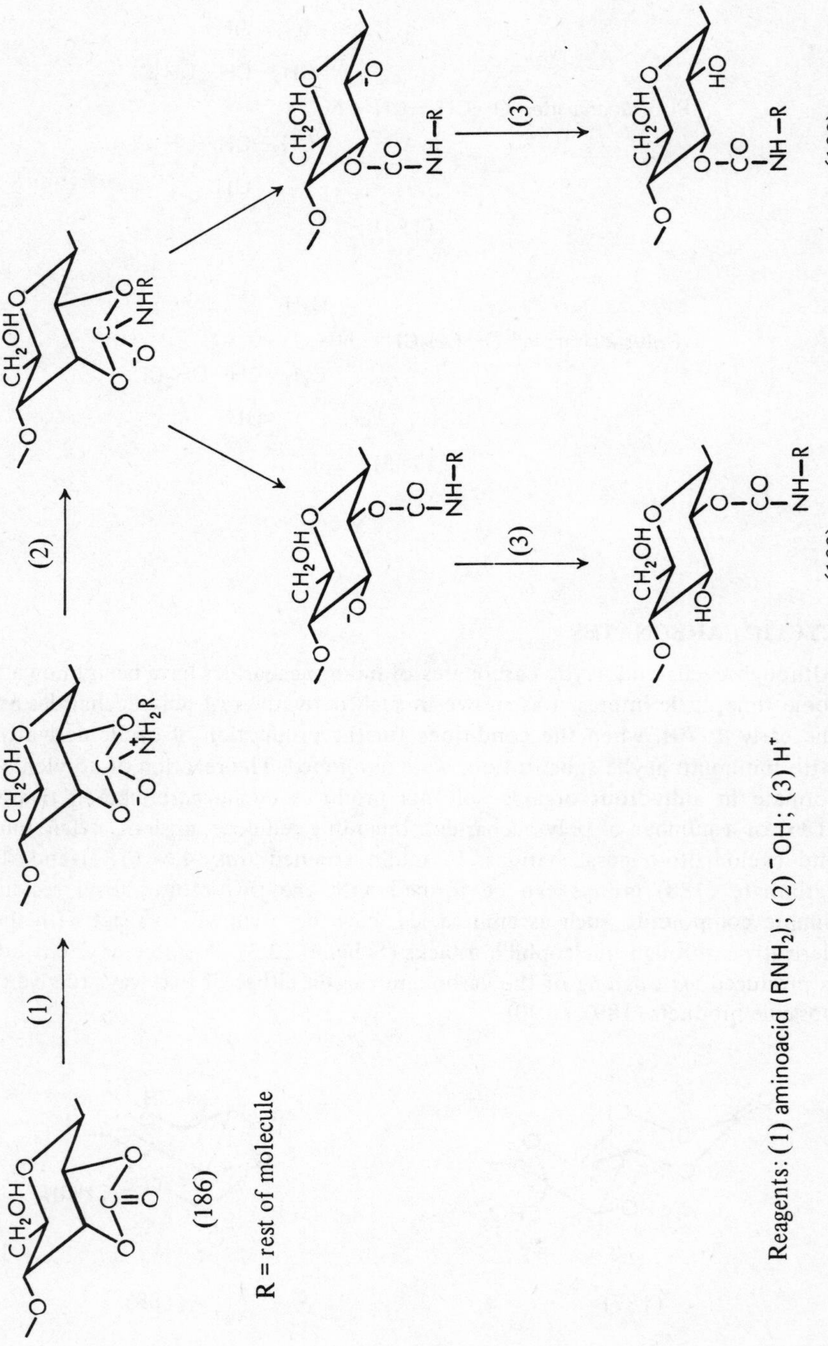

(186) (189) (190)

R = rest of molecule

Reagents: (1) aminoacid (RNH_2); (2) ^{-}OH; (3) H^+

Scheme 13.3

CYCLIC IMIDOCARBONATES

These compounds, which are closely related to cyclic carbonates, are traditionally called cyclic imidocarbonates although there is some evidence which indicates that the real situation is not as simple as originally thought. The action of cyanogen bromide on dextrans or cellulose produces a *trans*-2,3-imidocarbonate (191) whilst with agarose, the absence of vicinal hydroxyl groups prevents the formation of an energetically favourable imidocarbonate and the reaction has been shown, by the use of ^{13}C n.m.r. and specific chemical tests (Kohn and Wilchek, 1982), to involve cyanate esters (192). The cyclic imidocarbonates react in a similar manner to the cyclic carbonates with single molecules such as aminoacids (Scheme 13.4), whilst the cyanate esters react *via* the mechanism shown in Scheme 13.5 which gives an isourea derivative.

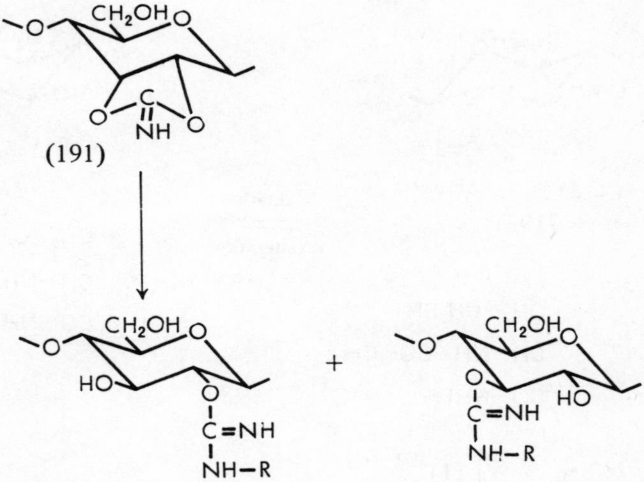

(191)

Scheme 13.4

Agarose —(1)→ Cyanate ester (192) —(2)→ Isourea derivative

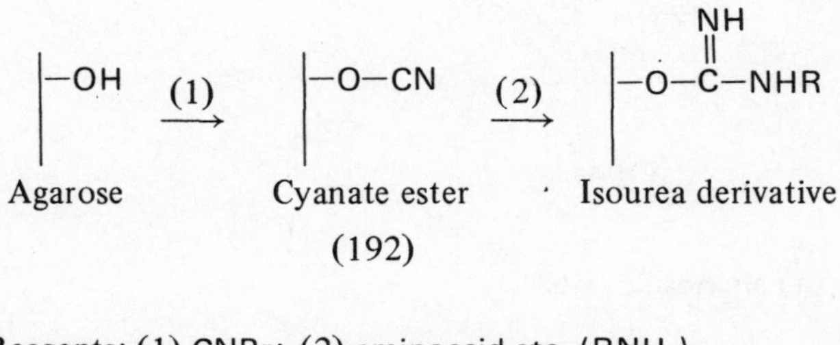

Reagents: (1) CNBr; (2) aminoacid etc. (RNH_2)

Scheme 13.5

XANTHATES

Xanthates are produced by the action of alkaline carbon disulphide on a polysaccharide, the most common derivative being cellulose xanthate (193). Xanthates do not react directly with amines and aminoacids but activation of these with *N*-acetylhomocysteine thiolactone (194) allows their reaction through the formation of disulphide linkages (Scheme 13.6). This process can be reversed by treatment with a thiol compound (usually L-cysteine) which regenerates the xanthate.

A number of other less common derivatives of polysaccharides have been reviewed, together with those mentioned herein (Kennedy, 1974a).

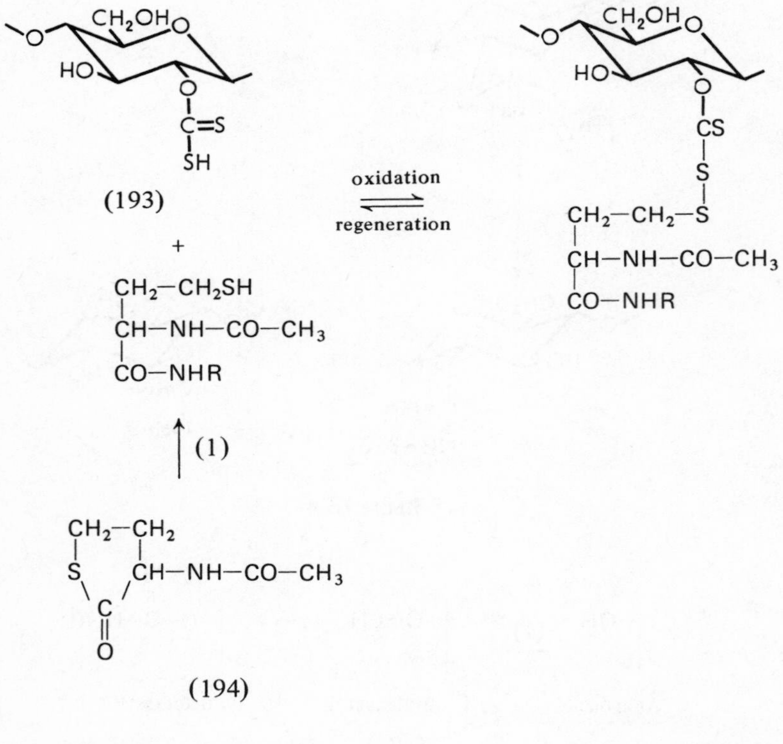

CH_2OH, HO, O, $C{=}S$, SH

(193)

\+

$CH_2{-}CH_2SH$

$CH{-}NH{-}CO{-}CH_3$

$CO{-}NHR$

↑ (1)

$CH_2{-}CH_2$

S, $CH{-}NH{-}CO{-}CH_3$

C=O

(194)

oxidation ⇌ regeneration

CH_2OH, HO, O, CS, S

$CH_2{-}CH_2{-}S$

$CH{-}NH{-}CO{-}CH_3$

$CO{-}NHR$

Reagents: (1) aminoacid (RNH_2)

Scheme 13.6

APPLICATIONS OF SYNTHETIC POLYSACCHARIDE DERIVATIVES

Chromatographic media

Cross-linking of soluble polysaccharide molecules gives an insoluble three-dimensional structure that swells in liquids, particularly in water, to give a gel. The most common method of cross-linking polysaccharides comes from their reaction with epichlorohydrin, and cross-linked dextrans, commercially available as the Sephadex® G series (Pharmacia), are well known. The degree of cross-linking is controlled in order to control the size of the pores of the gel; the more cross-linking, the smaller the pore size. Gels with larger pore sizes can be prepared by cross-linking agarose with 1,3-bis-(2,3-epoxypropoxy)-butane and again these are commercially available under the trade names of Bio-Gel® A series (Bio-Rad) and Sepharose® (Pharmacia). As such, these gels are used for the separation of molecules using the size/shape of the molecule as a basis for separation (gel filtration chromatography).

When these cross-linked polysaccharides are used in derivatisation reactions, the resulting insoluble derivatives are used for a variety of chromatographic methods. Carboxymethyl (CM) derivatives of cellulose or dextran are used as cation exchange supports whilst a range of anion exchange supports can be obtained by the formation of alkylaminoalkyl derivatives, with strongly basic anion exchange supports being quaternary ammonium derivatives (such as diethyl-(2-hydroxypropyl)-aminoethyl derivatives (195) and weakly basic anion exchange supports being primary, secondary or tertiary amino derivatives (most commonly diethylaminoethyl (DEAE) derivatives are used). Other derivatives of cross-linked polysaccharides are prepared, as a means of activating the cross-linked polysaccharide, to allow other molecules to be attached in order to produce, for example, affinity chromatographic supports (see later, p. 302).

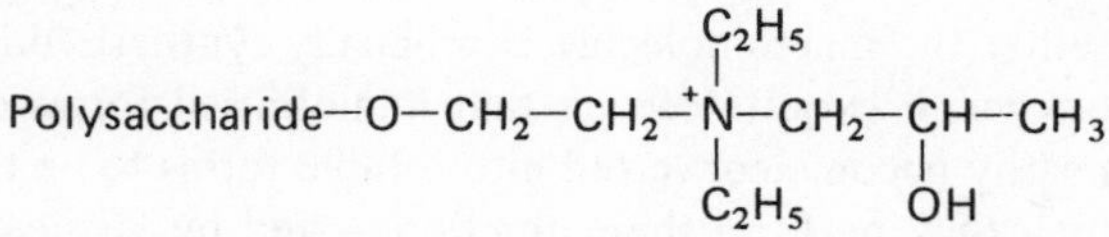

$$\text{Polysaccharide}{-}O{-}CH_2{-}CH_2{-}\overset{+}{N}(C_2H_5)_2{-}CH_2{-}CH(OH){-}CH_3$$

(195)

Immobilised biologically active molecules

Immobilised enzymes

Enzymes have been used by man for hundreds of years in the preparation of food, drink and clothing, but it was not until the beginning of the century, when the first individual enzyme (an amylase) was isolated, that full use could be made of the highly specific nature of most enzymes. Despite the tremendous advances that have been made in the isolation and purification of enzymes over the last few decades, the complex nature of the mixtures in which enzymes exist *in vivo* still makes purification of enzymes a lengthy and usually very expensive procedure. Re-isolation of enzymes, after use, is generally impractical because of the low concentrations usually employed, and consequently enzymes have not been used widely for industrial purposes.

In recent years, biotechnology has moved beyond the whole-cell phase, and the use of enzymes, immobilised on inert supports to aid stability and ease of recovery, has developed to an extent when it is no longer simply an aspect of microbiology; it has become a discipline in its own right, brought about by interactions between enzymologists, microbiologists and engineers. However, one must not be led by all the extensive literature which has now been published on the chemical production of such compounds to think that the principle of immobilisation is something new. The overall principle of attachment of a biologically active molecule to an insoluble matrix is simple and simulates the natural mode of action and environment of enzymes, antibodies, antigens, etc., which are carried on the surfaces or in the interiors of cells, or which are embedded in biological membranes and tissues. Indeed, as is often discovered, 'Nature was there first' and the greater proportion of the biologically active molecules in the human body exist at some time in an immobilised form. Although a precise biological activity has been identified for some of the molecules present in the body, many if not all of the others can be regarded as insoluble biological reactors of some description, for example, the proteoglycans in their tissue matrix-forming role. However, perhaps less attention has been given to the immobilised forms, rather than the soluble forms of such molecules since most chemical techniques, analyses, and manipulations are designed to be carried out in solution, and the chemistry of activity in the solid phase is less well developed.

In natural systems, the immobilisation of biologically active macromolecules such as enzymes and glycoprotein hormones may well be a reversible process, according to whether the macromolecule is originally synthesised in the solid or liquid phase. However, it is quite certain that immobilised forms of such active macromolecules easily become converted into soluble forms to be transported to a new site at which they perform their function – and by virtue of performing that function, they may once again become immobilised. In this respect the natural immobilised molecules differ markedly from those prepared in the laboratory. This is because synthetically-immobilised biologically active molecules are usually required to perform their biological function without being released into the surrounding solution and thereby contaminating it.

There are many applications of immobilised, biologically active molecules. Immobilised enzymes are principally used to effect the reaction catalysed by the free enzyme, but in a simplified form since the enzyme (insoluble) can be very easily and simply removed from the substrate and products (soluble) by filtration or centrifugation, whereas use of the soluble enzyme in the conventional fashion requires subsequent laborious separation of the enzyme from the products by, for example, gel filtration and ion-exchange chromatography. Further advantages of immobilised enzymes are that: the enzyme becomes stabilised to decomposition in storage and to heat on immobilisation, the reaction of the enzyme may be rapidly terminated without the addition of foreign substances, the enzyme may be packed into a column and used for continuous conversion processes, the products of the enzymic action are not contaminated with any unwanted biologically active material, and the immobilised enzymes can be easily re-used. Also, changes in stability and kinetic properties are sometimes found upon immobilisation, and these may be put to good use. Uses include: simplification of reactors, industrial processes and clinical analyses, employment in analytical chemistry and bio-chemistry – that is, enzyme electrodes, etc. (see Carr and Bowers, 1980), sequence analysis and synthesis, separation techniques, isolation of compounds related to enzymes, and use in membrane and chromatographic column forms.

A number of methods of immobilisation of biological molecules exist (Fig. 13.1), and no one method is perfect for all molecules or purposes. When attaching a biologically active molecule to an insoluble support, it is important to avoid a mode of attachment that reacts with or disturbs the active site(s) of the molecule, as otherwise a loss of activity will result on binding. It is also important to avoid overloading the matrix when binding molecules, since overloading leads to overcrowding and hence reduced activity, by reason of steric hindrance of approach of the substrate etc. molecules to the active sites of the bound molecules. However, it does not follow that limited loading of the enzyme molecules on the matrix surface will be successful, since hydrogen bonding and hydrophobic forces may occur between immobilised enzyme and 'free' molecules, thus causing the latter to block up the spaces between the immobilised molecules. Attention to the way in which the macromolecule can be immobilised and the choice of matrix is also a matter of importance. A number of matrix types and techniques have been used in the field of immobilisation, and it can be concluded that there is no ideal or universal support matrix or immobilisation technique. For specific applications many of the support matrices available can be discounted because of their characteristics. In a medical application, for example, where an immobilised enzyme is in contact with blood or tissue a material which evokes an immune response (wool support) or clotting reaction (glass support) should not be chosen, a support based on methacrylate or silicone rubber, which are inert in these aspects, being preferable. On the other hand, for continuous-flow reactors, materials with poor dimensional stability (such as cellulose, starch gels and dextrans) should be avoided with preference being given to, for example, controlled-pore glass or ceramic material.

The choice of support matrix is also influenced by its effect on the characteristics of the enzyme and its substrate. Comparison of the activity of an enzyme bound to a matrix with the activity of a freely dissolved enzyme at various pH levels has shown that if the enzyme is attached to a negatively charged matrix,

(a) Adsorption

(b) Ion Exchange

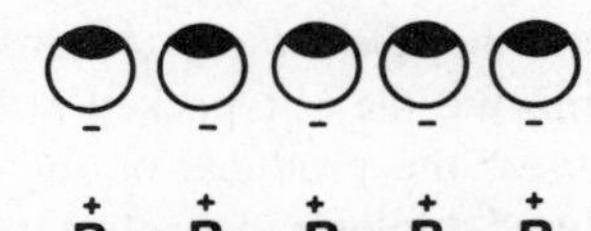

(c) Entrapment*

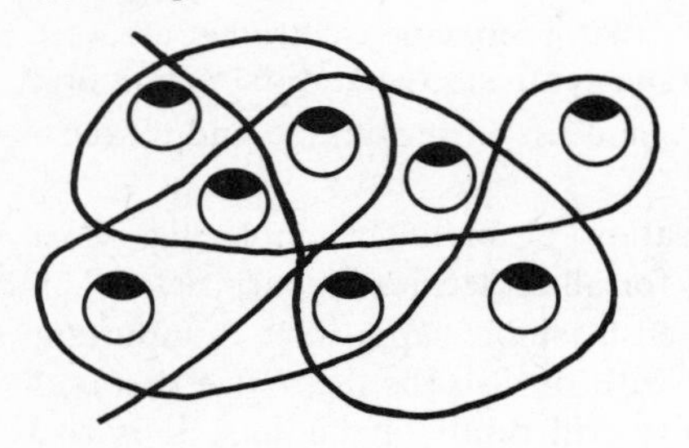

(d) Microencapsulation*

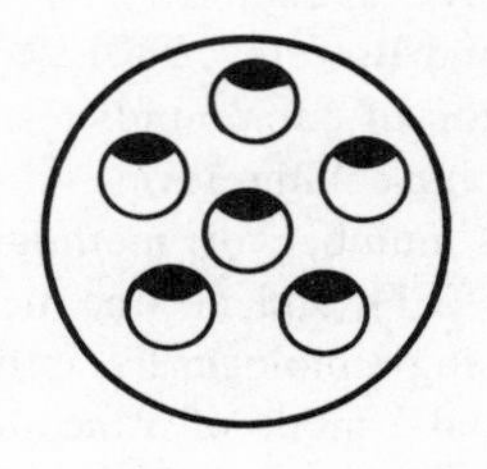

(e) Cross-linking

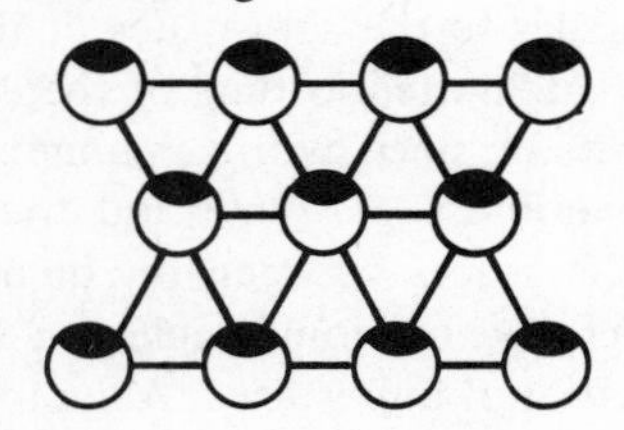

(f) Copolymerisation*

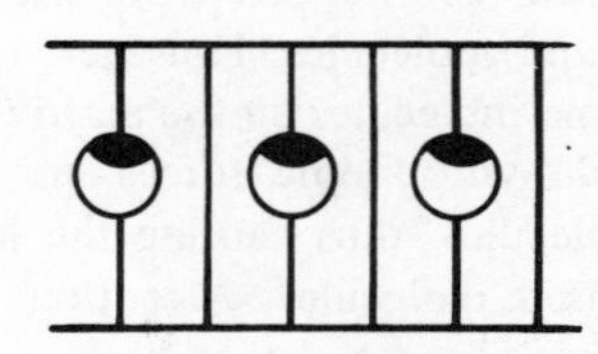

(g) Entrapment and Cross-linking*

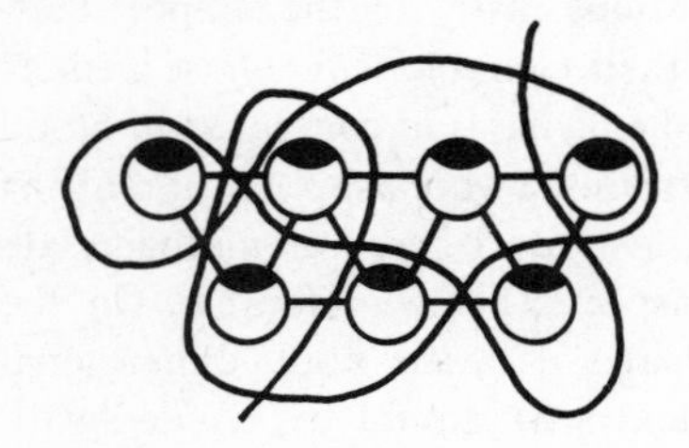

(h) Covalent attachment

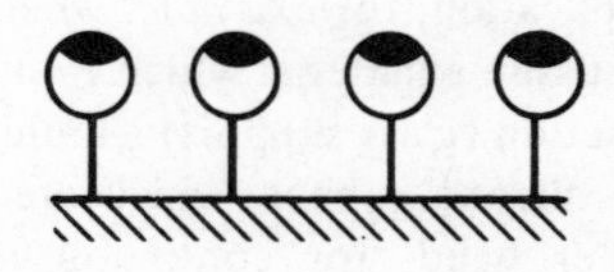

Fig. 13.1 – Diagramatic methods of enzyme immobilisation.
◓ Enzyme molecule, * Diffusion controlled.

for example DEAE-cellulose, the pH optimum is shifted towards the alkaline side: the immobilised enzyme reaches maximum activity at apparently higher alkalinity. This effect is due to the negatively charged groups of the matrix attracting a thin 'film' of positive hydrogen ions, thereby creating a micro-environment for the bound enzyme that has a higher hydrogen ion concentration (lower pH) than the concentration in the surrounding solution where the pH is actually measured. Similarly, for a positively charged matrix, for example CM-cellulose, the apparent shift in optimum pH is to the acid side (Fig. 13.2). The characteristics of the support matrix can be such that the substrate for the enzyme is attracted to the matrix, whilst the products of enzyme action may be unaffected. It is also possible to prepare support matrices which will allow easy removal of the immobilised enzyme from the reaction medium. Examples which have already been used are magnetic supports and film-forming materials, both methods allowing the recovery of the immobilised enzyme from reaction media containing other insoluble matter which makes separation by filtration impossible. The ideal support for a given application is one which would increase substrate binding, decrease product inhibition, shift the apparent pH optimum to the desired value, discourage microbial growth and could be readily recovered for reuse.

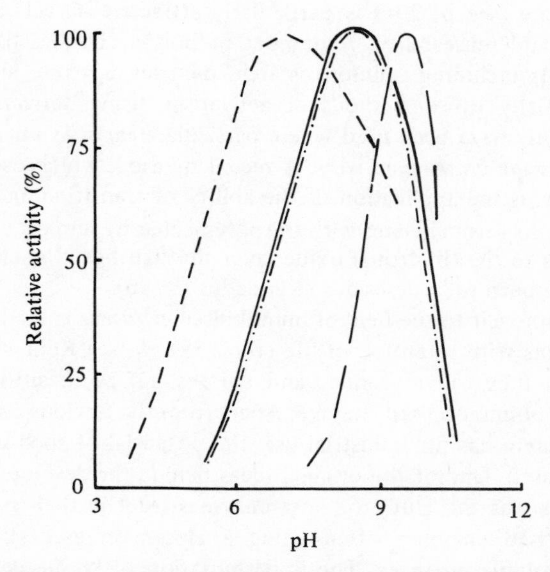

Fig. 13.2 – The effect of microenvironment of the pH-activity profile of an immobilsed enzyme. ——— soluble enzyme; – – – – – immobilised enzyme, cationic support; – · – · – immobilised enzyme, neutral support; —— —— immobilised enzyme, anionic support.

Full discussions of all the methods available and their suitability to particular applications can be found in the many books and reviews on the subject, (for example, Zaborsky, 1972; Wiseman, 1975; Trevan, 1980, and Kennedy and Cabral, 1982a) and in an ongoing series of literature surveys (Sturgeon and Kennedy, 1979 onwards). A review of a number of materials, including non-carbohydrate materials suitable for solid supports is also available (White and Kennedy, 1980).

The use of polymeric carbohydrate derivatives for enzyme immobilisation has received more emphasis than hydrophobic supports, since the latter may lead to the destabilisation of the enzyme. The major advantage of the use of polysaccharidic material is that the solid state residual hydroxyl groups provide a protective hydrophobic environment for the attached macromolecule. Immobilisation within the pores of macroporous polysaccharides gives added protection from exposure to destabilising influences whilst overcoming the disadvantage of poor diffusion of higher molecular weight substrates associated with microcrystalline polysaccharides.

In order to attach the enzyme chemically to a polysaccharide, the latter must first be activated in such a way that the enzyme can subsequently be attached under conditions which will not cause its inactivation. The most common method, for laboratory scale use, is the use of the so-called cyclic *trans*-2,3-imidocarbonate (see p. 293), whilst for large scale industrial uses cellulose xanthate (see p. 294) is particularly attractive on account of its ease of production and reuseability. A number of polysaccharides have been used as solid supports including cellulose, starch, dextran, agarose, alginic acid and carrageenans. Many other methods of activation, using derivatives described above, and others have been used where particular reactions must or must not be used to preserve enzyme activity. A recent method, which uses particularly mild conditions, is the application of the ability of transition-metal oxides and hydrous oxides to form chelates with the polysaccharide support (Fig. 13.3). The enzyme couples to the (hydrous) oxide layer through ligand exchange mechanisms which have been fully described (Kennedy, 1979b).

A novel approach to the field of immobilised enzymes is the immobilisation of microbial cells with retention of life (for a review, see Kennedy and Cabral, 1982b) so that they can reproduce and thereby act as an automatically self-renewing form of immobilised enzyme. Apart from the obvious advantage which this type of system has for industrial use, the potential of such systems has yet to be fully realised. One of the original ideas behind the development of immobilised enzymes was the study of how enzymes react in their natural environment, immobilised enzymes often being a closer representation of natural systems than soluble enzymes. The immobilisation of whole cells means that enzymes can be readily studied in situations which resemble their natural environment. With the development of more immobilised-cell systems it may become possible to apply this technology to the investigation of enzyme deficiency disorders by studying comparatively the action of particular enzymes in affected cells and in normal cells. A development of this could lead to more-accurate diagnosis and ultimate treatment (by implantation of immobilised whole normal cells into the affected areas) of enzyme deficiency disorders.

There are two major disadvantages in the use of polysaccharides for support materials, namely their nonspecific adsorption properties and their susceptibility to microbial attack. Nonspecific adsorption of protein can be avoided or overcome on a laboratory scale by washing the final product with a high ionic strength buffer, but this process can inactivate an enzyme and, for an industrial process, is frequently too expensive. The second disadvantage could possibly be overcome by immobilisation of antibiotics (see below, p. 303) but has not, as yet, been attempted successfully.

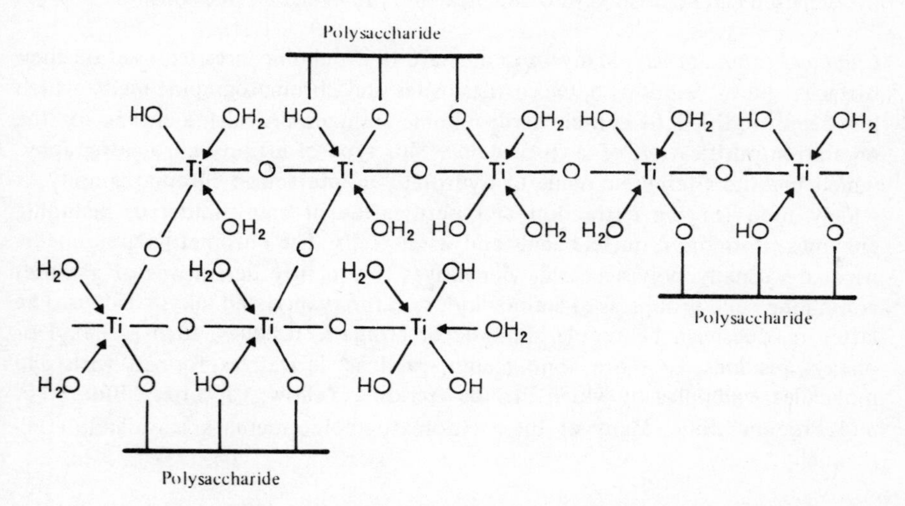

Fig. 13.3 – Schematic representation of polysaccharides chelated with hydrous titanium(IV) oxide.

Immunoadsorbents

Immobilised antibodies are principally useful for the purification of homologous antigens, usually by a type of column chromatography (immunoadsorption) in which the solution of impure antigen is passed through a bed of immobilised antibody: the specific antigen is adsorbed by the antibody whilst impurities are washed through the column. Subsequently, the antigen may be desorbed from the column in pure form. Thus the lengthy conventional techniques of various types of column, etc., chromatography are short-circuited. Immunoadsorption can of course also be applied in the reverse sense, using insolubilised antigen to purify an antibody. Immunoadsorption is a very versatile technique since many macromolecules are antigenic and therefore antibodies can be raised to them; but an important prerequisite is, of course, that the antigen which is to be immobilised can be obtained in pure form or that the antibody to be insolubilised can be obtained in pure and/or highly specific form. Immobilised antigens and antibodies are also of use in radioimmunoassay techniques.

Carbohydrate polymers, particularly Sepharose® *trans*-2,3-imidocarbonate, have been used extensively to prepare immunoadsorbents.

Affinity chromatography media

Affinity chromatography is a technique in which substances are separated on the basis of the differing strengths of their interactions with the support material which has been modified to contain groups with specific interactions. The interactions used can be divided into chemical and biological interactions.

Chemical interactions. Many proteins have hydrophobic sites exposed on their surfaces and interactions between these sites and chromatographic media which has been modified to contain hydrophobic residues provides a means for the separation/purification of such proteins. This type of affinity chromatography, which has the alternative name of hydrophobic interaction chromatography, is widely used for the extraction and purification of many materials including enzymes, hormones, nucleic acids and whole cells. The chromatographic media used are usually polysaccharide derivatives (frequently derivatives of agarose) containing such groups as ω-aminoalkyl, ω-carboxyalkyl and alkyl residues. The latter residues can be simple aliphatic or aromatic residues, such as hexyl or phenyl residues, or more complicated, such as derivatives formed with dye molecules, examples of which include Acridine Yellow, Cibacron® Blue 3G-A and Procion® Blue. Many of these chromatographic media are available commercially.

Biological interactions. In this type of affinity chromatography the residue attached to the chromatographic support is usually one of low molecular weight but one for which the macromolecule to be purified has a specific affinity. Thus for the purification of carbohydrate-directed enzymes, an immobilised carbohydrate, for which the enzyme is active, is used. Many such products are available under the trade name Selectins® (Pierce). Carbohydrates, either in the form of polysaccharides or as carbohydrate-containing macromolecules such as glycoproteins, etc., can be purified by immobilisation of lectins (see Table 13.1 and Chapter 9, p. 209) to retain the selected carbohydrate macromolecule until the elution conditions are altered to disrupt the lectin–carbohydrate affinity. A number of lectins, etc., are commercially available bound to carbohydrate supports, including agarose (for example, ConA-Sepharose®, immobilised concanavalin A from Pharmacia; and Glycosylex® A and Glycaminosylex®, immobilised concanavalin A and wheat germ agglutinin, respectively, from Miles).

Table 13.1 Carbohydrate specificity of some lectins

Lectin	Carbohydrate residue specificity
Concanavalin A	α-D-glucosyl α-D-mannosyl
Dolichos biflorus agglutinin	terminal 2-acetamido-2-deoxy-α-D-galactosyl group
Helix pomatia lectin	2-acetamido-2-deoxy-D-galactosyl
Jimson weed lectin	2-acetamido-2-deoxy-β-D-glucosyl
Kidney bean agglutinin	2-acetamido-2-deoxy-D-galactosyl
Lens culinaris lectin	α-D-mannosyl
Lentil lectin	α-D-glucosyl α-D-mannosyl
Lima bean agglutinin	2-acetamido-2-deoxy-D-galactosyl
Limus polyphenus lectin	5-acetamido-3,5-dideoxy-D-*glycero*-D-*galacto*-2-nonulopyranosylonic acid
Lotus tetragonolubus agglutinin	α-L-fucosyl
Pea tree agglutinin	2-acetamido-2-deoxy-D-galactosyl D-galactosyl
Potato lectin	2-acetamido-2-deoxy-D-glucosyl
Ricinus communis lectin	2-acetamido-2-deoxy-D-galactosyl D-galactosyl
Soya bean agglutinin	D-galactosyl
Ulex europeus agglutinin	
(form I)	α-L-fucosyl
(form II)	2-acetamido-4-*O*-(2-acetamido-2-deoxy-β-D-glucopyranosyl)-2-deoxy-D-glucosyl
Vicia graminea lectin	5-acetamido-3,5-dideoxy-D-*glycero*-D-*galacto*-2-nonulopyranosylonic acid
Wheat germ agglutinin	2-acetamido-2-deoxy-D-glucosyl

Immobilised antibiotics

The most recent innovation in the field of immobilisation has been the preparation of immobilised antibiotics. Where an anti-bacterial surface is required (for example, water storage tanks, industrial membranes, chromatographic columns), such surfaces could be realised by using cellulose-based paints, membranes, etc., and insolubilisation of the antibiotic by covalent attachment. In such cases, loss of the antibiotic would be minimal. Other applications, which have come to light, include provision of selective protection against microbial attack of paper and legal documents, of canvas and chromatographic media based on cellulosic materials, and of cellulose-based packings of cooling towers. The techniques also provide a novel form of sterility for sheets and other cotton-based fabrics and gauze, and for treating infected root canals in teeth before root filling. Also, where it is required to have a slow continual release of antibiotics but a higher initial release (for example, bandages and surgeons' thread), immobilised antibiotics can be expected to be of use. Since cellulose and other polysaccharides and their derivatives are used extensively in a number of forms as accessories to life, active immobilised antibiotics could well be of great use in a number of other areas (for example, food packaging materials).

Immobilised polysaccharides

A number of polysaccharides and carbohydrate-containing macromolecules have been immobilised by reaction with imidocarbonate derivatives of polysaccharides. These derivatives have been used to study the interactions between carbohydrate-containing macromolecules and other macromolecules. Examples include the interaction of glycosaminoglycans with lipoproteins and the interaction of carbohydrates with carbohydrate-directed enzymes. The latter example has been used as a method of purifying such enzymes (by affinity chromatography).

Insoluble derivatives of carbohydrates such as the dye-derivatives and immobilised glycosaminoglycans have been used as solid phase substrates for enzymes including α-amylase (EC 3.2.1.1), β-amylase (EC 3.2.1.2) and hyaluronidases (hyaluronate 3- and 4-glycanohydrolase, EC 3.2.1.36 and 3.2.1.35).

Immobilised nucleic acids

Polysaccharides, particularly cellulose, have been used almost exclusively for the immobilisation of nucleic acids and polynucleotides. These solvent-insoluble derivatives are used for the fractionation and purification of other nucleic acids, nucleotides, etc., isolation of single stranded nucleic acids by base pairing, analysis of base-sequences, affinity supports for nucleic acid-directed enzymes and nucleic acid-binding proteins.

CHAPTER 14

Technological aspects and applications

TRADITIONAL APPLICATIONS

Currently, the industrial uses of polysaccharides and carbohydrate-containing macromolecules are dominated by the plant polysaccharides, starch and cellulose, which are used and have been used in some cases for hundreds of years, in the polymeric or depolymerised and derivatised forms in a number of industries. The traditional uses of carbohydrates have been reviewed (Stacey, 1973).

Construction and packaging uses

Wood, the major source of cellulose, in which the polysaccharide exists in association with lignin, oils, resins, minerals and pigments, etc., is one of the most useful and versatile structural and building materials in use. Technology has advanced its usefulness from both the production and usage points of view. The improvement of wood by the introduction of preservatives, plastic fillers, fire retardants, etc., to give plywoods, fibreboards, blockboards and veneers, although now commonplace, has been quite remarkable. A major use for soft-woods lies in pulp for the paper industry for newsprint, books, packaging and boardmaking, etc., the demand for which increases at an evergrowing rate with the demand from the newer industries of computing, copying and convenience goods adding the needs for specialist products to those for the more traditional papers, etc.

Other forms of cellulose include cotton (the purest form in which cellulose occurs naturally), flax, jute, sisal, hemp and various straws and grasses, all of which are used in the construction and packaging industries to produce papers, (cotton, straws and grasses) and rope (sisal and hemp).

Derivatives of cellulose, particularly esters and ethers, have for a long time held a special place in the packaging industry for the production of transparent wrappings and films. Cellulose acetate film can be laminated to itself, to foil or the plastic films (particularly polythene films which produce a tough heat seal barrier film useful for packing meat and cheese).

Starch and its derivatives find many uses in the paper industry as sizes, stiffeners and adhesives. The more common derivatives include esters, ethers, phosphates, oxidized starches, hydroxyalkyl starches and cationic starches.

Food and brewing uses

Starch (and its components) is the major food polysaccharide with the main sources being maize, wheat, rice, potato, tapioca and sago. These various starches are used for a variety of foodstuffs which depend on the nature of the starch granule and their associated constituent waxes and oils. Maize is used whole for animal feedstuffs, wheat for bread, and barley for the brewing industry. The various polysaccharides used in the food industry are the subject of a recent book (Blanshard and Mitchell, 1979).

Sucrose, the household commodity, obtained from sugar cane and sugar beet, is used in the food industry in large quantities as a sweetening agent and preservative and as a raw material for the fermentation industry where it is used in forms which are by-products of sugar refining such as molasses, treacles and syrups.

A number of other oligosaccharides are used in the food industry, but to a much lesser extent than sucrose, and these include maltose and cellobiose (degradation products of starch and cellulose) and lactose (which is obtained as a by-product from milk after removal of the curd in cheese production). Lactose can be used in fermentation processes, but it required the development of yeasts which had been modified to lactose fermentation since normal yeasts have no action on lactose. Uses of these oligosaccharides have been reviewed (Lee, 1980).

Textile uses

Cellulose has been used for many years as a material for the production of textiles with cotton being one of the world's major fibres. With the introduction of man-made fibres, the position of cotton has declined but the developments of cellulose based man-made fibres (for example, rayon and triacetate) have ensured that carbohydrate-based textiles will retain their position for some time yet, especially for use and wear in hot climates. The dyeing of cellulose fabrics has been discussed in Chapter 13 (p. 288).

Pharmaceutical and cosmetic uses

Many carbohydrates are used in the pharmaceutical and cosmetic industry to bind preparations or to act as 'inert' bulking agents for small quantities of active ingredients. Among those used are D-glucose, sucrose, mannitol and α,α-trehalose with starch (and its derivatives) being used for its adhesive properties. Starch is also used as a talc substitute. Dextran and lower molecular weight fractions are used as plasma substitutes, which restore blood volume in patients who have lost considerable amounts of blood or are in a state of shock. Other fractions improve blood flow in capillaries and, as the sulphate derivatives, have anticoagulant, antilipemic and anti-ulcer activity.

Other uses

Derivatives of cellulose (mainly the acetate-butyrate, propionate and acetate-phthalate) are used in the plastics industry to make sheets, films, coatings and backing materials whilst fatty acid, alkyl or alkoxy esters and ethers of non-reducing disaccharides, including sucrose, α,α-trehalose and maltitol, form a group of surface active agents (surfactants) with an extraordinary range of applications, extending from industrial detergents and surface coatings to food emulsifiers and antimicrobial agents (see Lee, 1980, for a review).

NEW APPLICATIONS AND FUTURE TRENDS

With the development of microbial technology, that is, the multidisciplinary science of microbiology applied to production of industrially important materials, the value of producing polysaccharides has increased. The production and isolation of microbial (bacterial) polysaccharides has the advantages, compared to production by and from growing plants, of assured production and quality which is unaffected by marine pollution, tides, weather, war, famine, drought, etc., which affect the production of plant polysaccharides, whilst production can be geared to market trends and located such that convenient, cheap substrates can be utilised.

The uses to which microbial polysaccharides are put arise from the natural biological functions of polysaccharides, including thickening agents in joint fluids, protectants from desiccation and lubricants. Whilst the time taken between the initial reporting of xanthan gum to its pilot scale production was of the order of twelve years, our present understanding of how structural features affect the properties of polysaccharides and our ability to modify bacteria through genetic engineering (see Chapter 11, p. 263) has resulted in greater investment in time (and money) for this area of biotechnology with a number of polysaccharides (see Table 14.1) being developed more rapidly as possible industrially important materials (see reviews by Berkely *et al.*, 1979, Bull *et al.*, 1979, Sandford, 1979, and Davidson, 1980).

Oil industry

Perhaps the greatest single potential market for microbial polysaccharides is that provided by the oil industry for enhanced oil recovery and drilling muds. Enhanced oil recovery requires polymers which improve water flooding techniques by increasing efficiency of contact with, and displacement of, oil. Alternatively, the polymer must reduce the flow capacity of the solution in the rock system, either by increasing viscosity of the solution or by decreasing the permeability of the system. A prerequisite for the polymer is that it must be unaffected by the salt concentrations and temperatures found in oil wells. Drilling muds are aqueous suspensions, containing clays and colloidal materials, which lubricate drill heads and counterbalance the upward pressure of oil. As a prerequisite of these uses, the polymers used in drilling muds must have high viscosities to maintain the colloidal properties of the mud. This viscosity must again be relatively insensitive to temperature but pseudoplastic properties are essential. Xanthan and scleroglucan have been compared favourably with the petroleum based polymers traditionally used because of their greater insensitivity to temperatures, etc., but the prices are, at present, a major drawback.

Table 14.1 Polysaccharides, with (potential) commercial importance, from microbial and fungal origins

Polysaccharide	Organism	Trade name
Alginate	*Azotobacter vinelandii, Pseudomonas aeruginosa*	
—[a]	*Arthrobacter viscosus*	PS B-1797, PS B-1973
PS-7	*Azotobacter indicus*	PS-7
Baker's yeast glycan	*Saccharomyces cerevisiae*	BYG
—[a]	*Bacillus polymyxa*	
—[a]	*Chromobacterium violaceum*	
—[a]	*Cryptococcus laurentii* variant *flavescens*	
Curdlan (succino-glucan)	*Alcaligens faecalis* variant *myxogenes*	
Dextran	*Leuconostoc, Acetobacter* and *Klebsiella* species	Various
Elsinan	*Elsinoe leucospila*	
Erwinia polysaccharides	*Erwinia* species	Zanflo
—[a]	*Hansenula* species	
—[a]	*Pseudomonas* species	
Pullulan	*Aureobasidium pullulans*	
—[a]	*Rhinocladiella elatior*	
—[a]	*Rhinocladiella mansonii*	
Scleroglucan	*Sclerotium* species	Actigum CS Polytran F.S.
—[a]	*Tremella mesenterica*	
Xanthan	*Xanthomonas campestris*	Keltrol, Kelzan, Rhodigel 23

[a] No specific name is used for the polysaccharide.

Food industry

Polysaccharides are included into foods to function, among other purposes, as suspending agents, thickeners, gelling agents and ice crystal formation controllers. Plant polysaccharides such as starch, alginate, guar gum and locust bean gum have been used in various amounts, but the increase in demand for instant and processed foods has led to the use of microbial polysaccharides for many processes (see Table 14.2). New uses are continually being developed, with proposals to include these polysaccharides in bread to improve the texture and water retention properties. The inclusion of microbial polysaccharides into new food formulations is probably easier than replacing plant polysaccharides in existing formulations, due to the multifunctional nature of these new polysaccharides, since it is difficult to adjust the balance of wetting agents, thickening agents and stabilisers to that of the accepted product.

Table 14.2 Functions and potential food uses of microbial polysaccharides

Function	Polysaccharide	Potential use
Stabiliser of low pH gels	Xanthan	Milk shakes
Gelling agent	Curdlan, xanthan, scleroglucan, alginate	Jellies, custards, gravies, sauces, instant dry desserts, pie fillings
Colloidal stabiliser	Xanthan, alginate, scleroglucan	Ice cream, milk shakes
Non-caloric materials	Curdlan, pullulan	Additives for diabetic and dietetic foods (desserts, dressings, etc.)
Film and fibre former	Curdlan, pullulan	Edible films and fibres
Water-retention agent	Curdlan, pullulan	Sausages, ham, starchy jellies
Binding agent	Curdlan	Meat products (e.g. burgers), pastas, jellies
Pseudoplastic thickeners/stabilisers	Curdlan, xanthan, alginate	Sauces, dressings, gravies spreads, bakery fillings
Deodorant	Curdlan	Boiled rice
Coating agent	Xanthan, scleroglucan	Sauces and gravies for pasta and meat products
Crystallisation inhibitor	Xanthan, alginate	Ice cream, sugar syrups
Foam stabiliser	Alginate	Whipped toppings

The major drawback to development of these polysaccharides for food uses is the cost of the materials, and their approval for food use, but some of this can be offset by using existing, approved gels in mixtures which have properties different to the materials in isolation. The use of polysaccharide mixtures as new gelling agents has been reviewed (Dea, 1979).

A related area of carbohydrate uses in the food industry which has resulted from the biotechnological advances of recent years is the production of single cell protein (SCP) and other fermentation products from carbohydrate feedstocks which are often waste products from other food manufacturing processes (alcohol can be produced from waste whey which contains lactose, see p. 136).

Pharmaceutical and cosmetic industries

The pharmaceutical and cosmetic industries use a number of polysaccharides, such as those used in the food industry to form gels, colloids, thixotropic solutions, extrudable pastes and creams, etc., and to form films to bind tablets and produce protective creams. There are also a number of specific uses of carbohydrates in the pharmaceutical industry which have come about as a result of developments in genetic engineering and microbial technology.

Production of carbohydrate-containing antibodies on a commercial scale has revolutionised the treatment of infections, etc., and genetic manipulation and engineering is producing new or modified antibodies or increasing the yields obtained for existing ones. Similar production of enzymes, antigens, glycoprotein hormones, and other biologically active carbohydrate macromolecules can be expected to bring about similar revolutions in clinical treatment. Recently purified polysaccharides of bacterial origins have been prepared for use as antigenic vaccines against meningococcal and pneumococcal infections and, due to the ability of certain of these polysaccharides to cross react with other antisera, they may also provide immunity against other infections. The advantages of using purified polysaccharide vaccines is the low risk of adverse reactions compared to vaccines containing whole cells (either living or dead) and such vaccines can give immunity to infection caused by antibiotic-resistant bacteria.

Reviews of the developments in this field can be found in the ongoing series of biennial conference reports (Various Editors, 1974 onwards) and the series by Wiseman (1977 onwards).

Other uses

Production of carbohydrates by biotechnological processes has resulted in quantities of materials being available as raw materials for other industries, which have traditionally been made from oil, but with the ever-decreasing supply of oil, renewable carbohydrate sources may hold the key to the economic future of the industrialised countries. One advantage which carbohydrates have over many oil-based chemicals is that the correct stereochemical arrangements of the product can be present in the carbohydrate which makes them suitable starting products for the chemical synthesis of other natural products which require the correct stereochemical arrangement for their natural function (Harman, 1979).

NEW PRODUCTION TECHNOLOGY

The major drawback to the use of fermentation technology for the industrial production of, or utilisation of, carbohydrate materials is the cost of production. At present batch culture techniques are used, in which the fermenter is charged and the reaction allowed to proceed for a given time, after which the contents are removed in order to harvest the product. The down-time inherent in this method (caused by harvesting, sterilisation and recharging) could be reduced, and almost eliminated, if continuous culture methods were to be adopted. This, together with the ease of control, greater uniformity, reduced reactor volumes, etc., which are associated with continuous operation, would increase production and lower the final cost of the product. Things are not quite as simple as they appear and the problems of strain stability, low product concentration, and contamination problems have so far thwarted many attempts to improve the efficiency of production.

The future holds many opportunities for engineers, geneticists, microbiologists and biochemists in the rapidly expanding science of biotechnology to develop and improve production until economically viable products can be obtained. This gives mankind a future which is not locked into the ever-dwindling supplies of oil and oil-based products.

References

Aminoff, D., Binkley, W. W., Schafer, R. and Mowry, R. W. (1970), in *The Carbohydrates, Chemistry and Biochemistry*, W. Pigman and D. Horton (eds.), Academic Press, New York, 2nd edn., **IIB**, Chap 45.

Angyal, S. J. and James, K. (1970), *Chem. Comm.*, **320**.

Ashwell, G. and Morell, A. G. (1974), *Advances Enzymol.* **41**, 99.

Atherton, K. T., Byrom, D. and Dart, E. C. (1979), in *Microbial Technology: Current State, Future Prospects*, A. T. Bull, D. C. Ellwood and C. Ratledge (eds.), Cambridge University Press, Cambridge, p. 379.

Baddiley, J. (1972), *Essays Biochem.* **8**, 35.

Bayne, S. and Fewster, J. A. (1956), *Advances Carbohydr. Chem.* **11**, 43.

Bender, M. L. and Komiyama, M. (1978), *Cyclodextrin Chemistry*, Springer, Berlin.

Berkeley, R. C. W., Gooday, G. W. and Ellwood, D. C. (eds.) (1979), *Microbial Polysaccharides and Polysaccharases*, Academic Press, London.

Björndahl, H., Hellerqvist, C. G., Lindberg, B. and Svensson, S. (1970), *Angew. Chem. Internat. Edn.* **9**, 610.

Blackburn, G. M. (1979a), in *Comprehensive Organic Chemistry*, E. Haslam (ed.), Pergamon Press, Oxford, **5**, Chap 22. 4.

Blackburn, G. M. (1979b), in *Comprehensive Organic Chemistry*, E. Haslam (ed.), Pergamon Press, Oxford, **5**, Chap 22. 1.

Blackburn, G. M. (1979c), in *Comprehensive Organic Chemistry*, E. Haslam (ed.), Pergamon Press, Oxford, **5**, Chap 22. 5.

Blanshard, J. M. V. and Mitchell, J. R. (1979), *Polysaccharides in Food*, Butterworths, London.

Bull, A. T., Ellwood, D. C. and Ratledge, C. (eds.) (1979), *Microbial Technology: Current State, Future Prospects*, Cambridge University Press, Cambridge.

Butt, W. R., Lynch, S. S. and Kennedy, J. F. (1972), in *Structure-Activity Relationships of Protein and Polypeptide Hormones*, M. Margoulies and F. C. Greenwood (eds.), Internat. Cong. Ser. 241, Exerpta Medica, Amsterdam, Part 2, p. 355.

Cabib, E. and Shematek, E. M. (1981), in *Biology of Carbohydrates*, V. Ginsburg and P. Robbins (eds.), Wiley, New York, **1**, p. 51.

Candy, D. J. (1980), *Biological Functions of Carbohydrates*, Blackie, Glasgow.

Carr, P. W. and Bowers, L. D. (1980), *Immobilized Enzymes in Analytical and Clinical Chemistry*, Wiley, New York.

Cook, G. M. W. and Stoddart, R. W. (1973), *Surface Carbohydrates of the Eukaryotic Cell*, Academic Press, London.

Coxon, B. (1980), in *Developments in Food Carbohydrates–2*, C. K. Lee (ed.), Applied Science, London, p. 351.

Davidson, R. L. (ed.) (1980), *Handbook of Water-Soluble Gums and Resins*, McGraw Hill, New York.

Dawson, G. and Tsay, G. C. (1977), in *Research to Practice in Mental Retardation*, P. Mittler (ed.), University Park Press, Baltimore, **III**, p. 157.

Dea, I. C. M. (1979), in *Polysaccharides in Food*, J. M. Blanshard and J. R. Mitchell (eds.), Butterworths, London, p. 229.

Dutton, G. G. S. (1973), *Advances in Carbohydr. Chem. Biochem.* **28**, 11.

Dutton, G. G. S. (1974), *Advances in Carbohydr. Chem. Biochem.* **30**, 9.

Ferrier, R. J. (1976), in *Carbohydrates*, G. O. Aspinall (ed.), M.T.P. Internat. Rev. Sci. Organic Chem., Ser. 2, Butterworths, London, **7**, Chap 2.

Ferrier, R. J. (1980), in *The Carbohydrates, Chemistry and Biochemistry*, W. Pigman and D. Horton (eds.), Academic Press, New York, 2nd edn., **IB**, Chap 19.

Fukuda, M., Kondo, T. and Osawa, T. (1976), *J. Biochem. (Tokyo)* **80**, 1223.

Gilbert, W. and Villa-Komaroff, L. (1980), *Scientific American* **242(4)**, 68.

Gottschalk, A. (ed.) (1972), *Glycoproteins*, Elsevier, Amsterdam, 2nd edn.

Gurr, M. I. and James, A. T. (1980), *Lipid Biochemistry: An Introduction*, Chapman and Hall, London, 3rd edn.

Hakomori, S.-I. (1964), *J. Biochem. (Tokyo)* **55**, 205.

Hakomori, S.-I. (1976), in *Carbohydrates*, G. O. Aspinall (ed.), M.T.P. Internat. Rev. Sci. Organic Chem., Ser. 2, Butterworths, London, **7**, Chap 7.

Hakomori, S.-I. and Kobata, A. (1974), in *The Antigens*, M. Sela (ed.), Academic Press, New York, **2**, p. 80.

Hall, L. D. and Morris, G. A. (1980), *Carbohydr. Res.* **82**, 175.

Hanessian, S. and Haskell, T. H. (1970), in *The Carbohydrates, Chemistry and Biochemistry*, W. Pigman and D. Horton (eds.), Academic Press, New York, 2nd edn., **IIA**, Chap 31.

Harmon, R. E. (ed.) (1979), *Asymmetry in Carbohydrates*, Dekker, New York.

Hascall, V. C. (1981), in *Biology of Carbohydrates*, V. Ginsburgh and P. Robbins (eds.), Wiley, New York, **1**, p. 1.

Hassid, W. Z. (1970), in *The Carbohydrates, Chemistry and Biochemistry*, W. Pigman and D. Horton (eds.), Academic Press, New York, 2nd edn., **IIA**, Chap 34.

Heidelberger, M., Dische, Z., Neeley, W. B. and Wolfrom, M. L. (1955), *J. Amer.*

Chem. Soc. **77**, 3511.

Holum, J. R. (1978), *Organic and Biological Chemistry*, Wiley, New York.

Horton, D. and Wander, J. D. (1980a), in *The Carbohydrates, Chemistry and Biochemistry*, W. Pigman and D. Horton (eds.), Academic Press, New York, 2nd edn., **1B**, Chap 16.

Horton, D. and Wander, J. D. (1980b), in *The Carbohydrates, Chemistry and Biochemistry*, W. Pigman and D. Horton (eds.), Academic Press, New York, 2nd edn., **1B**, Chap 18.

Hough, L. and Richardson, A. C. (1972), in *The Carbohydrates, Chemistry and Biochemistry*, W. Pigman and D. Horton (eds.), Academic Press, New York, 2nd edn., **IA**, Chap 3.

Hough, L. and Richardson, A. C. (1979), in *Comprehensive Organic Chemistry*, E. Haslam (ed.), Pergamon Press, Oxford, **5**, Chap 26. 1.

Ivatt, R. J. and Gilvarg, C. (1979), *J. Biol. Chem.* **254**, 2759.

Jennings, H. J. and Smith, I. C. P. (1978), in *Methods in Enzymology*, G. Ginsburg (ed.), Academic Press, New York, **50**, p. 39.

Jermyn, M. A. and Yeow, Y. M. (1975), *Aust. J. Plant Physiol.* **2**, 501.

Kennedy, J. F. (1971–1981), in *Carbohydrates Chemistry – Specialist Periodical Reports*, J. S. Brimacombe (ed.), The Chemical Society, London, **4–12**, part II.

Kennedy, J. F. (1974a), *Advances in Carbohydr. Chem. Biochem.* **29**, 305.

Kennedy, J. F. (1974b), *Biochem. Soc. Trans.* **2**, 54.

Kennedy, J. F. (1976), *Advances Clin. Chem.* **18**, 1.

Kennedy, J. F. (1979a), *Proteoglycans – Biological and Chemical Aspects in Human Life*, Elsevier, Amsterdam.

Kennedy, J. F. (1979b), *Chem. Soc. Rev.* **8**, 221.

Kennedy, J. F. and Cabral, J. M. S. (1982a), in *Solid Phase Biochemistry: Analytical and Synthetic Aspects*, W. H. Scouten (ed.), Wiley, New York, (in press).

Kennedy, J. F. and Cabral, J. M. S. (1982b), in *Applied Biochemistry and Bioengineering*, L. B. Wingard and I. Chibata (eds.), Academic Press, New York, **4**, (in press).

Kennedy, J. F. and Fox, J. E. (1980), in *Methods in Carbohydrate Chemistry*, R. L. Whistler and J. N. BeMiller (eds.), Academic Press, New York, **8**, p. 13.

Kiss, J. (1970), *Advances Carbohydr. Chem.* **24**, 382.

Ko, A. M. Y. and Somers, P. J. (1974), *Carbohydr. Res.* **34**, 57.

Kohn, J. and Wilchek, M. (1982), *Enzyme Microb. Technol.* **4**, 161.

Kornfeld, R. and Kornfeld, S. (1976), *Ann. Rev. Biochem.* **45**, 217.

Lee, C. K. (ed.) (1980), *Developments in Food Carbohydrate–2*, Applied Science, London.

Lee, E. Y. C. and Whelan, W. J. (1966), *Arch. Biochem. Biophys.* **116**, 162.

Lennarz, W. J. (1975), *Science* **188**, 986.

Li, Y. T. and Li, S.-C. (1976), in *Methods in Carbohydrate Chemistry*, R. L.

Whistler and J. N. BeMiller (eds.), Academic Press, New York, **7**, p. 221.
Lis, H. and Sharon, N. (1977), in *The Antigens*, M. Sela (ed.), Academic Press, New York, **4**, p. 429.
Lis, H., Sharon, N. and Katchalski, E. (1964), *Biochim. Biophys. Acta* **83**, 376.
Lonngren, J. and Svensson, S. (1974), *Advances Carbohydr. Chem. Biochem.* **29**, 41.
Manners, D. J. and Matheson, N. K. (1980), *Carbohydr. Res.* **90**, 99.
Mathews, M. B. (1976), in *Methods in Carbohydrate Chemistry*, R. L. Whistler and J. N. BeMiller (eds.), Academic Press, New York, **7**, p. 116.
McArthur, H. A. I. (1981), *British Polymer J.* **13**, 111.
McKibbin, J. M. (1970), in *The Carbohydrates, Chemistry and Biochemistry*, W. Pigman and D. Horton (eds.), Academic Press, New York, 2nd edn., **IIB**, Chap 44.
McKusick, V. A. (1972), *Heritable Disorders of Connective Tissue*, C. V. Mosby, St. Louis, 4th edn.
Morris, H. R. (1980), *Nature* **286**, 447.
Muir, H. and Hardingham, T. E. (1975), in *Biochemistry of Carbohydrates*, W. J. Whelan (ed.), M.T.P. Internat. Rev. Sci. Biochem., Ser. 1, Butterworths, London, **5**, Chap 4.
Munson, R. S. and Glaser, L. (1981), in *Biology of Carbohydrates*, V. Ginsburg and P. Robbins (eds.), Wiley, New York, **1**, p. 91.
Palmer, T. (1981), *Understanding Enzymes*, Ellis Horwood, Chichester.
Perlin, A. S. (1959), *Advances Carbohydr. Chem.* **14**, 9.
Preiss, J. and Walsh, D. A. (1981), in *Biology of Carbohydrates*, V. Ginsburg and P. Robbins (eds.), Wiley, New York, **1**, p. 199.
Ramachandran, G. N., Ramakrishman, C. and Sasisekharan, V. (1963), in *Aspects of Protein Structure*, G. N. Ramachandran (ed.), Academic Press, New York, p. 121.
Ramachandran, G. N. and Reddi, A. H. (eds.) (1976), *Biochemistry of Collagen*, Plenum, New York.
Rauvala, H., Finne, J., Krusius, T., Kärkkäinen, J. and Järnefelt, J. (1981), *Advances Carbohydr. Chem. Biochem.* **38**, 389.
Recommendations (1970), Abbreviations and Symbols for Nucleic Acids, Polynucleotides and their Constituents (1970), *Eur. J. Biochem.* **15**, 203; and (1972), *Eur. J. Biochem.* **25**, 1.
Recommendations (1973), Nomenclature of Cyclitols (1975), *Eur. J. Biochem.* **57**, 1.
Recommendations (1976), Nomenclature of Lipids (1977), *Eur. J. Biochem.* **79**, 11.
Recommendations (1978), Enzyme Nomenclature (1979), Academic Press, New York; Supplement 1 (1980), *Eur. J. Biochem.* **104**, 1; Supplement 2 (1981), *Eur. J. Biochem.* **116**, 423; and Supplement 3 (1982), *Eur. J. Biochem.* **125**, 1.
Recommendations (1980a), Conformational Nomenclature for Five and Six-

Membered Ring Forms of Monosaccharides and their Derivatives (1980), *Eur. J. Biochem.* **111**, 295.

Recommendations (1980b), Abbreviated Terminology of Oligosaccharide Chains, *J. Biol. Chem.* **257**, 3347.

Recommendations (1980c), Polysaccharide Nomenclature, *J. Biol. Chem.* **257**, 3352.

Recommendations (1980d), Nomenclature of Branched-Chain Monosaccharides (1981), *Eur. J. Biochem.* **119**, 5.

Recommendations (1980e), Nomenclature of Unsaturated Monosaccharides (1981), *Eur. J. Biochem.* **119**, 1.

Recommendations (1981a), Symbols for Specifying the Conformation of Polysaccharides, *Eur. J. Biochem.* (in press).

Recommendations (1981b), Abbreviations and Symbols for the Description of Polynucleotide Chains, *Eur. J. Biochem.* (in press).

Reden, J. and Dürckheimer, W. (1979), *Topics in Current Chemistry* **83**, 105.

Rees, D. A. (1977), *Polysaccharide Shapes*, Chapman and Hall, London.

Rees, D. A. and Welsh, E. J. (1977), *Angew. Chem. Internat. Edn.* **16**, 214.

Rinehart, K. L., Jnr. and Suami, T. (eds.) (1980), *Aminocyclitol Antibiotics*, A.C.S. Symp. Ser. 125, Amer. Chem. Soc., Washington.

Rodén, L. and Schwartz, N. B. (1975), in *Biochemistry of Carbohydrates*, W. J. Whelan (ed.), M.T.P. Internat. Rev. Sci. Biochem., Ser. 1, Butterworths, London, **5**, Chap 3.

Rodén, L., Forsee, T. W., Jensen, J., Feingold, D. S., Prihar, H., Bäckström, M., Höök, M., Jacobsson, I., Lindahl, U., Riesenfeld, J. and Malmström, A. (1980), in *Mechanism of Saccharide Polymerisation and Depolymerisation*, J. J. Marshall (ed.), Academic Press, New York, p. 395.

Ryle, M., Chaplin, M. F., Gray, C. J. and Kennedy, J. F. (1970), in *Gonadotrophins and Ovarian Development*, W. R. Butt, A. C. Crooke and M. Ryle (eds.), Livingstone, Edinburgh, p. 98.

Saenger, W. (1980), *Angew. Chem. Internat. Edn.* **19**, 344.

Sammes, P. G. (ed.) (1977 onwards), *Topics in Antibiotic Chemistry*, Ellis Horwood, Chichester, **1–5**.

Sandford, P. A. (1979), *Advances Carbohydr. Chem. Biochem.* **36**, 265.

Sandhu, J. S., Hudson, G. J. and Kennedy, J. F. (1981), *Carbohydr. Res.* **93**, 247.

Schleifer, K. H. and Kandler, O. (1972), *Bacteriol. Rev.* **36**, 407.

Schwarzenbach, R. (1979), in *Biological and Biomedical Applications of Liquid Chromatography* **2**, G. L. Hawk (ed.), Chromatographic Science Ser. **12**, Dekker, New York, p. 193.

Serianni, A. S., Nunez, H. A. and Barker, R. (1979a), *Carbohydr. Res.* **72**, 71.

Serianni, A. S., Clark, E. L. and Barker, R. (1979b), *Carbohydr. Res.* **72**, 79.

Setlow, J. K. and Hollaender, A. (eds.) (1979), *Genetic Engineering, Principles and Methods*, Plenum, New York, **1**.

Setlow, J. K. and Hollaender, A. (eds.) (1980), *Genetic Engineering, Principles and Methods*, Plenum, New York, **2**.

Sharon, N. (1975), *Complex Carbohydrates, Their Chemistry, Biosynthesis and Functions*, Addison-Wesley, Massachusetts.

Sharon, N. (1980), *Scientific American* **243(5)**, 80.

Sharon, N. and Lis, H. (1979), *Biochem. Soc. Trans.* **7**, 783.

Spiro, R. G. and Bhoyroo, V. D. (1974), *J. Biol. Chem.* **249**, 5704.

Stacey, M. (1973), *Chem. and Ind.*, **222**.

Stewart-Tull, D. E. S. (1980), *Ann. Rev. Microbiol.* **34**, 311.

Sturgeon, C. M. and Kennedy, J. F. (1979 onwards), *Enzyme Microb. Technol.* **1**, 53, 129, 210 and 290; **2**, 66, 155, 244 and 318; **3**, 76, 160, 260 and 367; and **4**, 50, 118, 198, 276 and 356.

Stryer, L. (1981), *Biochemistry*, W. H. Freeman, San Francisco, 2nd edn.

Sutherland, I. W. (ed.) (1977), *Surface Carbohydrates of the Prokaryotic Cell*, Academic Press, London.

Sutherland, I. W. (1979), *Trends Biochem. Sci.* **4**, 55.

Svennerholm, L. (1963), *J. Neurochem.* **10**, 613.

Sweeley, C. C. (ed.) (1980), *Cell Surface Glycolipids*, A.C.S. Symp. Ser. 128, Amer. Chem. Soc., Washington.

Szarek, W. A. (1973), in *Carbohydrates*, G. O. Aspinall (ed.), M.T.P. Internat. Rev. Sci., Organic Chem., Ser. 1, Butterworths, London, **7**, Chap 3.

Tentative Rules (1969), Carbohydrate Nomenclature, Part 1 (1971), *Eur. J. Biochem.* **21**, 455; and (1972), *Eur. J. Biochem.* **25**, 4.

Theander, O. (1980), in *The Carbohydrates, Chemistry and Biochemistry*, W. Pigman and D. Horton (eds.), Academic Press, New York, 2nd edn., **IB**, Chap 23.

Trevan, M. D. (1980), *Immobilized Enzymes, An Introduction and Applications in Biotechnology*, Wiley, London.

Umezawa, S. (1976), in *Carbohydrates*, G. O. Aspinall (ed.), M.T.P. Internat. Rev. Sci., Organic Chem., Ser. 2, Butterworths, London, **7**, Chap 5.

Various Authors (1968 onwards), in *Carbohydrate Chemistry – Specialist Periodical Reports*, The Chemical Society, London.

Various Editors (1974 onwards), *Enzyme Engineering*, Plenum, New York; **2**, (1974) E. K. Pye and L. B. Wingard Jnr. (eds.), **3** (1978) E. K. Pye and H. H. Weetall (eds.), **4** (1978) G. B. Broun, G. Manecke and L. B. Wingard Jnr. (eds.), and **5** (1980) H. H. Weetall and G. P. Royer (eds.).

Watson, J. D. and Crick, F. H. C. (1953a), *Nature* **171**, 737.

Watson, J. D. and Crick, F. H. C. (1953b), *Nature* **171**, 964.

Wells, W. W. and Eisenberg, F. Jnr. (eds.) (1978), *Cyclitols and Phosphoinositides*, Academic Press, New York.

Weigandt, H. (1973), *Hoppe-Seylers Z. Physiol. Chem.* **354**, 1049.

Whistler, R. L. and BeMiller, J. B. (1958), *Advances Carbohydr. Chem.* **13**, 289.

Whistler, R. L. and Kosik, M. (1971), *Arch. Biochem. Biophys.* **142**, 106.

Whistler, R. L. and Others (eds.) (1962 onwards), *Methods in Carbohydrate Chemistry*, Academic Press, New York, **1** (1962), **2** (1963), **3** (1963), **4** (1964), **5** (1965), **6** (1972), **7** (1976) and **8** (1980).
White, C. A. and Kennedy, J. F. (1980), *Enzyme Microb. Technol.* **2**, 82.
White, C. A. and Kennedy, J. F. (1981), *Techniques Carbohydr. Metabolism*, **B312**, 1.
White, C. A., Corran, P. H. and Kennedy, J. F. (1980), *Carbohydr. Res.* **87**, 165.
Williams, N. R. and Wander, J. D. (1980), in *The Carbohydrates, Chemistry and Biochemistry*, W. Pigman and D. Horton (eds.), Academic Press, New York, 2nd edn., **IB**, Chap 17.
Wiseman, A. (ed.) (1975), *Handbook of Enzyme Biotechnology*, Ellis Horwood, Chichester.
Wiseman, A. (ed.) (1977 onwards), *Topics in Enzyme and Fermentation Biotechnology*, Ellis Horwood, Chichester, **1–5**.
Wolfrom, M. L. and Schumacher, J. N. (1955), *J. Amer. Chem. Soc.* **77**, 3318.
Wu, R. (ed.) (1979), *Methods in Enzymology*, Academic Press, New York, **68**.
Zaborsky, O. R. (1973), *Immobilized Enzymes*, C.R.C. Press, Cleveland.
Zadrazil, S. and Sponar, J. (eds.) (1980), *DNA – Recombination, Interactions and Repair*, Pergamon, Oxford.

Index

E

F

H

U

V

W

X